শিমাংস (কি) ।
বদ্বোত্তর ।
উত্তোর ।
১৯৮৫

Dr. Sunil R. Das
University of Ottawa
1985

MORPHOMETRICS

The Multivariate Analysis of Biological Data

Richard A. Pimentel

California Polytechnic State University
San Luis Obispo, California

KENDALL/HUNT PUBLISHING COMPANY
2460 Kerper Boulevard,
Dubuque, Iowa 52001

This book is dedicated to
Jerome C. R. Li.

I hope he would have approved.

Contents

Preface

Multivariate analysis as presented in most text books provides trivial results for biological data. Neither the mathematical nor applied treatments are of much use to the scientist who has good data and wishes to gain their full implications. Existing applied texts contain powerful techniques for interpreting data for the fields of education, psychology and social sciences. Since workers in those fields often must use indirect measurements of items of study and must use multivariate analysis to unravel the data, existing texts emphasize means for interpreting such data. On the other hand when data are direct measures of phenomena, results can and should be interpreted differently. Since procedures for such interpretation are scattered in the literature, this book was written. Although useful to anyone having good data that describe the form of a phenomenon, it will be obvious that this book was written by a biologist interested in field biology.

This book has developed over seven years of teaching a course in multivariate biometry. Its approach was finalized on the basis of the standard, inadequate mathematical and statistical background of students. Also, placing most of the terminology of multivariate analysis in one section of one chapter was found to be the best way to overcome the terminology problem with biology students well conditioned to sessions of memorizing terms. More important, of course, are the concepts gained from the language of multivariate analysis.

The book was planned to be used in the sequence presented; however, the chapter on the multivariate analysis of covariance could be omitted. If one wished further restriction of the material to be considered, factor analysis and canonical correlation analysis could be omitted; but this would restrict the kinds of problems that could be studied.

Many individuals influenced the preparation of this book. Students, while taking courses in multivariate biometry and morphometrics, helped formalize the manner of presentation. Later, while working on individual problems, students further disclosed pitfalls and gaps in discussions. Three students, Eric H. Castain, James O. Oliver, and William L. Weigle, presented page-by-page editorial comments that extended over several weekly sessions. Dr. Rhonda L. Riggins-Pimentel also influenced extensive revisions. Most helpful was the encouragement and time she provided. Drs. C. Dennis Hynes, Kingston L. H. Leong and Aryan I. Roest also gave beneficial criticisms.

Several other colleagues wrote lengthy critiques or brief comments that resulted in drastic additions or revisions in the original manuscript. Dr. Robert E. Blackith, University of Dublin, caused the addition of sections on Principal Coordinate Analysis and Correspondence Analysis. Dr. C. David McIntyre, Oregon State University, brought about the treatment of ecological ordination and data influence on ordination. Dr. James E. Mosimann, National Institute of Health, and Dr. Pierre Jolicoeur, Université de Montréal, made various suggestions that improved the chapter on Principal Component Analysis. Dr. James D. Smith, California State University at Fullerton, helped by utilizing one of the computer programs designed in conjunction with the book and by demonstrating what was needed in text support to clarify the output. Many others made comments that were incorporated and benefited the book. The latter individuals hopefully forgive their omission here.

I am especially indebted to the personnel of the California Polytechnic State University at San Luis Obispo Computer Center for aid in developing the computer programs discussed in the appendix. Miss Lila Young, Mr. Parker Gillespie and Mr. Dana Freiburger helped at various times. Mr. Luther Bertrando contributed throughout programs development.

To all these people, I am very grateful.

Chapter 1
Introduction

Biologists study complex phenomena but generally use one, two or three measurements in an attempt to describe any phenomenon. Most of us would agree that no three measurements would adequately describe a species, or an ecological situation, or a physiological complex. However, we tend to use few measurements because we find it difficult to analyze more than one, two or three measurements simultaneously. We can analyze the measurements singly, or in pairs or triplets. Such simple analyses, as well as those involving graphing, certain multiple comparison methods, and/or conventional statistical analyses, generally fall far short of unraveling the full relationships of the measurements. Such conventional analyses do not enable us to comprehend either the independent patterns of variation, or the synergistic contrasting patterns of variation that are defined by the total relationships of a realistic number of measurements.

Fortunately, there are analyses that provide a single, repeatable, meaningful description of biological phenomena as a complex of independent and/or contrasting patterns of variation. Such descriptions result from a branch of statistics called *multivariate analysis*. Multivariate analysis differs from conventional statistical analyses in that two or more measurements, each varying among individuals, are considered as a single set of variables to describe the form of a sample of individuals. Multivariate analysis that unravels the patterns of variation describing the form of a phenomenon is aptly termed *morphometrics*.

In any given morphometric study the major areas of methodology often are intertwined. Three broad areas of methodology are recognized. One is the study of variation within a population and is accomplished by principal component analysis of a single set of data. Another also pertains to single populations but examines relationships between two different sets of data, such as environmental and morphological data, and is accomplished by canonical correlation analysis. The third area of methodology is the consideration of similarities and differences between two or more populations, and is known as discriminant analysis. The latter, in part, amounts to an extension of the univariate analysis of variance to the multivariate case. However, discriminant analysis goes so far beyond the univariate analysis of variance as to hardly be comparable. Related to the third area in purpose but starting from a single group rather than two or more populations are ordination and cluster analysis.

To my knowledge, the major multivariate methods and various statistics used here as interpretive aids of results have not previously been proposed as a single interrelated unit. In spite of this, each of the various analyses and interpretive aids are well-documented, sound statistical procedures. In fact, many of the methods and aids are widely used and can be found in other references on multivariate analysis. Partly because the methods are well documented, but also because of the philosophy behind this book, mathematical proof of methods occurs secondarily, if at all, in descriptions of what a particular analysis does.

A Contrast.   With conventional statistical analysis of variation and relationships such as multiple regression and correlation, if $p = 6$ variables are measured on each individual of a sample, tests of hypotheses and estimates could be calculated for

p means,
p variances,
$p (p - 1)/2$ covariances,
$p (p - 1)/2$ intercorrelations and
$p - 1$ regression coefficients.

A confusing array of 47 statistics results. These 47 statistics often lead to poor definitions of a biological system.

With multivariate analysis of variation and relationships among six variables, only six basic statistics are calculated. The statistics are either variances or correlation coefficients. Although related to the conventional statistics, those of multivariate analysis are not the same. The basic device for obtaining the six multivariate statistics is transformation of the original data. The transformations are very different from the logarithmic, the square root, or other transformations of data applied to satisfy statistical assumptions of normality or equality of variances. The transformations rotate original variable axes into new positions, creating new variables from the old. The statistics of multivariate analysis apply to the new axes.

When $p = 6$, each of the six statistics has associated with it a set of six coefficients (one for each variable) also obtained as part of the process of data transformation. Each set of coefficients pertains only to its associated statistic because each set is independent of the others; hence, the six statistics are independent of each other. The exact methodology of obtaining the six independent statistics will be presented in Chapter 2.

With multivariate analysis, each of the six statistics is subject to a test of hypothesis. For each test that is statistically significant, the associated coefficients can be used to interpret an association of the variables that helps define the biological system. Even if all six tests are significant, only six interpretations are made. Therefore, multivariate analysis provides a simpler and more concise means of analyzing biological data.

An Example.   The only purposes here are to clarify the procedure of a simple morphometric analysis and to show the independent patterns of variation extracted from six bone measurements (skull, humerus, ulna, femur and tibia lengths and skull width) from each of 276 leghorn chickens (Wright, 1954). These data also serve to indicate what a particular multivariate method, called principal component analysis, can do for the biologist. Do not become concerned about any aspect of the analysis per se.

Principal component analysis extracts independent patterns of variation from a sample—in this case a sample of 276 chickens. In the example, patterns of variation are extracted from a correlation matrix (Table 1.1). For those not familiar with matrices, note that each column and each row pertains to a chicken bone measurement and that the measurements are in the same sequence of skull length to tibia length in both the columns and the rows. The diagonal elements of the matrix all have values of one and represent the correlation of a variable with itself which always is perfect and unity. Numbers not on the diagonal are correlation coefficients between pairs of variables, which pair being indicated by the variable label for the particular row and column in which the correlation coefficient occurs.

Table 1.1 Correlations between six chicken bone measurements

	skull		wing		leg	
	length	width	humerus	ulna	femur	tibia
length	1.00	.58	.62	.60	.57	.60
width	.58	1.00	.58	.53	.53	.56
humerus	.62	.58	1.00	.94	.88	.88
ulna	.60	.53	.94	1.00	.88	.89
femur	.57	.53	.88	.88	1.00	.92
tibia	.60	.56	.88	.89	.92	1.00

Also note that the matrix is symmetric, corresponding rows and columns having the same sequence of values for the correlation coefficients. Now accept that the matrix also measures variation.

It was said that principal component analysis based upon p measurements provides p statistics and p sets of p coefficients. Therefore, six chicken bone measurements provide six statistics (six variances) and six sets (components) of coefficients (Table 1.2). Each component has six coefficients with numeric values for each bone dimension. Each successive component accounts for a smaller proportion of the total variation (the sum of the six original variances) in the sample than does any previous component.

For the purposes of simplification, the first step in the analysis can be considered a test of the hypothesis that each of the six variances of the population is equal to zero. If we assume that all six population variances are statistically significantly different from zero, each of the six components can be interpreted as an independent pattern of variation. The patterns of variation can be read without any additional statistical tests or estimates.

Table 1.2 Principal components of six chicken bone measurements

variables	components					
	1	2	3	4	5	6
skull						
length	.35	.53	.76	−.05	.04	.00
width	.33	.70	−.64	.00	.00	−.04
wing						
humerus	.44	−.19	−.05	.53	.18	.67
ulna	.44	−.25	.02	.48	−.15	−.71
leg						
femur	.43	−.28	−.06	−.51	.65	−.13
tibia	.44	−.22	−.05	−.48	−.69	.17
variance	4.57	.71	.41	.17	.08	.06
% total variance	76.1	11.9	6.9	2.9	1.3	.9

The interpretation of components can be quite simple from a statistical point of view. Pearce (1965b) stresses the concept of components representing independent patterns of variation (responses) from causal stimuli. In the sense that principal component analysis examines variation in a sample and components of a set are mutually independent, each component is said to represent an independent pattern of variation. The stimuli acting as causal agents are assumed to be genetic or environmental in the broadest sense.

The actual interpretation or "reading" of components can follow definite criteria. Although the basis must be clarified later, Jolicoeur and Mosimann (1960) demonstrated that any component having coefficients all of the same sign was indicative of size variation and any component having both positive and negative coefficients was indicative of shape variation. In a less specific sense, size components often are termed general components and shape components, bipolar components. Rao (1964) provides formal proof that components can be recognized as size or shape factors.

In the sense that organismal growth generally is accepted as an increase in size, a growth component might be restricted to a size component. This appears to be a matter of personal preference. Actually all components derived from measurements made on a sample of organisms might be said to represent patterns of growth since both size and shape changes are involved in growth. Here reference shall be to patterns of variation and a growth component shall be a size component.

Returning to the example, the first component accounts for 76.1% of the total variation in bone dimensions. All coefficients of the first component are positive numbers and all wing and leg dimensions have coefficients of like magnitude. However, the skull dimensions have coefficients of only slightly smaller magnitude. At this point we can say that the first component describes a pattern of variation in which all bone dimensions increase together (they all have positive coefficients). Naturally this pattern is consistent with general growth of bones in chickens, or any other growing organism.

The second component is more complex. Since its coefficients have both positive and negative signs, shape changes are implied. Naturally the negative signs for wing and leg bones cannot imply that the pattern involves bone shrinkage. Rather, the implication is that in terms of all patterns of variation operating simultaneously, this 11.9% of the sample's variation is explained by an increase in skull dimensions and a decrease in leg and wing dimensions. In reference to all components, any negative sign for a coefficient does indicate a decrease in magnitude of a dimension. This can also be considered skull growth at the expense of relationship with leg and bone relationship. Biologically, this is consistent with known differential growth between skull and appendages.

The third component accounts for 6.9% of the variation and emphasizes only skull dimensions, since coefficients for both wing and leg bones have numeric values approximating zero. The pattern of variation is one of shape in which skull length increases at the expense of its relationship with skull breadth. Biologically, this amounts to differential growth in skull dimensions.

The remaining components account for successively lesser variation and all have near-zero values for the coefficients of skull dimensions. Component four contrasts wing and leg. Component five contrasts proximal and distal appendage bones but mostly in the leg. Component six contrasts proximal and distal wing bones but also distal and proximal leg bones. For a biological interpretation, these components probably pertain to epiphyseal growth. The implication is that ulna epiphyseal growth is associated with like growth of the humerus. However, some humerus growth is associated with that of the femur and some with the tibia. Finally, although some very small epiphyseal growth of femur and tibia might be associated (component four), much of this growth is independent.

The above cursory analysis demonstrates the potential power of any multivariate analysis and the morphometric approach. The six independent components, or patterns of variation, summarize much that is known about the independent aspects of bone growth. What is truly remarkable is that a single principal component analysis, involving only six bone measurements, unraveled many of the facets of bone growth.

The accomplishments of the component analysis might be appreciated even more if contrasted with a more conventional statistical analysis, correlation. Table 1.1 presents the intercorrelations between the various bone dimensions. The results of correlation analysis would not surprise a biologist. The correlation coefficients are generally high and positive which can be expected since all bone dimensions increase as the chicken grows. The coefficients serve to show that the bones are associated. However, for multivariate analysis these correlations are data, the source of variation, from which the six components were extracted.

Prerequisites.   This book is written with the firm conviction that an understanding of morphometrics is too important to allow unnecessary barriers to prevent its use. A possible barrier is terminology. The language of any scientific discipline is very important; approximately three hundred terms of multivariate analysis are introduced, a large proportion of them in Chapter 2. I feel that terminology should be no great problem since the reader probably is already well indoctrinated to the value of language in any field of science. Another possible barrier is that of mathematical sophistication. Judging from experience, mathematical sophistication for many biologists amounts to little more than a dim acquaintance with college algebra and trigonometry. Even biologists with more formal training in mathematics often have difficulty in relating formulae to the interpretation of results. I believe most books on multivariate analysis are considered incomprehensible by the average biologist because of terminology and expected mathematical background. Terms are important and can be learned, but mathematical background for this book is minimal.

Most of the language of multivariate analysis has precise mathematical meaning, but much of this mathematics is beyond the scope of the book. Therefore, instead of being presented in a manner that would satisfy a statistician, most terms are defined in the sense of presenting their essential qualities rather than their precise limits. Such definitions frequently are developed gradually as one proceeds through the text. In a few cases full appreciation of the implications of a term is late in the text. This approach has proven satisfactory; but, at the insistence of my students, a glossary has been added. Although the nature of definitions is the same, the glossary represents attempts to make definitions independent of the text, more precise, concise, and self explanatory. Another goal was to include all terms necessary to provide a conceptual image of multivariate analysis—some terms not in the text occur in the glossary.

A rigorous attempt has been made to minimize the need for mathematical prerequisites. However, formulae for all analyses and interpretive aids are included. But, for the formulae, their implications, concepts, use and ''proofs'' are developed through discussion of what they may do, can do, and how they can be interpreted. All the mathematics is linked to simple biological examples that are intended to expose the approach to morphometric analysis. ''Proofs'' are empirical, often including graphic representation of the effects of using a particular formula.

What then are the mathematical prerequisites? The critical requirements are the abilities gained from most one- or two-quarter courses in applied statistics. Such courses should include the necessary algebra, fixed one-way analysis of variance, linear regression, and some knowledge of multiple regression. Also, the reader should be aware of basic trigonometric functions. Additional mathematical background, mostly matrix algebra, is included in the text. The latter material is intentionally brief, but rich in terminology that must be mastered, since it is commonplace in papers using multivariate analysis and in this book. In fact, the entire short section of matrix algebra must be mastered to comprehend the remainder of the book.

Although the prerequisites are true as stated, some readers will probably disagree. One statistician wondered if biologists might not be insulted by ''the deemphasis of mathematical rigor and of previous statistical knowledge.'' On the other hand, a number of biologists with minimal to moderate background in mathematics or statistics considered the treatment challenging. This dichotomy of opinion amounts to an unstated prerequisite: at least some familiarity with mathematical approaches and formulae.

References.   References are kept at a minimum, since experience indicates to me that literature compendia act more as a barrier rather than an aid to learning. The references stress those works basic to verifying methodology and the morphometric approach, plus certain general works on multivariate analysis. For those interested in the origin of multivariate analysis, the first papers of Fisher (1936), Hotelling (1931, 1933, 1935, 1936a b) and Mahalanobis (1930, 1936) are cited. They also demonstrate that although multivariate analysis might be considered new, the framework of methodology predates computers hence application to complex problems.

There are several excellent volumes on multivariate analysis. The first major treatment was that of Rao (1952). Although Rao directed his presentation toward biological applications, many biologists find the mathematics too difficult. The same difficulty exists with another book emphasizing biological applications (Seale, 1964). The manual of Lee (1971) is somewhat mathematical but does contain several excellent biological examples. The works of Atchley and Bryant (1975) and Bryant and Atchley (1975) are compendia of papers that examine historical perspectives and biological examples. Mather (1976) provides the most detailed critique of computer algorithms in conjunction with multivariate procedures. Anderson (1958) considered multivariate analysis in great mathematical detail. Other current, excellent, comprehensive, and mathematical treatments include those of Morrison (1967), Dempster (1969), Press (1972) and Overall and Klett (1972).

Less mathematical, but stressing data analysis in the social and behavioral sciences, are Tatsuoka (1971) and Harris (1975). Still less mathematical, but in many ways more comprehensive, is Cooley and Lohnes (1971). Even less mathematical, but often frustratingly so, since methodology often is not clear, is Hope (1969).

The basic literature on morphometric methods is scattered, and frequently specific analyses are poorly defined. The best summary of the approach, and the best source of basic references, is the contribution of Blackith (1965). Certain papers by Blackith (1957, 1960, 1962), Jolicoeur (1959, 1963a), and Jolicoeur and Mosimann (1960) are most helpful. Blackith and Reyment (1971) provides an extensive bibliography and many examples of multivariate data analysis, but their presentation of morphometrics and methodology is less detailed than here. However, the book is excellent for indicating possible areas of study.

The interpretative aids used here include statistics discussed in some detail by Cooley and Lohnes (1971). Other aids are discussed in Hope (1969) and in the above works by Blackith; however, few of these are linked to basic mathematical formulae.

Multivariate Data Analysis.   Multivariate analyses developed very early in the 20th century and rapidly reached refinement well before the advent of the high speed computers that made their full application possible. Application has been mostly to the behavioral sciences, especially psychology and education; but is seen more and more in other areas including biology (Blackith and Reyment, 1971). From inception methods often were used more to develop theoretical constructs about the nature of phenomena than to emphasize statistical analyses, a so-called data analysis approach. Morphometrics also has developed with an heuristic approach but is more statistical than heuristic.

Current interpretation of the terms data analysis and data analyst are the consequence of a convincing discussion by a renowned mathematical statistician, John W. Tukey (1962). His "Future of Data Analysis" both justifies and encourages the data analyst, the non-statistician whose primary concern is analyzing and interpreting multivariate studies. Tukey characterized the data analyst as being willing to "seek for scope and usefulness rather than security," and, "willing to err moderately often in order that inadequate evidence shall more often *suggest* the right answer." Also, his recommended method for the analyst was to combine scientific and mathematical judgments, stressing the former without ignoring the latter. In essence, Tukey provided both credence and respectability to the "indication procedures" of multivariate heuristics that place secondary the "conclusion procedures" of statistical estimations and tests of hypotheses (Cooley and Lohnes, 1971).

Morphometrics. Multivariate procedures for analyzing the form of phenomena have been both biologically and statistically oriented. Rao's (1952) summary of early work is quite mathematical and relies little upon heuristics. Even the development of methodology for extending conclusions beyond those resulting from tests of hypotheses and estimations of parameters have relied almost entirely upon mathematical theory (Blackith, 1960, 1961, 1965; Jolicoeur, 1959, 1963a, 1963b; Jolicoeur and Mosimann, 1960; and Mosimann, 1958, 1975a, 1975b). Somewhat less mathematical, although still dependent upon statistical assumptions being satisfied, are the interpretations proposed by Pearce (1965b) and the reification proposed by Blackith (1965).

An attempt will be made to distinguish between the conservative conclusions justified by statistics and the interpretations derived from heuristics. For heuristics, reification shall be emphasized. Blackith (1965) applied the term *reification* to describe morphometric methods for drawing biological and/or physical conclusions from multivariate data. Here reification is formalized into a more detailed sequence of procedures that act as checks and balances on, as well as aids to, interpretation. The procedures will be developed gradually and are finally integrated in chapter ten.

Perhaps obviously, detailed biological and/or physical conclusions are not possible for many studies. The reasons for this are analytical and biological. Under analytical reasons are those of data, sampling, statistical analysis, and other items to be developed later. Under biological reasons are such items as insufficient knowledge and incorrect premises about the phenomenon under study. Reification is not a black box procedure that replaces biology or the biologist. Actually, the potential of any reification must reflect the detail of biological knowledge.

In a sense the checks and balances features are most important when biological implications are most difficult or even impossible to determine. Then, reification serves to prevent unwarranted conclusions and to restrict a physical description of relationships.

In spite of all this, the heuristic nature of reification cannot be denied. One never should conclude that determinations from reification are proven—a definite potential for error exists. Therefore, although the methodology of reification will be pursued in some detail, its possible shortcomings must be kept in mind.

Chapter 2
Matrix Algebra and Multivariate Methods

To understand the methods and procedures of multivariate analysis, matrix algebra must be understood. Once mastered, matrix algebra allows visualization of methodology. The present treatment of matrix algebra is both concise and brief; however, it is sufficient to understand the statistical models of the multivariate methods developed later.

The first section of this chapter includes all the mathematical operations and the associated terminology of matrix algebra used in this book. Section two briefly describes the multivariate normal distribution and some of its parameters. Section three is an abbreviated presentation of basic multivariate statistics. Section four presents general criteria for good data. The final section summarizes the multivariate methods derived from a specific matrix algebra operation, the characteristic equation.

2.1 Vectors and Matrices

Any operation in matrix algebra can be written in familiar scalar algebraic form. Any single number is a scalar. In scalar algebra, all individual values of variables and constants are indicated in formulae. Matrix algebra provides a shorthand that is less confusing than scalar algebra, e.g., in vectors and matrices individual values often are not indicated in formulae.

Vectors and matrices are the basic operational units for addition, subtraction, and multiplication in matrix algebra, and are the individual items in the multivariate models.

Basic Definitions. A *matrix* is a rectangular row by column arrangement (array) of numbers or terms, the array written between brackets, parentheses, or double lines on either side. For example, a data table

measurement	individual 1	2
1	7	2
2	9	4
3	8	7
4	4	5

can be written as a matrix, a data matrix,

$$\begin{bmatrix} 7 & 2 \\ 9 & 4 \\ 8 & 7 \\ 4 & 5 \end{bmatrix}$$

Any single number or term in a matrix is an *element*. The *order* or *dimension* of a matrix is the number of elements in the rows by the number of elements in the columns, symbolically $r \times c$, read "r by c". Note that a matrix is identified first by its number of rows and second by its number of columns. For example, a $r \times c$ matrix **A** has r rows and c columns and can be symbolized

$$\mathbf{A} = \begin{bmatrix} a_{11} & a_{12} & a_{13} & \ldots & a_{1c} \\ a_{21} & a_{22} & a_{23} & \ldots & a_{2c} \\ a_{31} & a_{32} & a_{33} & \ldots & a_{3c} \\ \ldots & \ldots & \ldots & \ldots & \ldots \\ a_{r1} & a_{r2} & a_{r3} & \ldots & a_{rc} \end{bmatrix}$$

The elements also are identified by their row and column, e.g., element a_{23} is in the second row and the third column. In fact, an entire matrix can be defined in the form of elements of a_{ij}

$$\mathbf{A} = (a_{ij}) \text{ where } i = 1, 2, \ldots, r; \text{ and } j = 1, 2, \ldots, c.$$

The *transpose* of a matrix is obtained by interchanging the rows and columns of a $r \times c$ matrix to form a $c \times r$ matrix. For example, the transpose of the above matrix **A** is symbolized **A'**, and is

$$\mathbf{A}' = \begin{bmatrix} a_{11} & a_{21} & a_{31} & \ldots & a_{r1} \\ a_{12} & a_{22} & a_{32} & \ldots & a_{r2} \\ a_{13} & a_{23} & a_{33} & \ldots & a_{r3} \\ \ldots & \ldots & \ldots & \ldots & \ldots \\ a_{1c} & a_{2c} & a_{3c} & \ldots & a_{rc} \end{bmatrix}$$

A *column matrix* has a single column ($r \times 1$) and is also called a *vector*. A *row matrix* has a single row ($1 \times c$) and here always is the transpose of a vector. In many applications elsewhere, either a column matrix or a row matrix is considered to be a vector.

Certain kinds of matrices and the properties of some of them are of special interest in multivariate analysis. A *null matrix* has all zero elements. A *square matrix* has the same number of rows and columns, i.e., its order is $r \times r$ or $c \times c$, since $r = c$. The order of a square matrix often is simplified to being r (or c). A *symmetric matrix* is a square matrix equal to its transpose, $\mathbf{A} = \mathbf{A}'$. A *rectangular matrix* has an unequal number of rows and columns ($r \neq c$).

Symmetric matrices are very common in multivariate analysis. Any correlation matrix is a symmetric matrix containing the bivariate correlations between pairs of variables (see Table 1.2 and accompanying discussion). In the case of p variables, a correlation matrix is a $p \times p$ matrix having all p diagonal elements of unity (the intracorrelations of each of the p variables with itself) and all off diagonal elements are intercorrelations between pairs of variables. A variance-covariance matrix also is a symmetric matrix. The diagonal elements of the latter matrices are variances and the off diagonal elements are covariances.

The *diagonal* or *principal diagonal* of a $p \times p$ square matrix consists of all elements from the upper left-hand corner through the lower right-hand corner. The *trace* is the sum of all diagonal elements, $i = 1, 2, \ldots, p$.

$$\mathbf{tr} \ (\mathbf{A}) = \sum a_{ii}.$$

A *diagonal matrix* has non-zero elements only on the diagonal (off diagonal elements all are zero). An *identity matrix* is a diagonal matrix having all diagonal elements of unity and symbolically is represented by $\mathbf{I}$. One use of an identity matrix is to place elements of a scalar in the diagonal of a matrix, creating a diagonal matrix, e.g., the product of the scalar k and a 2×2 identity matrix is

$$k\mathbf{I} = k\begin{bmatrix} 1 & 0 \\ 0 & 1 \end{bmatrix} = \begin{bmatrix} k & 0 \\ 0 & k \end{bmatrix}$$

Also, the product of any matrix $\mathbf{A}$ and a conforming identity matrix is equal to the original matrix, and the opposite is true,

$\mathbf{AI} = \mathbf{A}$ and $\mathbf{IA} = \mathbf{A}$.

An *inverse matrix* exists for most square matrices. The inverse of $\mathbf{A}$ is symbolized $\mathbf{A}^{-1}$. An inverse matrix is a unique matrix that satisfies the relation

$\mathbf{A}^{-1}\mathbf{A} = \mathbf{A}\mathbf{A}^{-1} = \mathbf{I}$.

Since the result is an identity matrix, multiplying $\mathbf{A}$ by $\mathbf{A}^{-1}$ is equivalent to dividing $\mathbf{A}$ by itself. Inverse matrices are used in lieu of division, a procedure that is not possible in matrix algebra. The equivalent of dividing $\mathbf{A}$ by $\mathbf{B}$ is $\mathbf{AB}^{-1}$. Such a procedure often is called conditioning $\mathbf{A}$ by $\mathbf{B}$.

An *orthogonal matrix* is a square matrix having the special property that when multiplied by its transpose produces an identity matrix, i.e.,

$\mathbf{AA}' = \mathbf{I}$.

Orthogonal matrices are commonplace in multivariate analysis.

Elementary Operations.
For the addition, subtraction or multiplication of two matrices, the elements of both matrices must correspond according to fixed conditions. When two matrices satisfy the conditions necessary for an arithmetic procedure, the matrices are said to *conform* for that operation.

Matrices are added and subtracted by adding or subtracting corresponding elements, so both the number of rows and columns of the two matrices must be equal for the matrices to conform. Matrix multiplication provides new elements from the sum of the products of corresponding elements of the rows of the first (premultiplier) matrix and the columns of the second (postmultiplier). Therefore, to multiply two matrices, only the number of elements in the rows of the first $r_1 \times c_1$ matrix must equal the number of elements in the columns of the second $r_2 \times c_2$ matrix, i.e., the matrices must conform in that the number of columns in the first matrix must equal the number of rows in the second matrix ($c_1 = r_2$).

For an example, consider two matrices, $\mathbf{A}$ and $\mathbf{B}$, and their transposes, $\mathbf{A}'$ and $\mathbf{B}'$:

$$\mathbf{A} = \begin{bmatrix} 1 & 2 & 3 \\ 4 & 5 & 6 \end{bmatrix} \quad \mathbf{B} = \begin{bmatrix} 7 & 10 \\ 8 & 11 \\ 9 & 12 \end{bmatrix} \quad \mathbf{A}' = \begin{bmatrix} 1 & 4 \\ 2 & 5 \\ 3 & 6 \end{bmatrix} \quad \mathbf{B}' = \begin{bmatrix} 7 & 8 & 9 \\ 10 & 11 & 12 \end{bmatrix}$$

Since **A** and **B**′ are 2 × 3 matrices and **B** and **A**′ are 3 × 2 matrices, neither **A** + **B**, nor **A** − **B** is possible; but **A**′ + **B**, **A**′ − **B**, **A** + **B**′ and **A** − **B**′ are possible, e.g.,

$$\mathbf{A}' + \mathbf{B} = \begin{bmatrix} 1 + 7 & 4 + 10 \\ 2 + 8 & 5 + 11 \\ 3 + 9 & 6 + 12 \end{bmatrix} = \begin{bmatrix} 8 & 14 \\ 10 & 16 \\ 12 & 18 \end{bmatrix}$$

In multiplication, **AB** and **A**′**B**′ are possible, but **AB**′ and **A**′**B** do not conform for multiplication. In the case of **AB**,

$$\mathbf{AB} = \begin{bmatrix} (1)(7) + (2)(8) + (3)(9) & (1)(10) + (2)(11) + (3)(12) \\ (4)(7) + (5)(8) + (6)(9) & (4)(10) + (5)(11) + (6)(12) \end{bmatrix}$$

$$= \begin{bmatrix} 50 & 68 \\ 122 & 167 \end{bmatrix}$$

Note that the product of a matrix of order $r \times p$ and a matrix of order $p \times c$ is possible because they conform in p and is equal to an $r \times c$ matrix. In the example, the product of a 2 × 3 and 3 × 2 matrix is possible because they conform in 3, and the result is a 2 × 2 matrix.

Since vectors are column matrices, addition and subtraction of two vectors requires that both have the same number of elements, r. Addition or subtraction of one vector and the transpose of another vector is not possible, but the transposes of two vectors can be added or subtracted if both have the same number of elements. For multiplication, two vectors must conform in the same manner as any other matrix.

$$\text{If } \mathbf{a} = \begin{bmatrix} 7 \\ 9 \\ 8 \\ 4 \end{bmatrix} \text{ and } \mathbf{b} = \begin{bmatrix} 2 \\ 4 \\ 7 \\ 5 \end{bmatrix}$$

ab cannot be calculated because there is one element in the first row of **a** and four elements in the column of **b**. But **a**′**b** can be calculated:

$$\mathbf{a}' = [7 \ 9 \ 8 \ 4] \text{ and}$$

$$\mathbf{b} = \begin{bmatrix} 2 \\ 4 \\ 7 \\ 5 \end{bmatrix}$$

$$\mathbf{a}'\mathbf{b} = (7 \times 2) + (9 \times 4) + (8 \times 7) + (4 \times 5)$$
$$= 14 + 36 + 56 + 20$$
$$= 126.$$

Therefore, **a**′**b** is a scalar.

Also, **ab**′ can be calculated since there is one element in each row of **a** and each column of **b**′.

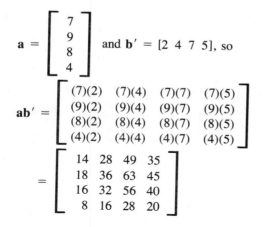

$$\mathbf{a} = \begin{bmatrix} 7 \\ 9 \\ 8 \\ 4 \end{bmatrix} \quad \text{and } \mathbf{b}' = [2 \ 4 \ 7 \ 5], \text{ so}$$

$$\mathbf{ab}' = \begin{bmatrix} (7)(2) & (7)(4) & (7)(7) & (7)(5) \\ (9)(2) & (9)(4) & (9)(7) & (9)(5) \\ (8)(2) & (8)(4) & (8)(7) & (8)(5) \\ (4)(2) & (4)(4) & (4)(7) & (4)(5) \end{bmatrix}$$

$$= \begin{bmatrix} 14 & 28 & 49 & 35 \\ 18 & 36 & 63 & 45 \\ 16 & 32 & 56 & 40 \\ 8 & 16 & 28 & 20 \end{bmatrix}$$

Therefore, $\mathbf{ab}'$ is a matrix of new elements.

Permissible Matrix Manipulations. Matrix operations are similar to those for scalars. Aside from division of matrices being impossible, the main point of concern is that different orders of multiplication must be recognized. This can be reduced to a few simple rules:

1. For two matrices to be added or subtracted, they must conform by having the same number of rows and columns $(r \times c)$. Then,

    $\mathbf{A} \pm \mathbf{B} = (a_{ij} \pm b_{ij})$ for $i = 1,2,\ldots,r$ and $j = 1,2,\ldots,c$.

2. The sequence of addition and/or subtraction of conforming matrices is immaterial. Thus,

    $\mathbf{A} - \mathbf{B} + \mathbf{C} = \mathbf{C} - \mathbf{B} + \mathbf{A} = -\mathbf{B} + \mathbf{A} + \mathbf{C}$, etc.

3. The product of two matrices is possible if the number of column elements of the first equals the number of row elements of the second, c. Thus,

    $$\mathbf{AB} = \sum_{k=1}^{c} a_{ik} b_{kj} \text{ for } k = 1,2,\ldots,c.$$

4. The product $\mathbf{AB}$ provides $\mathbf{C}$ with row elements equal to the number of rows in $\mathbf{A}$ (p) and with column elements equal to the number of columns in $\mathbf{B}$ (q). Thus,

    $\mathbf{C} = (c_{ij})$ for $i = 1,2,\ldots p$ and $j = 1,2,\ldots,q$.

From (3) and (4), note in the process $\mathbf{AB} = \mathbf{C}$ amounts to

$$\sum_k a_{ik} b_{kj} = c_{ij},$$

which is possible because a and b conform in k.

5. If $\mathbf{B}$ and $\mathbf{C}$ conform for addition and subtraction (are $r \times c$) and $\mathbf{A}$ conforms as a premultiplier for both $(p \times r)$,

$$\mathbf{A(B + C)} = \mathbf{AB} + \mathbf{AC} = \mathbf{D},$$

where $\mathbf{D}$ is $p \times c$.

6. If three matrices conform for multiplication ($\mathbf{A}, m \times n$; $\mathbf{B}, n \times p$; and $\mathbf{C}, p \times q$), then

$$\mathbf{ABC} = \mathbf{(A)(BC)} = \mathbf{(AB)(C)}$$

and the result is a $m \times q$ matrix.

7. The transpose of the product of two matrices is equal to the product of their transposes,

$$\mathbf{(AB)'} = \mathbf{A'B'}.$$

8. Except for special cases such as a matrix and its inverse where

$$\mathbf{AA^{-1}} = \mathbf{A^{-1}A} = \mathbf{I},$$

the order of matrix multiplication is critical,

$$\mathbf{AB} \neq \mathbf{BA}.$$

9. The equality of two equations is unaffected by premultiplication or postmultiplication by the same matrix. If

$$\mathbf{A} = \mathbf{B}, \text{ then } \mathbf{CA} = \mathbf{CB} \text{ and } \mathbf{AC} = \mathbf{BC}.$$

Linear Dependence and Rank.   An important characteristic of matrices has to do with *linear dependence* of rows and columns. In multivariate analysis the only source of such dependency is the original variables. Also, if variables are linearly dependent so are the rows and/or columns of any matrix derived from those variables, e.g., a correlation matrix. All linear dependence implies is that from part of a vector of different variables the remaining variable(s) can be derived. For example, if all p variables are percentages of a whole, any $p - 1$ variables provide the value of the missing variable since the p variables must sum to 100%. More formally, a set of p vectors, each of a single variable, is linearly dependent if a linear combination of any two or more vectors equals zero.

$$k_1\mathbf{X}_1 + k_2\mathbf{X}_2 + \ldots = 0.$$

The constants k_1, k_2, etc. may be equal or unequal but at least one of the constants must be nonzero. For example, the vectors $\mathbf{X}_1' = [1\ \ 4]$ and $\mathbf{X}_2' = [3\ \ 12]$ are linearly dependent since $-3\mathbf{X}_1 + 1\mathbf{X}_2 = 0$. In the case of derived matrices linear dependence can be more complex. For example, $\mathbf{v}_1' = [-1\ \ 5\ \ -2]$, $\mathbf{v}_2' = [1\ \ 2\ \ -1]$, and $\mathbf{v}_3' = [3\ \ -1\ \ 0]$ are linearly dependent since $-1\mathbf{v}_1 + 2\mathbf{v}_2 - 3\mathbf{v}_3 = 0$. The absence of linear dependence is *linear independence*. Removal of one or more variables from a vector of linearly dependent variables can result in the remaining variables being linearly independent. The largest possible number of linearly independent rows or columns of a matrix is the *rank* of a matrix. In multivariate analysis this amounts to independent measurements. A matrix having all possible independent rows or columns is of *full rank* and all others are of *reduced rank*. Many matrices of multivariate analysis must be of full rank. Full rank is accomplished by deleting the minimum number of variables possible from the original p variables to produce the largest possible set of linearly independent variables.

Determinants. A determinant is a single scalar number associated with any square matrix. This represents a unique function of the numbers in a matrix and is symbolized: $|\mathbf{A}|$. Understanding the formal definition or any mathematically meaningful definition of a determinant requires some sophistication in matrix algebra, but is unnecessary for understanding multivariate analysis. Calculation of the determinant of a large matrix requires complex procedures. However, the formal definition leads to the following direct solutions of 2×2 and 3×3 matrices:

$$\begin{vmatrix} a_{11} & a_{12} \\ a_{21} & a_{22} \end{vmatrix} = a_{11}a_{22} - a_{12}a_{21}$$

$$\begin{vmatrix} a_{11} & a_{12} & a_{13} \\ a_{21} & a_{22} & a_{23} \\ a_{31} & a_{32} & a_{33} \end{vmatrix} = a_{11}a_{22}a_{33} + a_{12}a_{23}a_{31} + a_{13}a_{21}a_{32} - a_{13}a_{22}a_{31} - a_{11}a_{23}a_{32} - a_{12}a_{21}a_{33}$$

For example, the determinant of

$$\begin{vmatrix} 7 & 5 \\ 4 & 8 \end{vmatrix} = 56 - 20 = 36;$$

and the determinant of

$$\begin{vmatrix} 7 & 6 & 3 \\ 4 & 5 & 2 \\ 2 & 1 & 8 \end{vmatrix} = \begin{aligned} &(7)(5)(8) + (6)(2)(2) + (3)(4)(1) - \\ &(3)(5)(2) - (7)(2)(1) - (6)(4)(8) \end{aligned}$$
$$= 280 + 24 + 12 - 30 - 14 - 192$$
$$= 80.$$

A *singular matrix* is any square matrix whose determinant is zero. A *nonsingular matrix* is any square matrix whose determinant is non-zero. In multivariate analysis determinants often are calculated for a correlation or variance-covariance matrix partly because the matrices often must be nonsingular.

The determinant of a variance-covariance matrix is a single numeric value called the *generalized variance* and that of a correlation matrix, the *generalized correlation*.

Characteristic Equation.   Like having a determinant and perhaps an inverse, each square matrix has associated with it an extremely important mathematical function, a *characteristic equation*. The characteristic equation of the square matrix $\mathbf{D}$ of order p is the determinant of the difference of $\mathbf{D}$ minus the scalar λ times an identity matrix $\mathbf{I}$ of order p set equal to zero,

$$\mathbf{D} - \lambda\mathbf{I} = 0.$$

The product $\lambda\mathbf{I}$ is used to create a diagonal matrix of order p with all diagonal elements of λ.

$$\lambda\mathbf{I} = \begin{bmatrix} \lambda & 0 & . & . & 0 \\ 0 & \lambda & . & . & 0 \\ . & . & . & . & . \\ 0 & 0 & 0 & . & \lambda \end{bmatrix}$$

This is done to cause $\lambda\mathbf{I}$ to conform with $\mathbf{D}$ for subtraction. For a p-order matrix, up to p different values for λ might exist, i.e., λ is a scalar variable. Parenthetically this means that $\lambda\mathbf{I}$ can result in different values for λ along the principal diagonal. The values for the lambdas are called *eigenvalues*, a reason the characteristic equation is also called the *eigenvalue function*. Some synonyms for eigenvalue are latent root, characteristic root and lambda root. An *eigenvector* is associated with each eigenvalue and some synonyms for eigenvector are latent vector and characteristic vector. Each eigenvector is selected to satisfy the *eigenvector function*

$$(\mathbf{D} - \lambda_i\mathbf{I})\mathbf{a}_i = 0 \text{ or } \mathbf{D}\mathbf{a}_i = \lambda_i\mathbf{a}_i$$

where λ_i is one of the p eigenvalues and $\mathbf{a}_i$ is one of p corresponding eigenvectors.

Numerical Example.   Solution of the characteristic equation and eigenvector function usually is beyond our capability. For any matrix larger than 3×3 a computer program normally is used. However, since the solution of these two functions is an integral part of all multivariate techniques, a simple numerical example of a 2×2 matrix might prove useful. The application will be considered in reference to principal component analysis.

The characteristic equation satisfies the purpose of principal component analysis, to transform original data vectors (measurements that are correlated) to vectors of component scores (variables that have all zero bivariate correlations). Therefore, in the bivariate case two original variables, X_1 and X_2, are transformed to two new variables, y_1 and y_2. We shall not be concerned here with X_1 and X_2 or y_1 and y_2, but rather with the original variance-covariance matrix, $\mathbf{S}^2$, of X_1 and X_2 and the transformation of $\mathbf{S}^2$ to $\lambda\mathbf{I}$, the variance-covariance matrix of y_1 and y_2. It will be seen later that $\lambda\mathbf{I}$ will consist of two variances, λ_1 and λ_2, and zero covariances. Therefore, since calculation of a correlation coefficient between two variables involves their covariance and in $\lambda\mathbf{I}$ the covariance of y_1 and y_2 is zero, the correlation between y_1 and y_2 must be zero.

If we consider the 2×2 matrix below to be a variance-covariance matrix, the example will include the basic calculations of principal component analysis.

If

$$\mathbf{S}^2 = \begin{bmatrix} 4 & \sqrt{3} \\ \sqrt{3} & 2 \end{bmatrix}$$

in solving the characteristic equation, $\mathbf{S}^2 - \lambda\mathbf{I} = 0$, only λ is unknown, we can solve

$$\lambda\mathbf{I} = \lambda\begin{bmatrix} 1 & 0 \\ 0 & 1 \end{bmatrix} = \begin{bmatrix} \lambda & 0 \\ 0 & \lambda \end{bmatrix}$$

and

$$\mathbf{S}^2 - \lambda\mathbf{I} = \left| \begin{bmatrix} 4 & \sqrt{3} \\ \sqrt{3} & 2 \end{bmatrix} - \begin{bmatrix} \lambda & 0 \\ 0 & \lambda \end{bmatrix} \right| = \left| \begin{matrix} 4 - \lambda & \sqrt{3} \\ \sqrt{3} & 2 - \lambda \end{matrix} \right|$$

Then, since the determinant of a 2×2 matrix is

$$\left| \begin{matrix} a & c \\ d & b \end{matrix} \right| = ab - cd,$$

$|\mathbf{S}^2 - \lambda\mathbf{I}| = 0$ is equal to

$$\begin{aligned} (4 - \lambda)(2 - \lambda) - (\sqrt{3})(\sqrt{3}) &= 8 - 6\lambda + \lambda^2 - 3 = 0 \\ &= \lambda^2 - 6\lambda + 5 = 0 \\ &= (\lambda - 5)(\lambda - 1) = 0 \end{aligned}$$

where λ is a scalar variable that has two solutions,

$$\lambda = 5,1.$$

Since in component analysis, successive eigenvalues must be the maximum possible values,

$\lambda_1 = 5$ and $\lambda_2 = 1$ rather than
$\lambda_1 = 1$ and $\lambda_2 = 5$.

For a $p \times p$ symmetric matrix, in general there are p such solutions, each a variance for one of the p variables y_i.

It is now convenient to examine $\lambda\mathbf{I}$

$$\begin{bmatrix} 5 & 0 \\ 0 & 1 \end{bmatrix}$$

as a transformation of the original variance-covariance matrix. Note that the sum of eigenvalues, 6, is equal to the sum of the original variances. Also note that the product of eigenvalues, 5, is equal to the determinant of the variance-covariance matrix, i.e., the generalized variance, but the covariances of $\lambda\mathbf{I}$ are zero.

Knowing the values of λ_1 and λ_2, in the eigenvector functions only the eigenvectors $\mathbf{a}_1$ and $\mathbf{a}_2$ are unknown,

$$(\mathbf{S}^2 - \lambda_1\mathbf{I})\mathbf{a}_1 = 0 \text{ and } (\mathbf{S}^2 - \lambda_2\mathbf{I})\mathbf{a}_2 = 0.$$

Solving for $\mathbf{a}_1$:

$$(\mathbf{S}^2 - \lambda_1\mathbf{I}) = \begin{bmatrix} 4 & \sqrt{3} \\ \sqrt{3} & 2 \end{bmatrix} - \begin{bmatrix} 5 & 0 \\ 0 & 5 \end{bmatrix} = \begin{bmatrix} -1 & \sqrt{3} \\ \sqrt{3} & -3 \end{bmatrix} \text{ and}$$

$$(\mathbf{S}^2 - \lambda_1\mathbf{I})\mathbf{a}_1 = \begin{bmatrix} -1 & \sqrt{3} \\ \sqrt{3} & -3 \end{bmatrix} \begin{bmatrix} a_{11} \\ a_{12} \end{bmatrix} = 0, \text{ so}$$

$-a_{11} + \sqrt{3}a_{12} = 0$ and
$\sqrt{3}a_{11} - 3a_{12} = 0,$

both equations leading to

$$a_{11} = \sqrt{3}a_{12}$$

which cannot be solved. However, as is the case in principal component analysis, if we set

$a_{11}^2 + a_{12}^2 = 1$ and substitute for a_{11}^2
$(\sqrt{3}a_{12})^2 + a_{12}^2 = 1,$ then
$a_{12}^2 = 1/4.$ Therefore,
$a_{12} = 1/2$ and $a_{11} = \sqrt{3/4}.$

Next, solving for $\mathbf{a}_2$:

$$(\mathbf{S}^2 - \lambda_2\mathbf{I}) = \begin{bmatrix} 4-1 & \sqrt{3} \\ \sqrt{3} & 2-1 \end{bmatrix} = \begin{bmatrix} 3 & \sqrt{3} \\ \sqrt{3} & 1 \end{bmatrix} \text{ and}$$

$$(\mathbf{S}_2 - \lambda_2\mathbf{I})\mathbf{a}_2 = \begin{bmatrix} 3 & \sqrt{3} \\ \sqrt{3} & 1 \end{bmatrix} \begin{bmatrix} a_{21} \\ a_{22} \end{bmatrix} = 0, \text{ so}$$

$3a_{21} + \sqrt{3}a_{22} = 0$ and
$\sqrt{3}a_{11} + a_{22} = 0,$

both equations leading to

$$a_{22} = -\sqrt{3}a_{21}.$$

Setting

$a_{21}^2 + a_{22}^2 = 1$ and substituting for a_{22}
$a_{21}^2 + 3a_{21}^2 = 1,$ then
$a_{21}^2 = 1/4.$ Therefore,
$a_{21} = 1/2$ and $a_{22} = -\sqrt{3/4}.$

Note that $\mathbf{a}_1'\mathbf{a}_2 = 0.$

The characteristic equation and eigenvector function provide the basic statistics shown in the chicken bone example. Each eigenvalue is a variance and the eigenvector, associated with each eigenvalue, is the set of coefficients, a component, that is read as a pattern of variation as was done in the chicken bone example.

Eigenvalue and Eigenvector Properties.

Certain properties of the eigenvalues and eigenvectors of *any square matrix,* **A,** of order p are very important in multivariate analysis. Some of these properties were mentioned before and all will be developed in further detail later. The properties are as follows:

1. The sum of the p eigenvalues is equal to the sum of the diagonal elements of the original matrix,

    $$\sum \lambda_i = \mathbf{tr(A)}.$$

    The reader can verify that the sum of chicken bone eigenvalues (Table 1.2) equals the trace of their correlation matrix (Table 1.1), i.e., both equal six.

2. The product of the p eigenvalues of **A** is equal to the determinant of **A,**

    $$\Pi \lambda_i = \mathbf{A}.$$

    This was shown to be the case in the numerical example for the characteristic equation.

3. The rank of **A** is equal to the number of non-zero eigenvalues of **A.** This property is especially important since direct computation of the number of linear independent rows or columns is both a side issue to multivariate analysis and time consuming even with a computer program.

4. In the methods to be used, a negative eigenvalue is not possible, values will be zero or greater. This is the case since determinants are calculated only for variance-covariance or correlation matrices. Therefore, $\Pi \lambda_i = \mathbf{A} \geq 0$. When potential $\mathbf{A} \geq 0$, the matrices are termed *positive semidefinite;* but, when potential $\mathbf{A} > 0$, the matrices are termed *positive definite.* Therefore, positive semidefinite matrices are singular or nonsingular, but all positive definite matrices are nonsingular. Although multivariate analysis matrices potentially are positive semidefinite for good biological data the matrices must be positive definite, hence of full rank defining p-dimensional space. A dispersion or correlation matrix generally is positive definite unless one or more variance are zero, bivariate correlation coefficients are unity, or two or more rows (or columns) are linearly dependent.

5. If $\mathbf{A} = 0$, $\mathbf{A}^{-1}$ does not exist, so any multivariate method requiring $\mathbf{A}^{-1}$ is not possible. Two methods to be discussed later, canonical correlation analysis and canonical analysis of discriminance, require inverses so in section 2.4 conditions contributing to a $\mathbf{A} = 0$ are considered.

6. Given the p eigenvectors of **A,** $\mathbf{v}_i$, the sum of the products of any two eigenvectors, $\mathbf{v}_i$ and $\mathbf{v}_j$, is zero, $\mathbf{v}_i'\mathbf{v}_j = 0$. This was shown to be the case in the above numerical example.

7. If an eigenvalue is non-zero, the associated eigenvector cannot be a null vector.

Nature of Determinants.

Determinants are commonplace in multivariate analysis. Although they are applied in a variety of ways, they are often influenced by linear dependence, being zero when there is linear dependence and increasing in magnitude relative to decreasing correlation between variables. For this reason, the generalized variance and generalized correlation are not reliable measures of overall variation and/or association.

2.2 Multivariate Normal Distribution

The methods to be presented generally depend upon the populations following the multinormal or multivariate normal distribution (hereafter often *mnd*). The sampling unit is the *individual* and each individual is represented by a p-element data vector of *p-observations* or *items*. Since each data vector contains elements which vary from individual to individual and since each is drawn at random from a *mnd*, the data vectors are *vector random normal variables,* or *vector normal variates.*

Any *mnd* is defined by its vector of means for each observation, the *mean vector* or *centroid,* and by a measure of variation or dispersion, a *variance-covariance* or *dispersion matrix.* The population centroid μ is the average of all elements of all (theoretically an infinite number) N data vectors in the population

$$\mu = \frac{1}{N} \sum \mathbf{X}_i$$

where $\mathbf{X}_i$ is any one of the N data vectors. The population dispersion $\sum^2$ is

$$\sum^2 = \frac{1}{N} \sum_{i=1}^{N} (\mathbf{X}_i - \mu)(\mathbf{X}_i - \mu),$$

a $p \times p$ symmetric matrix. If of interest, the density function of a *mnd* is

$$f(\mathbf{X}) = Ke^{-.5(\mathbf{X} - \mu)' \sum^{-2}(\mathbf{X} - \mu)}$$

where K is limited in that the integral over the p-element Cartesian space is unity. For any $\mathbf{X}_i$ following the *mnd* and having centroid μ and dispersion $\sum^2$,

$$K = (2\pi)^{-.5} \sum^2 {}^{-.5}.$$

Further features of the *mnd* include three major types of derived distributions. A *marginal distribution* is the distribution for any single element of a vector variable. If the vector variable of p-elements follows the *mnd,* each element, a measurement, follows the normal distribution. Parenthetically, just because all elements of a vector variable follow a normal distribution does not mean that the vector variable follows the *mnd.*

A *conditional distribution* is the predicted distribution of one element of a vector variable by the remaining elements. If a vector variable follows the *mnd,* every conditional distribution it can generate is normal. An example of a conditional distribution is that of the predicted variable in multiple regression.

A *component distribution* can be generated from any linear function of the vector variables. If the vector variables follow the *mnd,* so does the component distribution. We shall limit our detailed consideration of component distributions to principal components.

Bivariate Normal Distribution.   Visualization of a *mnd* is simplest in the bivariate case. This requires three dimensions. Two dimensions portray the axes of the two variables and the third dimension indicates the frequency of individuals at each coordinate of the axes of variables $\mathbf{X}_1$ and $\mathbf{X}_2$ (Figure 2.1). However, using three dimensions to portray the distribution of individuals based upon

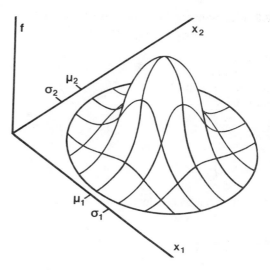

Figure 2.1 Bivariate normal distribution of X_1 and X_2 with the vertical dimension representing the frequency of individuals in the distribution.

two variables is inconvenient. For this reason, a set of nested frequency contours can be used to indicate individual frequency (Figure 2.2). No information is lost in the second figure. Both figures indicate the relative positions of the means, μ_1 and μ_2, and the inequality of variances $\sigma_1^2 > \sigma_2^2$. If the two variances were equal, the frequency contours would be circles rather than ellipses.

For most purposes, diagrams of a *mnd* do not require frequency contours. A single ellipse or circle works just as well. This is the case in the remainder of the book.

Figure 2.2 Bivariate normal distribution of X_1 and X_2 with contours representing the frequency of individuals in the distribution. The mean of both variables is represented by the central dot and each contour is based upon one standard deviation of each variable.

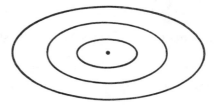

2.3 Basic Multivariate Statistics, an Example

We will now use matrix algebra to develop the basic statistics of multivariate analysis. The basic statistics will be based upon p-elements, or measurements, relating to each individual and upon a sample of N such individuals. Usually, we shall assume that any sample was drawn from a *mnd*.

A data vector, **X,** consists of a column of p items (observations or measurements) on a single individual,

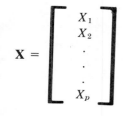

$$\mathbf{X} = \begin{bmatrix} X_1 \\ X_2 \\ \cdot \\ \cdot \\ \cdot \\ X_p \end{bmatrix}$$

All N individuals, $\mathbf{X}_i$, can be displayed in N columns in a $p \times N$ *data matrix*.

The *sample centroid,* is a vector containing means for each element in the N data vectors and is defined as

$$\overline{\mathbf{X}} = \frac{1}{N} \sum_{i=1}^{N} \mathbf{X}_i.$$

Upon subtraction of the sample centroid from each data vector, *deviation (score data) vectors* are derived,

$$\mathbf{x} = (\mathbf{X} - \overline{\mathbf{X}}).$$

The latter can be arranged in a $p \times N$ *deviation (scores) matrix,* also symbolized by $\mathbf{x}$, in which each column is a deviation vector. Then,

$$\mathbf{x}\,\mathbf{x}'$$

provides a $p \times p$ *sum of squares and cross products matrix, SSCP,* whose p diagonal elements are sums of squares of deviations from the mean and $p^2 - p$ off diagonal elements are sums of cross products between pairs of deviations from the mean. Then, when each element of the *SSCP* is multiplied by the scalar $1/(N - 1)$, the degrees of freedom, the result is the $p \times p$ *sample dispersion* or *variance-covariance* matrix,

$$\mathbf{S}^2 = \frac{1}{N - 1} \, \mathbf{x}\,\mathbf{x}',$$

an unbiased estimate of the population dispersion. Finally, the *sample correlation matrix* is computed from

$$\mathbf{R} = \mathbf{S}^2 \, (1/s_j s_k),$$

where s_j and s_k are scalar standard deviations of the jth and k^{th} variables and $(1/s_j s_k)$ is a scalar variable with a value specific for the jth row and k^{th} column.

It is frequently useful to use standardized normal scores (z-scores) in analyses,

$$\mathbf{z}_i = \mathbf{x}_i(1/s_j) = (\mathbf{X}_i - \overline{\mathbf{X}}) \, (1/s_j)$$

for $i = 1, 2, \ldots, N$ individuals, which through rules of uniform changes in observations provides a mean of $z = 0$ and standard deviation of unity for each variable. Also, the dispersion matrix of z-scores is

$$\mathbf{S}_z^2 = \frac{1}{N - 1} \mathbf{z}\,\mathbf{z}' = \frac{1}{N - 1} \mathbf{x}\,\mathbf{x}' \, (1/s_j s_k) = \mathbf{S}^2(1/s_j s_k)$$

which is equal to the correlation matrix $\mathbf{R}$. Much will be made of the fact that $\mathbf{S}^2$ is the dispersion of raw variables and $\mathbf{R}$ is the dispersion of z-scores or standardized variables.

Numerical Example. Given the simple 3×7 data matrix of 3 measurements on each of 7 individuals,

$$\mathbf{X} = \begin{bmatrix} 4 & 3 & 0 & 5 & 6 & 5 & 5 \\ 2 & 1 & 1 & 0 & 1 & 3 & 6 \\ 2 & 2 & 2 & 3 & 2 & 3 & 0 \end{bmatrix}$$

Summing each row and dividing by the sample size of $N = 7$, the sample centroid is

$$\overline{\mathbf{X}} = \begin{bmatrix} 4 \\ 2 \\ 2 \end{bmatrix}$$

Subtracting the centroid from each column of the data matrix provides the deviation score matrix

$$\mathbf{x} = \begin{bmatrix} 0 & -1 & -4 & 1 & 2 & 1 & 1 \\ 0 & -1 & -1 & -2 & -1 & 1 & 4 \\ 0 & 0 & 0 & 1 & 0 & 1 & -2 \end{bmatrix}$$

The reader can verify that

$$\mathbf{x}\mathbf{x}' = \begin{bmatrix} 24 & 6 & 0 \\ 6 & 24 & -9 \\ 0 & -9 & 6 \end{bmatrix} ; \text{ and that}$$

$$\mathbf{S}^2 = \left(\frac{1}{N-1} \right) \mathbf{x}\mathbf{x}' = \begin{bmatrix} 4 & 1 & 0 \\ 1 & 4 & -1.5 \\ 0 & -1.5 & 0 \end{bmatrix} \text{ so}$$

$$\mathbf{s} = \begin{bmatrix} 2 \\ 2 \\ 1 \end{bmatrix} \text{ and}$$

$$\mathbf{R} = \mathbf{S}^2(1/s_j s_k) = \begin{bmatrix} \dfrac{4}{2 \times 2} & \dfrac{1}{2 \times 2} & \dfrac{0}{2 \times 1} \\ \dfrac{1}{2 \times 2} & \dfrac{4}{2 \times 2} & \dfrac{-1.5}{2 \times 1} \\ \dfrac{0}{1 \times 2} & \dfrac{-1.5}{1 \times 2} & \dfrac{1}{1 \times 1} \end{bmatrix}$$

$$= \begin{bmatrix} 1 & .25 & 0 \\ .25 & 1 & -.75 \\ 0 & -.75 & 1 \end{bmatrix}$$

Turning to the z-score model, also verify that

$$\mathbf{z} = \mathbf{x}(1/s_j) = \begin{bmatrix} 0 & -.5 & -.2 & .5 & 1 & .5 & .5 \\ 0 & -.5 & -.5 & -1 & -.5 & .5 & 2 \\ 0 & 0 & 0 & 1 & 0 & 1 & -2 \end{bmatrix}$$

$$\mathbf{zz'} = \begin{bmatrix} 6 & 1.5 & 0 \\ 1.5 & 6 & -4.5 \\ 0 & -4.5 & 6 \end{bmatrix}$$

$$\mathbf{R} = \left(\frac{1}{N-1} \right) \mathbf{zz'} = \begin{bmatrix} 1 & .25 & 0 \\ .25 & 1 & -.75 \\ 0 & -.75 & 1 \end{bmatrix}$$

The reader may also wish to verify that

$$\mathbf{R} = 3/8$$

2.4 Criteria for Good Data Vectors

Good data are a prime concern throughout this book. Now, certain fundamental criteria for any good data vector can be defined. It is important to realize that these standards for judgment pertain to any set of data vectors no matter how they are analyzed, i.e., data vectors do not create problems that are unique to multivariate analysis. The only difference is that in other analyses the biologist might not be aware that poor data are influencing the results.

As steps to defining criteria, first the generalized variance and generalized correlation are reexamined, and then the computation of determinants is considered.

Generalized Variance and Correlation. Since $\mathbf{S}^2$ and $\mathbf{R}$ are positive semidefinite, the precise bases for a $\mathbf{S}^2 = 0$ or $\mathbf{R} = 0$ must be understood, again because neither $\mathbf{S}^{-2}$ nor $\mathbf{R}^{-1}$ exist in such situations. Some reasons for a determinant being zero can be appreciated in the bivariate case.

If the raw data dispersion formula for variables X_1 and X_2 is

$$\mathbf{S}^2 = \begin{bmatrix} s_1^2 & s_{12} \\ s_{21} & s_2^2 \end{bmatrix},$$

where s_1^2 and s_2^2 are variances of X_1 and X_2 respectively and $s_{12} = s_{21}$ is the covariance between X_1 and X_2, then the correlation coefficient between X_1 and X_2 is

$$r_{12} = s_{12}/s_1 s_2,$$

where s_1 and s_2 are standard deviations of the two variables. For later application, another form of the covariance, s_{12}, is required. If the covariance is multiplied by $s_1 s_2/s_1 s_2$ its value is unchanged, but

$$s_{12} = \frac{s_1 s_2}{s_1 s_2} s_{12}$$

$$= s_1 s_2 \frac{s_{12}}{s_1 s_2}$$

$$= s_1 s_2 r_{12},$$

so a covariance is a function of the standard deviation of each variable and the correlation coefficient between both variables. This leads to better appreciation of the nature of the determinant, since the determinant of

$$\mathbf{S}^2 = \begin{bmatrix} s_1^2 & s_1 s_2 r_{12} \\ s_1 s_2 r_{12}, & s_2^2 \end{bmatrix}$$

is equal to

$$\mathbf{S}^2 = s_1^2 s_2^2 - (s_1 s_2 r_{12})(s_1 s_2 r_{12})$$
$$= s_1^2 s_2^2 - s_1^2 s_2^2 r_{12}^2$$
$$= s_1^2 s_2^2 (1 - r_{12}^2)$$

which allows the following statements about the nature of $\mathbf{S}^2$ and also $\mathbf{R}$ by extension.

1. Since the possible values of r_{12} range from -1 through $+1$, the maximum possible value of $\mathbf{S}^2$ is $s_1^2 s_2^2$ and occurs when $r_{12} = 0$.
2. The minimum value of $\mathbf{S}^2$ is zero and occurs when $s_1^2 = 0, s_2^2 = 0$, and/or $r_{12} = \pm 1$. Recall that reflects the degree of independence of the two variables. Note that $\mathbf{S}^2$ increases as r_{12} approaches
3. Since a singular matrix neither possesses an inverse, nor is subject to certain transformations, it follows that multivariate analysis must in fact deal with variables ($s_j^2 \neq 0$) and the variables cannot be perfectly correlated ($r_{12} \neq \pm 1$).
4. It also follows that the magnitude of $\mathbf{S}^2$, although influenced by the magnitudes of s_1^2 and s_2^2, reflects the degree of independence of the two variables. Note that $\mathbf{S}^2$ increases as r_{12} approaches zero and decreases as r_{12} approaches ± 1. Therefore, the generalized variance is not a measure of the essence of variation per se in a sample.
5. It also follows that since

$$\mathbf{R} = 1 - r_{12}^2,$$

$\mathbf{R}$ is purely a measure of independence rather than association— $\mathbf{R}_{12} = 0$ when $r_{12} = \pm 1$ and $\mathbf{R} = 1$ when $r_{12} = 0$.

Explanation of what physically is measured by $\mathbf{S}^2$ might also be helpful. To do this requires a slight modification of the original covariance, some properties of a vector of a single variable, and the cosine of an angle.

The kind of vector needed is one of deviations of each observation of a single variable from its mean. In the above example, this would be the individual rows, transposes of $\mathbf{x}_1$ and $\mathbf{x}_2$ of N elements from a deviation score matrix. In previous notation,

$$\mathbf{x}' = [\mathbf{x}_1' \ \mathbf{x}_2'] = \begin{bmatrix} x_{11} & x_{21} \\ x_{12} & x_{22} \\ \ldots & \ldots \\ x_{1N} & x_{2N} \end{bmatrix}$$

Therefore, $\mathbf{x}_1'$ and $\mathbf{x}_2'$ are deviation score data "vectors" of single measurements from a sample of N individuals. Also, we must introduce the absolute value of such a vector, here both $|\mathbf{x}_1'|$ and $|\mathbf{x}_2'|$ (not determinants), which is defined as the square root of the sum of squares of its N elements

$$\mathbf{x}_1' = \sqrt{\sum x_{1i}^2} \text{ and } \mathbf{x}_2' = \sqrt{\sum x_{2i}^2} .$$

The absolute value of a vector of a single variable also is defined as being the length of the vector, and the length of a vector is related to the standard deviation. In the bivariate case

$$\mathbf{x}_1' = \sqrt{N-1} \ s_1 \text{ and } s_1 = \mathbf{x}_1' / \sqrt{N-1}, \text{ and}$$
$$\mathbf{x}_2' = \sqrt{N-1} \ s_2 \text{ and } s_2 = \mathbf{x}_2' / \sqrt{N-1},$$

so s_1 and s_2 are the lengths of $\mathbf{x}_1'$ and $\mathbf{x}_2'$ scaled by $\sqrt{N-1}$.

Next, since r_{12} could be shown to be the cosine of the angle θ between the variable vectors of deviations, $\mathbf{x}_1'$ and $\mathbf{x}_2'$, the $\mathbf{S}^2$ can be re-evaluated:

$$\begin{aligned} \mathbf{S}^2 &= s_1^2 s_2^2 \ (1 - r_{12}^2) \\ &= s_1^2 s_2^2 \ (1 - \cos^2 \theta) \\ &= s_1^2 s_2^2 \ (\sin^2 \theta) \\ &= (s_1 s_2 \sin \theta)^2. \end{aligned}$$

Since s_1 and s_2 are scaled lengths of $\mathbf{x}_1'$ and $|\mathbf{x}_2'|$, and θ provides the angle between the two vectors, the latter form of the determinant amounts to defining the square of the area of a parallelogram as is shown in Figure 2.3. In the multidimensional p-variate case, $\mathbf{S}^2$ is the square of the volume of a multidimensional extension of a parallelogram defined by p-vectors, each of a length equal to the standard deviation of a variable. It also can be seen that $\mathbf{S}^2$ measures the independence between

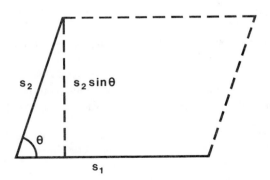

Figure 2.3 Relationships of $s_1 s_2 \sin \theta$, the square root of the determinant of a 2 × 2 dispersion matrix, S_2. Since s_1 defines the base and $s_2 \sin \theta$ the altitude of a parallelogram, $s_1 s_2 \sin \theta$ provides the area of the parallelogram.

variables as well as the variance of the variables since independent variables are 90° apart and dependent variables are 0°.

Although the 2-variable conditions causing $\mathbf{S}^2 = 0$ hold for p-variables, the relationships become more complex. For example, in the 3-variable case $\mathbf{S}^2$ reduces to

$$\mathbf{S}^2 = s_1^2 s_2^2 s_3^2 (1 + 2 r_{12} r_{13} r_{23} - r_{12}^2 - r_{13}^2 - r_{23}^2).$$

Therefore, $\mathbf{S}^2 = 0$ if

1. any variance equals zero,
2. all bivariate correlation coefficients have absolute values of unity (two can be negative or all must be positive), or
3. a complex relationship causes

$$2r_{12}r_{13}r_{23} - r_{12}^2 - r_{13}^2 - r_{23}^2 = -1,$$

i.e., linear dependency exists among the original variables.

Computation of Determinants. In computer programs, a determinant generally is calculated as part of a procedure that also provides the inverse of the matrix. In practice, matrix inversion is accomplished without any difficulty. On the other hand, the values of determinants are prone to error. Part of the error is data dependent, slight changes in the values of elements of a matrix can change the value of a determinant. Also, rounding errors in a computer program can alter the value of a determinant.

In practice, determinants pertain mostly to two aspects of multivariate analysis. The first is linear independence and the need for a $\mathbf{A} > 0$. It may be found that a determinant calculated during matrix inversion is zero or even a theoretically impossible value of less than zero. At the same time, but in another computation, it may be found that $\Pi\lambda_i > 0$ and that a computational check on a calculated $\mathbf{A}^{-1}$ shows that $\mathbf{AA}^{-1} = \mathbf{I}$ or closely approximates $\mathbf{I}$. This represents an inaccurate computation of $\mathbf{A}$. Such inaccurate $\mathbf{A} \leq 0$ are most likely in large matrices that approach but do not reach the conditions of linear dependence. Therefore, when all λ_i are positive ignore any $\mathbf{A} \leq 0$. In the second case inaccurate determinants are more important. Determinants are involved in the test of equality of group dispersions in the multivariate analysis of variance and some inaccuracy is almost always involved in the calculating of these determinants.

Also pertinent here is the fact that a $\mathbf{A} = 0$ might check with $\Pi\lambda_i = 0$ because the size of a sample, N, is equal to or less than the number of variables, p. Since the maximum possible rank of derived matrices then is $N - 1$, be sure that the minimum N is $p + 1$.

The foregoing problems in inverting a matrix become most extreme in so-called ill-conditioned matrices. Such matrices approach being singular and their inverses can be determined accurately only if computations involve many significant digits (perhaps more than found in a particular computer), or if special algorithms are implemented. In essence, this means that slight errors in measurement can produce erroneous results. Therefore, one should hesitate to stretch measurements beyond limits of accuracy.

Fortunately, although most matrices derived from biological data tend to display some features of ill-conditioned matrices, e.g., small values for their determinants, the extreme conditions should be encountered rarely. Moreover, such extremes generally become obvious in the form of drastic differences among the magnitudes of elements in the computed inverse. Also, the extreme values of such elements tend to represent the "problem" variable or variables contributing to ill-conditioning.

Data Vector Criteria.   The biological criteria for good data vectors are developed later. Now we must consider each measurement singly, so we are interested in bases for judging a set of vectors (each representing a single variable) that are linearly independent of one another. More appropriately, we wish to consider p-vectors, each of N elements of a single variable, that are linearly independent. In the present sense, the ultimate goal of course is to obtain positive definite matrices so their inverses exist, i.e., the aim is to obtain data vectors containing no redundant variables. Some kinds of variables always are linearly dependent, others well might be. The kinds of variables to avoid are

1. *ipsative measures,* a set of variables that sum to a constant. A relative minor characteristic of such variables is that they have inflated covariances and correlation coefficients. This is the case when for each individual a set of percentages sums to 100% or a set of proportions sums to unity. Inflated correlations also occur when variables of a set are coded to indicate presence or absence but only one feature can be present in any individual. For example, consider three localities, each represented by a variable coded zero for absence and one for presence. The localities conceptually should have zero correlation coefficients, but such coding causes any pair of localities to have a correlation coefficient of 1/3.

    The most important consequence of ipsative measures is that they reduce the rank of $\mathbf{x}\,\mathbf{x}'$, $\mathbf{S}^2$ and $\mathbf{R}$ from p to $p - 1$. The reduction in rank results from linear dependence since subtraction of the constant sum, e.g., 100%, from each element of each data vector causes the sum of elements, e.g., percentages, of any variable vector to equal zero.

2. *a measure and its component parts,* e.g., a total length and natural subunits that sum to the total length. If the parts are measured and the total is obtained by summing the measures of parts, linear dependence and a condition analogous to ipsative measures occurs. This is the case since all but one of the measures in the set (parts and total) automatically provide the remaining measure. When the total is measured separately, hence can deviate from the sum of parts, linear dependence still is likely but might not occur. In spite of possible linear independence, such totals should be avoided.

3. *a measure and some of its parts,* e.g., total length (composed of head, body and tail fractions) and tail length. Such variables tend to be highly correlated and to display similar degrees of correlation with other variables, a relationship that can cause linear dependence.

4. *geometrically redundant measures,* e.g., a set that overdefines a constant geometric figure. Measures beyond those necessary to define a triangle, rectangle, etc., create linear dependence.

5. *indices and their variables,* e.g., a ratio and one or both variables providing the ratio.

6. *related indices,* e.g., a set of indices that share one or more variables in their calculation.

Awareness of redundancy causing linear dependence usually leads to good data vectors, but the following procedure insures good data:

1. With redundancy in mind, select variables that outline the form of the phenomenon under study and collect the data.

2. Perform a principal component analysis of the data. If all eigenvalues are greater than zero, the requirement of linearly independent variables is satisfied.

3. If all eigenvalues are not greater than zero, the number of zero eigenvalues provides the number of redundant variables if $N > p$. Hopefully the investigator's expertise and the above guidelines will enable a quick solution to the problem. In many cases any variable can contribute to those being deleted. It should be realized that the original variables are the source that should be examined for linear dependency, because derived matrices often obscure sources of dependency. Parenthetically for those familiar with this problem, the device of using component scores as data vectors is not recommended since morphometric interpretation of the results is too difficult.

2.5 Characteristic Equation and Multivariate Methods

The use of the basic arithmetic operations of matrix algebra should be obvious. Also, the fundamental need for the characteristic equation was demonstrated. Both the arithmetic operations and characteristic equation are essential language for the multivariate methods of morphometrics. Since the characteristic equation involves a determinant, a function not having a counterpart in scalar algebra, the complete use of the characteristic equation will now be outlined. Quite naturally this can be confusing. The only important thing at this point is appreciating the fact that in providing so called canonical form, a characteristic equation results in simplification that aids in interpretation.

In this book, the three primary methods involving the characteristic equation are principal component analysis, canonical correlation analysis and canonical analysis of discriminance. A major matrix algebra operation of each method involves the transformation of a specific matrix to *canonical form* or *normal form*. The canonical form of a matrix is generally considered the simplest or most convenient transformed form of square matrices of a certain class. The canonical form to be developed for all of the above analyses, Jacobi canonical form, is a diagonal matrix where elements of the principal diagonal are eigenvalues, so the characteristic equation is the means of transformation.

Principal Component Analysis.   In principal component analysis, the variance-covariance matrix S^2 is transformed to a diagonal matrix L:

$$S^2 - \lambda_i I = 0,$$

where

$$L = \lambda_i I = \begin{bmatrix} \lambda_1 & 0 & . & . & . & 0 \\ 0 & \lambda_2 & . & . & . & 0 \\ . & . & . & . & . & . \\ 0 & 0 & . & . & . & \lambda_p \end{bmatrix}$$

Simplification occurs in L in that all $p\lambda_i$ are variances but there are all zero covariances. If eigenvectors a_i are calculated via the eigenvector function

$$(S^2 - \lambda_i I)a_i = 0,$$

original X variables are transformed to y_i variables by the linear function

$$y_i = a_i' (X - \overline{X})$$

and $\mathbf{y}_i$ has variance

$$\mathbf{a}_i'\mathbf{S}^2\mathbf{a}_i = \lambda_i$$

and zero covariances. Involved in this transformation are p original variables in each data vector $\mathbf{X}$, the sample centroid $\overline{\mathbf{X}}$, p eigenvalues or $\mathbf{y}$ variances λ_i, p components scores $\mathbf{y}_i$ for each individual data vector $\mathbf{X}$, and p eigenvectors or sets of component coefficients $\mathbf{a}_i$. Since the covariance between any two elements of $\mathbf{y}_i$ is zero, all pairs of $\mathbf{y}_i$ have zero correlations. Also, since any $\mathbf{a}_i'\mathbf{a}_j = 0$ $(i \neq j)$, they have zero correlation and define orthogonal axes.

Although all these statistics and their properties might be confusing at this point, this is indeed simplified form. More important is the fact that each $\mathbf{y}_i$ pertains to an $\mathbf{a}_i$ which consists of p elements, each representing an original variable. In essence, each $\mathbf{a}_i$ defines a relationship among variables that is independent of all other relationships. Therefore, the total set of p $\mathbf{a}_i$, the matrix $\mathbf{A}$, provides p independent sets of relationships among the original variables, that can be treated as p independent patterns of variation in terms of growth and/or form. Whether one is examining the morphology, physiology, ecology or ethology of organisms, the biological phenomenon examined can be separated into independent patterns of variation.

Canonical Correlation.   In chapter 3, much will be said about the problems of using the product moment correlation coefficient to measure the associations between any pair of variables. The problem with correlation coefficients becomes almost insurmountable if one wishes to examine associations between two sets of variables. Each set must consist of a logical grouping of variables, e.g., a morphological set vs. a physiological set.

Canonical correlation methodology starts by subdividing a total correlation matrix $\mathbf{R}$ of correlations within and between sets,

$$\mathbf{R} = \begin{bmatrix} \mathbf{R}_{11} & \mathbf{R}_{12} \\ \mathbf{R}_{21} & \mathbf{R}_{22} \end{bmatrix}$$

where $\mathbf{R}_{11}$ is the submatrix of intercorrelations between set one variables, $\mathbf{R}_{12} = \mathbf{R}_{21}'$ are the submatrices of intercorrelations between set one and set two variables, and $\mathbf{R}_{22}$ is the submatrix of intercorrelations between set two variables. Then, one calculates $\mathbf{R}^2$ the square of the intercorrelations between sets

$$\mathbf{R}^2 = \mathbf{R}_{11}^{-1}\mathbf{R}_{12}\mathbf{R}_{22}^{-1}\mathbf{R}_{21}.$$

If one considers the total matrix $\mathbf{R}$ a dispersion matrix of z-scores, the calculation of $\mathbf{R}^2$ is analogous to calculating the square of the product moment correlation between two variables. Also, some readers might recognize the relationship between the above $\mathbf{R}^2$ and the square of the multiple correlation coefficient in multiple regression.

The characteristic equation

$$\mathbf{R}_{11}^{-1}\mathbf{R}_{12}\mathbf{R}_{22}^{-1}\mathbf{R}_{21} - \lambda_i\mathbf{I} = 0$$

provides eigenvalues λ_i, each a maximum possible square of a so-called canonical correlation coefficient, that is a product moment correlation between sets. If there are p variables in set one and q variables in set two and $p \geq q$, there will be q values of λ_i. The number of eigenvalues is equal to the minimum (p,q).

As was the case for principal component analysis, the λ_i form a diagonal matrix $\mathbf{L}$. Since $\mathbf{R}_{11}^{-1}\mathbf{R}_{12}\mathbf{R}_{22}^{-1}\mathbf{R}_{21}$ is reduced to Jacobi canonical form, after this transformation, correlations between any two variables within a set is zero. Therefore, within either set there exists a condition equivalent to that resulting from principal component analysis.

The eigenvalue and eigenvector functions solve the model for canonical correlation; however, there are distinct differences from component analysis. Data for each individual in component analysis is a single data vector; in canonical correlation, two data subvectors $\mathbf{z}_1$ and $\mathbf{z}_2$ involve two vectors $\mathbf{c}$ and $\mathbf{d}$ rather than a single eigenvector. Finally, canonical correlation provides two sets of canonical variates, $\mathbf{x}$ and $\mathbf{y}$, rather than a single set of component scores. The model for canonical correlation is

$$\mathbf{x}_i = \mathbf{c}_i'\mathbf{z}_1 \quad \mathbf{y}_i = \mathbf{d}_i\mathbf{z}_2.$$

The eigenvector function

$$(\mathbf{R}_{11}^{-1}\mathbf{R}_{12}\mathbf{R}_{22}^{-1}\mathbf{R}_{21} - \lambda_i\mathbf{I})\mathbf{d}_i = 0$$

provides the set two canonical vectors, $\mathbf{d}_i$, and

$$\mathbf{c}_i = \mathbf{R}_{11}^{-1}\mathbf{R}_{12}\mathbf{d}_i/\lambda_i$$

provides the set one vectors, $\mathbf{c}_i$.

Therefore, canonical correlation provides q canonical correlations, $\sqrt{\lambda_i}$, and with each two vectors, $\mathbf{c}_i$ and $\mathbf{d}_i$. Since between any pair of $\mathbf{x}_i$ (or $\mathbf{y}_i$) there are zero correlations, it might seem that canonical correlation is just as appropriate for examining relationships within a set as is component analysis. Later we shall see that this is not the case.

Canonical Analysis of Discriminance.
Discriminant analysis includes two major procedures, multigroup discriminant function analysis and canonical analysis of discriminance, to analyze differences between groups. Multigroup discriminant function analysis does not involve the characteristic equation; canonical analysis of discriminance does. The latter operates on $\mathbf{W}^{-1}\mathbf{B}$ which is the inverse of the within group $SSCP$ times the among group $SSCP$. This is analogous to a univariate one-way analysis of variance F-test.

Later it will be shown that successive eigenvalues are maximum values representing differences between groups. Therefore, the characteristic equation

$$\mathbf{W}^{-1}\mathbf{B} - \lambda_i\mathbf{I} = 0$$

followed by the eigenvector function

$$(\mathbf{W}^{-1}\mathbf{B} - \lambda_i\mathbf{I})\mathbf{v}_i = 0$$

provides λ_i, indicating successive maximum differences among groups, and $\mathbf{v}_i$, indicating coordinate axes along which these differences exist. For example, the first two axes expose the maximum possible differences that occur and can be shown by graphing methods in two space. Again, inferences can be drawn from the coefficients of any $\mathbf{v}_i$.

Chapter 3
Multiple Regression and Correlation

In this chapter simple linear regression and correlation are extended to the multivariable case, multiple regression and multiple correlation. It is assumed that the reader is familiar with both. The primary purpose of this chapter is to convince the reader of the shortcomings of regression and correlation. There are legitimate applications of both methods.

Regression provides three kinds of results. First, regression can predict or estimate one variable from one or more other variables. The estimated variable is termed a predicted, criterion, or dependent variable. The one or more variables that estimate a predicted variable are termed predictors, covariates, or independent variables. Second, regression can determine the best formula for predicting some relationship. This function is not explored here. Finally, the success of a given regression can be ascertained. This usually amounts to measuring the overall relationship between the criterion and predictors by a multiple correlation coefficient.

In correlation, Pearson's product moment correlation coefficient is applied to measure the association between a pair of variables. In the simple case, the correlation coefficient is easily appreciated as a measure of goodness of fit of individuals to the line of regression. In multiple correlation, coefficients pairing a predictor with the criterion are stressed, but correlations between predictors often are examined.

Multiple regression and often multiple correlation are multivariable rather than multivariate techniques (Seal, 1964). The distinction is made upon the dependence structure of the variables. In multiple regression, a set of independent variables is applied to predict a single dependent variable which mathematically at least is the point of focus of the analysis. In multivariate analysis, the variables of the entire set are interdependent and are on an equal footing—one or more variables cannot be split off from the others and considered separately. In the latter sense, multivariate analysis is said to deal with a set of dependent variables (Kendall, 1957). In spite of the fact that multiple regression is not a multivariate technique, it is treated in books on multivariate analysis for a variety of reasons. Mathematically, multivariate analysis is a logical outgrowth and extension of multiple regression, which in turn is an extension of simple regression. Perhaps the common misconception that multiple regression is a multivariate method is another important factor. Also the widespread application of multiple regression cannot be ignored. This can be verified by usage of statistical computer programs. At the Regional Data Center of the California State University and Colleges, more than 90% of statistical usage is of a regression or correlation program. For all these reasons, both multivariable methods are included here.

There are biological situations in which regression and/or correlation can be helpful and should be utilized. However, predicting a variable, finding a formula, or demonstrating an association often are trivial results compared to the goals of most biologists. This, in part, is the reason for my attitude towards these methods; but my main concern comes from current usage. Correct applications of regression or correlation appear to be the exception rather than the rule. Interpretation commonly goes beyond the legitimate bounds of methodology. In fact, even the extremes of implying the relative merits of predictors or a cause and effect in associations are fairly common.

This entire chapter is directed towards biological goals that are not solved by regression or correlation. It is not intended that stressing the latter should be interpreted as a general condemnation of legitimate application of these techniques.

The focal point of this chapter is shortcomings of methodology. The treatment often parallels that of Snedecor and Cochran (1967) or Cooley and Lohnes (1971). The first section reviews the case of one predictor and one criterion. The second section extends procedures to the two or more predictor case. Additional discussion examines the merits of using the correlation coefficient as a measure of association. The third section, stepwise regression, criticizes another method for comparing the relative worth of predictors. Section four, multiple partial correlation, and section five, multiple part correlation, present methods that supposedly make the correlation coefficient a better measure of associations. The final critique is a concluding judgment of the statistical procedures presented in this chapter.

3.1 Simple Linear Regression and Correlation

In the case of one predictor, X, and one criterion, Y, the regression function is

$$\hat{Y}_j = a + b_X X_j$$

where $\hat{Y}_j$ is an estimate of the criterion Y for an X_j, and X_j is any one of a sample of N values of the predictor. The statistic a is the estimated value of the Y-axis intercept (the value of $\hat{Y}_j$ when X_j equals zero), and the statistic b_X is the estimated regression slope coefficient (the tangent of the angle between the X-axis and the estimated line of regression). The formula for the slope coefficient of X, b_X, could be shown to be

$$b_X = \frac{s_{12}}{s_{11}},$$

where s_{12} is the covariance between X and Y and s_{11} is the variance of X.

If one can assume that when $X = 0$, $Y = 0$, it is more convenient to change to a deviation score model,

$$\hat{y}_j = b_x x_j,$$

where the means of x and y both equal zero since both x_j and y_j are deviations from their respective means, i.e.,

$$x_j = X_j - \overline{X} \text{ and}$$
$$y_j = Y_j - \overline{Y}.$$

Therefore, $\hat{y}_j$ is the jth estimate of the deviation score y_j by x_j. The only effect of this model is to cause $a = 0$, since deviation scores of X and Y do not change the values of s_{12} and s_{11}, i.e.,

$$b_X = b_x = \frac{s_{12}}{s_{11}}$$

For many purposes, and especially when correlation is emphasized, the standardized normal score, or z-score, model is preferred. In this model, z_1 is the predictor and z_2 is the criterion so

$$\hat{z}_{2j} = b_z z_{1j}.$$

In accordance with transformation to z-scores,

$$z_{1j} = \frac{X_j - \overline{X}}{s_1} = \frac{x_j}{s_1}$$

and

$$z_{2j} = \frac{Y_j - \overline{Y}}{s_2} = \frac{y_j}{s_2}$$

where s_1 and s_2 are the standard deviations of X and Y respectively. Also, $\hat{z}_{2j}$ is the jth estimate of z_2, the standardized normal score of Y. In this model, the slope coefficient, b_z, takes on new meaning,

$$b_z = b_x \frac{s_1}{s_2} = \frac{s_{12}}{s_{11}} \frac{s_1}{s_2} = \frac{s_{12}}{s_1 s_2} = r_{12}$$

because to obtain

$$\hat{z}_{2j} = b_z z_{1j}$$

from

$$\hat{y}_j = b_x x_j,$$

both $\hat{y}_j$ and x_j are divided by their standard deviations,

$$\frac{\hat{y}_j}{s_2} = b_z \frac{x_j}{s_1}.$$

Therefore, to maintain the identity of the original formula, both sides are divided by s_2 and the right side is multiplied by s_1/s_1,

$$\frac{\hat{y}_j}{s_2} = \frac{b_x}{s_2} x_1 \frac{s_1}{s_1} , \text{ so}$$

$$\hat{z}_{2j} = b_x \frac{s_1}{s_2} z_{1j}, \text{ and}$$

$$b_z = b_x \frac{s_1}{s_2} .$$

Then, by substituting for b_x

$$b_z = \frac{s_{12}}{s_{11}} \frac{s_1}{s_2} = \frac{s_{12}}{s_1 s_2}$$

which by definition is equal to r_{12}, the correlation between X and Y, so

$$\hat{z}_{2j} = r_{12} z_{1j}.$$

3.2 Multiple Regression and Correlation

In the case of p-dimensional space, composed of p-1 predictors, $\mathbf{X}$, and a pth criterion variable, Y, the raw score model is

$$\hat{Y}_j = a + \mathbf{b}_X \mathbf{X}$$

and the deviation score model is

$$\hat{y}_j = \mathbf{b}_X \mathbf{x}$$

As was the case in simple linear regression, the weights for both models are the same; however, in multiple regression for both cases there are p-1 weights, a vector of p-1 coefficients, $\mathbf{b}_X$. The vector of **b**-weights are now computed from

$$\mathbf{b}_X = \mathbf{S}_{11}^{-1} \mathbf{S}_{12},$$

the product of the inverse of the variance-covariance matrix of the p-1 predictors and a vector of p-1 covariances, each between a predictor and the criterion. Note that the ith element of $\mathbf{b}_X$ would be the product of the ith row of $\mathbf{S}_{11}^{-1}$ and the total matrix (actually a vector) $\mathbf{S}_{12}$, so any b_{Xi} is influenced by the variance of the ith variable, the covariances of the ith predictor with all other p-2 predictors, and the p-1 covariances of all predictors with the criterion.

The z-score regression model is

$$\hat{\mathbf{z}}_{pj} = \mathbf{bz}.$$

Note that $\mathbf{b}$ has replaced the symbol $\mathbf{b}_z$ from simple linear regression, a convention to be followed hereafter. Since in the z-score model $\mathbf{S}_{11}$ becomes $\mathbf{R}_{11}$ and $\mathbf{S}_{12}$ becomes $\mathbf{R}_{12}$,

$$\mathbf{b} = \mathbf{R}_{11}^{-1}\mathbf{R}_{12}.$$

The relationship between $\mathbf{b}_X$ and $\mathbf{b}$ is more complicated now since each b_i represents the product of two vectors and not two scalars as in simple linear regression. However, in agreement with the above formula for $\mathbf{b}$, each

$$b_i = \frac{s_i}{s_y} b_{X_i},$$

rather than a correlation coefficient between an X and Y.

No attempt was made to follow the convention of some statisticians to differentiate between the notations, $\mathbf{b}$ for the raw or deviation score models. ($\mathbf{b}_X$ here), and $\boldsymbol{\beta}$ for the z-score model. ($\mathbf{b}$ here). Since the z-score model is the only one considered in further discussion, what will continue to be symbolized as $\mathbf{b}$ might be found to be $\boldsymbol{\beta}$ elsewhere.

The predictors, $\mathbf{z}_j$, locate a hyperplane of p-1 dimensions from which is estimated $\hat{z}_{pj}$, a single line of regression in the pth dimension. The $\mathbf{b}$-weights are called the standard partial regression coefficients and often are applied to evaluate the relative merits of their corresponding variables in predicting the criterion.

To compute the $\mathbf{b}$-weights one must first define partitions of the total correlation matrix as follows:

$$\mathbf{R} = \begin{bmatrix} \mathbf{R}_{11} & \mathbf{R}_{12} \\ \mathbf{R}_{21} & \mathbf{R}_{22} \end{bmatrix}$$

where $\mathbf{R}_{11}$ is the submatrix of intercorrelations among predictors. $\mathbf{R}_{12} = \mathbf{R}_{21}'$ is the vector of predictors-criterion intercorrelations, and $\mathbf{R}_{22} = 1$ is the scalar of intercriterion correlation. Matrix symbols, bold face type, are applied to the vector $\mathbf{R}_{12}$ and the scalar $\mathbf{R}_{22}$ for two reasons. First, the vector can be considered a $(p$-1$) \times 1$ matrix and the scalar a 1×1 matrix. Also, the symbols correspond to convention and to those appearing later in canonical correlation.

The above model of multiple regression often is preferred because, like its simple regression counterpart, it provides correlation coefficients for interpreting associations. However, additional interest is in evaluating the relative value of each independent variable as a predictor. Methods for such evaluation will be considered first.

Criterion Prediction. The square of the multiple correlation coefficient, R^2, classically has served as a singular scalar evaluator of the success of all independent variables as predictors,

$$R^2 = \mathbf{R}_{21}\mathbf{R}_{11}^{-1}\mathbf{R}_{12} = \mathbf{R}_{21}\mathbf{b}.$$

The multiple correlation coefficient, R, ranges in magnitude from zero to unity and is the product moment correlation coefficient between the predictor set $\mathbf{z}$ of p-1 variables and the estimated criterion $\hat{z}_p$.

R also is a measure of goodness of fit of the N individual z_{pj} to the line of regression defined by all $\hat{z}_{pj}$, i.e., the variance in terms of the line of regression, $s_{\hat{z}p}^2$. It is also the correlation between the observed and predicted criterion, and can be obtained from

$$R^2 = \left(\frac{1}{N-1} \right) \sum z_{pj} \hat{z}_{pj}$$

$$= \left(\frac{1}{N-1} \right) \sum z_{pj} (\mathbf{b}\mathbf{z}_j)$$

$$= \left(\frac{1}{N-1} \right) (\sum z_{pj} \mathbf{z}_j) \mathbf{b}$$

$$= \mathbf{R}_{21} \mathbf{b}.$$

When all z_{pj} are on the line of regression, $R = 1$. In the same sense as R measuring the goodness of fit to the regression line, R^2 termed the coefficient of determination, estimates the proportion of the total criterion variance that the variance of predictors can predict.

The estimates of predicted and deviation portions of regression can be used to test the null hypothesis that the population multiple correlation coefficient equals zero. Since R^2 is the predicted portion of regression the deviation from regression is $1.0 - R^2$, which can be translated as follows:

$$1.0 - R^2 = \mathbf{R}_{22} - \mathbf{R}_{21}\mathbf{b}$$
$$= \mathbf{R}_{22} - \mathbf{R}_{21}\mathbf{R}_{11}^{-1}\mathbf{R}_{12}.$$

The test of hypothesis utilizes

$$F = \frac{R^2(N - p)}{(1 - R^2)(p - 1)}$$

with $p - 1$ and $N - p$ degrees of freedom.

Having the explained portion of regression, R^2, and the unexplained portion, $1.0 - R^2$, it might seem a simple matter to evaluate the contribution of each predictor. However, this is not the case—the predictors interact in a very complex manner. This can be proven mathematically using two predictors, z_1 and z_2. In the two predictor case, the explained variance, R^2, is

$$R^2 = s_{\hat{z}_3}^2$$

which in scalar algebra form is

$$R^2 = \frac{1}{N-1} \sum \hat{z}_{3j}^2, \text{ but } \hat{z}_{3j} = b_1 z_{1j} + b_2 z_{2j} \text{ so}$$

$$R^2 = \frac{1}{N-1} \sum (b_1 z_{1j} + b_2 z_{2j})^2$$

$$= \frac{1}{N-1} (b_1^2 \sum z_{1j}^2 + b_2^2 \sum z_{2j}^2 + 2b_1 b_2 z_{1j} z_{2j})$$

$$= b_1^2 s_{z_1}^2 + b_2^2 s_{z_2}^2 + 2b_1 b_2 r_{12}$$

$$= b_1^2 + b_2^2 + 2b_1 b_2 r_{12}.$$

The final form of R^2 emphasizes why any attempt to judge the importance of X_1 and X_2 in estimating X_3 leads to difficulties. Judgment of importance often is on the basis either of b_1 and b_2, or of b_1^2 and b_2^2; but both bases ignore $2b_1 b_2 r_{12}$ that is a function of the amount of correlation between the two predictors. Other attempts to judge the importance of the ith predictor have involved $b_i r_{ip}$, since another way of calculating the multiple correlation coefficient is from

$$R^2 = \sum b_i r_{ip}.$$

When all b_i and r_{ip} are positive scalars they can be helpful; however, this naturally limits their usage in practical problems. On the other hand, when negative, individual $b_i r_{ip}$ can detect a suppressor variable, i.e., a predictor having almost no correlation with the criterion but having a large b_i owing to the predictor being correlated with other predictors. In essence, supressor variables decrease the magnitude of a multiple correlation coefficient.

Predictor-Regression Function Correlations. The importance of predictor variables might be evaluated by correlations between each predictor and the regression function. To calculate these correlations, some introductory relationships are necessary:

Since $\hat{z}_p$ follows the normal distribution with mean equal to zero and variance equal to R^2,

$$(1/R)\hat{z}_p$$

follows the normal distribution with mean equal to zero and standard deviation equal to unity, i.e., standardizes $\hat{z}_p$; but

$$\hat{z}_p = \mathbf{b}'\mathbf{z} \text{ so } \frac{1}{R} \hat{z}_p = \frac{1}{R} \mathbf{b}'\mathbf{z}.$$

Therefore, the correlations between the individual predictors $\mathbf{z}$ and the estimated line of regression $\hat{z}$ can be developed as follows:

$$\mathbf{r}_{zz_p} = \frac{1}{N-1} \sum \mathbf{z} \left(\frac{1}{R} \mathbf{b}'\mathbf{z} \right)$$

$$= \frac{1}{N-1} \sum \mathbf{z}\mathbf{z}'\mathbf{b} \frac{1}{R}$$

$$= \mathbf{R}_{11}\mathbf{b} \frac{1}{R}.$$

However,

$$\mathbf{b} = \mathbf{R}_{11}^{-1}\mathbf{R}_{12}, \text{ so}$$

$$\mathbf{r}_{z\hat{z}_p} = \mathbf{R}_{11}\mathbf{R}_{11}^{-1}\mathbf{R}_{12} \; \frac{1}{R}$$

$$= \mathbf{I}\mathbf{R}_{12} \; \frac{1}{R}$$

$$= \mathbf{R}_{12} \; \frac{1}{R} \; .$$

Therefore, the vector of predictor-regression function correlations, also termed regression structure, is obtained by dividing the predictor-criterion intercorrelations by the multiple correlation coefficient.

Cooley and Lohnes (1971) consider the above correlation coefficients better possible indicators of the relative importance of predictors, partly for the following reasons: The predictor-regression function correlations make little use of the vector of standard partial regression coefficients (**b**-weights). The **b**-weights only influence the scaling value $(1/R)$ which unless R is unity, increases the magnitude of regression structure over the $\mathbf{R}_{12}$ correlations. Since **b** plays only a small role, the structural correlations are much more reliable indicators of the relative importance of predictor variables.

Harris (1975) not surprisingly questions the merits of scaled bivariate correlations, structure, as predictor evaluators. Harris's question is in the framework of the purposes behind rating variables, either deciding to retain a variable in a regression function, or evaluating variable contributions to criterion scores. The former purpose is considered later under stepwise regression. In the latter and present purpose context, Harris aptly refers to the fact that any regression coefficient provides information about a predictor's contribution only within the highly specific context of the other p-1 predictors. Naturally, this is demonstrated in the above formula.

Harris represents the school of thought that prefers **b**-weights. However, one cannot ignore the extensive discussion of Cooley and Lohnes in this matter. They provide good reasons for questioning the merits of **b**-weights. First, **b**-weights show surprising variation from sample to sample taken from the same population. This variation appears to hold in samples approximating 200 individuals. Second, even arbitrarily assigned values of unity to each element of **b**, 65% of the time outperformed calculated elements of **b**. Third, when interpredictor correlations approach unity, **b**-weights are unreliable (Kendall, 1957). These three problems all reflect the fact that regression coefficients are influenced by all errors that exist in estimating the $(p-1)(p-2)/2$ correlation parameters. Therefore, as a final item, it should be no surprise that in a Monte Carlo study estimations from repeated samples of the same population revealed that structure outperformed regression coefficients 75% of the time (Cooley and Lohnes, 1971).

In my opinion the above constitutes most damning evidence against evaluation of predictors. Perhaps other Monte Carlo studies would produce different results. However, the above one implies that structural correlations are better. Since structural correlations are bivariate correlations scaled by a constant, the shortcomings of structure are those of $\mathbf{R}_{12}$.

Problems of Multiple Regression and Correlation.   Many biometricians are hesitant to recommend regression and correlation to biologists. This is the case only partly because investigators are not content with the legitimate use of these statistics. The hesitation of biometricians goes beyond this. The $p^2 + 2p - 1$ statistics (means, variances, covariances, intercorrelations, and regression weights) apply to any regression analysis. Even if one could place reliance in these statistics, their number alone and the fact that they are evaluated both individually and in concert makes interpretation extremely difficult. In either event, the valid application of these statistics is so limited that for practical purposes biologists are almost forced to reject them.

The fact that the correlation coefficient is related mathematically to the regression coefficient,

$$\mathbf{R}_{12} = \mathbf{R}_{11}\mathbf{b},$$

causes limitations of one coefficient to be limitations of the other. However, examples of situations leading to invalid coefficients are better appreciated as correlation coefficient difficulties. In such examples, it is well to recall that correlation coefficients are also z-score covariances. Therefore, whenever a z-score is in part a function of another z-score, covariances represent a most complex function of the covariates and other variables.

Some examples of how complex functions develop are as follows:

Many attributes of organisms are functions of increasing size and growth. For this reason, variable correlations might do no more than display the fact that most body parts increase together in growth. Another way of stating the same thing is that unrelated parts, in the hereditary sense, can display spurious correlations. Such correlations occur whenever two variables share a common multiplicative or additive element owing to an extraneous influence, again often growth.

Time is another agent of invalid correlations among parts. Since time is involved in growth and seasonal changes, either of the latter can bring about nonsense correlations between attributes if sampling extends over some period.

Chance alone can influence the value of a correlation coefficient. This is a problem of any statistical analysis. One must remember that even a sample that satisfies the assumption of being a random sample from a multivariate normal population can represent an extreme of that population. In terms of correlation coefficients, the set may contain misleadingly high and/or low values.

Perhaps an extension of the latter problem is the fact that given many measurements, high nonsense correlations are likely. Examples include the often reported fact that increased alcohol consumption and numbers of ministers is positively correlated. More remarkable is the fact that the correlation between U.S. pigiron production and Great Britain birthrate between 1875 and 1920 is $-.98$. The latter might be just a nonsense correlation but it could be influenced by the next item.

Correlation of two items with a third measured or unmeasured variable can modify correlations between the original two. Such spurious correlations might be joined to the same sources of error as involved in growth, but when associations are a function of density-dependent ecological factors another problem occurs. Let us consider this problem in more detail. A well documented instance involves adult weight, length of larval period, and the larval density in flour beetles as influenced by food supply. When larval density is low, the food supply is adequate, so larval period is positively correlated with adult weight. However, when larval weight decreases, larval period increases, and adult weight decreases. In the latter case larval period is negatively correlated with adult weight. Therefore, in an experiment considering only adult weight and length of larval period, the results might well be meaningless.

Finally is a statistical limitation. For correlations to be valid, they must be based upon linear functions between variables.

For future use, let us imagine the impossible—a set of biological variables in which all correlations between predictors are zero. A whole new marvelous situation exists. Only p means, p variances, and p predictor-criterion correlations need be considered. Since predictors are uncorrelated with one another, any predictor-criterion correlation is interpreted separately and unambiguously. Another added attraction is that regression coefficients equal predictor-criterion intercorrelations. Although

$$\mathbf{b} = \mathbf{R}_{11}^{-1}\mathbf{R}_{12},$$

$$\mathbf{R}_{11}^{-1} = \mathbf{I}^{-1} = \mathbf{I}, \text{ so}$$

$$\mathbf{b} = \mathbf{R}_{12}.$$

Also, the squared multiple correlation coefficient is simplified to the sum of squares of predictor criterion correlations,

$$R^2 = \mathbf{b}'\mathbf{R}_{12} = \mathbf{R}_{12}'\mathbf{R}_{12}.$$

Therefore, the total variance can be seen as a set of independent components of variance, each component being the variance due to one of the predictors. These conditions are analogous to the end products of principal component analysis.

3.3 Stepwise Regression

Another approach to evaluating predictors actually ends in selecting a subset of supposedly "best" predictors and discarding the other predictors.

The most thorough procedure, the exhaustive method, calculates the regression of Y on every possible subset of the k predictors ($k = p - 1$). In this method, each X singly, every pair, every triplet, etc. of predictors is used. Then, the subset with the smallest deviation mean square $(1 - R^2)$ is selected for further regression analysis. The drawback of the exhaustive method is the number of regressions that must be calculated, $2^k - 1$. For ten variables, 1023 calculations, a task to challenge even a computer, must be performed. Understandably the method is rarely used.

The primary substitute for the exhaustive method is stepwise regression. It is the most systematic of the more practical devices for obtaining a subset of the $p - 1$ predictors that supposedly give the best prediction of the criterion. Unfortunately stepwise regression might be the worst possible method for rating predictors. In general application are two methods, the step down and the step up methods.

The step down method starts with the regression of Y on all k X-variables. Then, the contribution of each ith predictor, X_i, to the reduction in the sum of squares of Y after fitting the other $\mathbf{X}$ variables presumably is measured by

$$b_i^2/s_{ii},$$

the square of the regression coefficient of a given predictor divided by the variance of the predictor. Then X_u, the predictor variable having the smallest such value is selected and perhaps omitted from the regression function according to some rule. The rule might be that the calculated value is less than unity or that b_u is not significant at some predetermined level of significance. If X_u is omitted, the same procedure is applied to the remaining $k-1$ predictors until no variable qualifies for deletion.

The step up method starts with the regression of Y on each X_i singly. Then the variable giving the greatest reduction in the sum of squares of Y is designated X_1 and all bivariate regressions involving X_1 and one of the other $k-1$ X-variables are calculated. The other variable giving the greatest reduction in the sum of squares of Y is selected and designated X_2. This procedure is continued until additional contribution is too small to satisfy some rule for inclusion.

The two stepwise methods do not necessarily select the same X-variables and neither method is likely to make the same selection as the exhaustive method. These facts alone should not be reasons for concern since different subsets might give equally good prediction if intercorrelations among predictors are high. However, there are three good reasons for being suspicious of stepwise regression methods. First, rules for including or deleting variables are more arbitrary than statistical. Second, in some cases a rule will include poor predictors and, in others, exclude good predictors. Third and most critical of all, both step up and step down methods are prone to select variables that just happen to predict well in a given sample. When the same method is applied to another sample, previously "good" predictors may become "poor" predictors.

3.4 Multiple Partial Correlation

The discussion of this method of correlation analysis shall be limited to a set of p predictors, z_1, and a set of q criteria, z_2. The purpose is to remove the influence of the z_1 from the z_2. The modified z_2 variables are residuals, or parts of z_2 not explained by z_1. The residuals, also called adjusted criteria, symbolically $\tilde{z}_2$, can provide intercorrelations independent of the predictors. If the predictors are such that they minimize correlation difficulties, the method is very useful. For example, in a given study, if the only sources of invalid correlations are time and growth and predictor variables account for temporal changes and increases in size, the residual correlations might provide valid indications of biological relationships. However, it is not likely that anyone regularly will know the sources of invalid correlations.

The procedure of multiple partial correlation is precisely that to be used later in the multivariate analysis of covariance.

In multiple partial correlation there are the following steps:

1. Transformation of data

   X to z, where $z' = (z_1' \; z_2')$

2. Compute and partition the correlation matrix as indicated before.
3. Form the standardized regression weights

   $$B = R_{11}^{-1}R_{12}.$$

Note that the regression weights form a $p \times q$ matrix B with each column containing the p weights for regressing one of the q elements of z_2 on all the elements of z_1.

4. Calculate the adjusted $\mathbf{z}_2$

   $\hat{\mathbf{z}}_2 = \mathbf{B}'\mathbf{z}_1$ which in algebraic form is
   $\hat{z}_{2ij} = B_{hi}z_{1hj}$ where
   $h = 1, 2, \ldots, p$ predictors
   $i = 1, 2, \ldots, q$ criteria
   $j = 1, 2, \ldots, N$ individuals.

5. Calculate the predicted regression matrix, the variance-covariance matrix, of $\hat{\mathbf{z}}_2$

   $$\hat{\mathbf{R}}_{22} = \mathbf{R}_{21}\mathbf{B} = \mathbf{R}_{21}\mathbf{R}_{11}^{-1}\mathbf{R}_{12}$$

6. Form the variance-covariance matrix of residuals,

   $$\tilde{\mathbf{z}}_2 = \mathbf{z}_2 - \hat{\mathbf{z}}_2,$$

   $$\tilde{\mathbf{R}}_{22} = \mathbf{R}_{22} - \hat{\mathbf{R}}_{22}.$$

7. Finally, form the intercorrelations of $\tilde{\mathbf{z}}_2$, the $q \times q$ multiple partial correlation matrix of all the mth and n^{th} adjusted criterion variables

   $$\mathbf{R}_{2.1} = (r_{2.1_{mn}}) = \frac{\tilde{r}_{22_{mn}}}{\sqrt{\tilde{r}_{22_{mm}}\tilde{r}_{22_{nn}}}}$$

   from which associations of criteria, independent of predictors, are interpreted.

3.5 Multiple Part Correlation

The purpose is to use a third set of variables to adjust only the predictor variables. In this model $\mathbf{z}_1$ is a vector of control variables, $\mathbf{z}_2$ is now a vector of predictors, and $\mathbf{z}_3$ is a vector of criteria. The procedure is to develop the following matrix:

$$\begin{bmatrix} \mathbf{R}_{2.1} & \mathbf{R}_{2.1,3} \\ \mathbf{R}_{3,2.1} & \mathbf{R}_{33} \end{bmatrix}$$

where $\mathbf{R}_{2.1}$ is the same multiple partial correlation matrix as calculated and defined above, $\mathbf{R}_{2.1,3} = \mathbf{R}_{3,2.1}'$ is the correlation between predictors adjusted by controls and the original criteria, and $\mathbf{R}_{33}$ is the original submatrix of intercorrelations among criteria.

Part correlation is applied when it is known that predictors are influenced by some extraneous factor but criteria are not. For example, if size differences warp predictors but not criteria, predictor-criteria intercorrelations would be reduced. Therefore, a set of controls might adjust predictors for size. This leads to the above matrix which is the basis for evaluating relationships between adjusted predictors, between adjusted predictors and criteria, and between criteria.

The computations are as follows:

1. Compute and partition the correlation matrix

$$\begin{bmatrix} \mathbf{R}_{11} & \mathbf{R}_{12} & \mathbf{R}_{13} \\ \mathbf{R}_{21} & \mathbf{R}_{22} & \mathbf{R}_{23} \\ \mathbf{R}_{31} & \mathbf{R}_{32} & \mathbf{R}_{33} \end{bmatrix}$$

2. Use the same partial correlation procedure shown in section 3.4 to obtain the multiple partial correlation matrix $\mathbf{R}_{2.1}$.

3. Compute the covariance of adjusted residuals and unadjusted criteria

$$\tilde{\mathbf{R}}_{23} = \mathbf{R}_{23} - \hat{\mathbf{R}}_{23} = \mathbf{R}_{23} - \mathbf{B}'\mathbf{R}_{13} = \mathbf{R}_{23} - \mathbf{R}_{21}\mathbf{R}_{11}^{-1}\mathbf{R}_{13}$$

4. Scale the covariance $\mathbf{R}_{23}$ into a correlation matrix of the hth adjusted predictor and i^{th} criterion

$$\mathbf{R}_{2.1,3} = \frac{\tilde{r}_{23_{hi}}}{\sqrt{\tilde{r}_{22_{hh}}}}$$

and since $\mathbf{R}_{3.2.1} = \mathbf{R}_{2.1,3}{}'$, all elements of the required matrix for interpreting associations are now provided.

3.6 Critique

Regression and correlation are very popular tools for analyzing data. I believe this is true for three reasons. First, the methods are excellent when little is known about a particular phenomenon. The methods provide relationships that can become focal points for further study. Second, the limitations of these statistics frequently either are not considered, or are minimized in texts and courses in applied statistics. Why, I do not know. The limitations are not minimized in Snedecor and Cochran (1967). Third, both statistics approximate human judgment and are allied to Aristotelean logic. For this reason, the statistics most definitely can lead to correct conclusions. In fact, the overwhelming majority of sound scientific knowledge stems from Aristotelean logic and frequently the logic is aided by regression and/or correlation. On the other hand, much knowledge has been gained only after denial or reevaluation of previous concepts. In fact, the history of biology would indicate that denial or reevaluation is the rule rather than the exception of developing principles. For this reason, as knowledge proceeds beyond the exploratory phase, I believe that methodology for evaluating data must change.

If you are convinced of the limitations of regression and correlation, there is no need for discouragement. In a general sense, when used inappropriately the methods examine variation and relationships between variables from a sample of a population. For such purposes Principal Component Analysis is suggested as a substitute primarily on the bases of validity of the method and more comprehensive results. The most exhaustive regression and correlation analysis for the purpose of examining variation does not match the potential outcome of a principal component analysis of the same data.

Chapter 4
Principal Component Analysis

Principal Component Analysis, frequently abbreviated PCA, is a unique method. Part of the uniqueness arises from the fact that it analyzes data that comprise a single sample of individuals from a single population and each individual is represented by a set of $p > 1$ measurements. Also, in component analysis, the original variables are transformed to variables that have zero intercorrelations. The transformation rotates the original axes but maintains the original relationship among data points. The new axes define independent patterns of variation (frequently recognized as size or shape of variables) that should typify the population being sampled. Univariate analysis of a like sample would lead to a sample mean and variance and to estimation of the population mean and variance. Using the sample variance as a measure of variation would hardly compare with the results of PCA. Even extension of the univariate case to simple or multiple regression within this framework is not much better.

This chapter is subdivided into five sections. The first section, the methods of principal axes examines basic methodology in some detail. Much of the material was presented previously; however, often from a different point of view. The second section, interpretation of principal components, introduces some kinds of interpretations. The third section, principal component analysis, an example, provides basic statistics for a new problem. The fourth section, interpretive aids of component analysis, examines a variety of statistical methods for evaluating a component analysis and considers other kinds of interpretation of results. The fifth section, critique of component analysis, evaluates the methodology when assumptions are approximated.

A final point of clarification. PCA can be performed upon the dispersion of variables as is emphasized throughout this chapter. This is termed an R-technique. On the other hand, a similar method, principal coordinate analysis, can operate on individuals, a so-called Q-technique. For example, the Q-technique of principal coordinate analysis might also involve a correlation matrix, but one of correlation between individuals rather than between variables. A Q-type analysis can also be applied to PCA.

4.1 Methods of Principal Axes

Two statistical models and their limitations summarize the properties and features of PCA. One model is for studying absolute variation among variables and the other model, standardized variation.

Absolute and standardized variation are used to indicate the dispersion matrix from which the components were extracted. The variance-covariance matrix is said to measure absolute variation, implying that this dispersion measures original variable variation and reflects their scales of measurement. The correlation matrix provides standardized variation in the sense of also being z-score dispersion with all variances of unity and covariances reflecting equality of variances. More precisely, the correlation matrix is independent of scales of measurement.

The model for studying absolute variation is

$$\mathbf{y} = \mathbf{A}'(\mathbf{X} - \overline{\mathbf{X}}) = \mathbf{A}'\mathbf{x}.$$

When p variables are involved, $\mathbf{y}$ is a vector of p component scores for an individual, $\mathbf{A}$ is a matrix of p eigenvectors, each $\mathbf{X}$ is a data vector for an individual, $\overline{\mathbf{X}}$ is the sample centroid, and each $\mathbf{x} = \mathbf{X} - \overline{\mathbf{X}}$ is a vector of deviations of a data vector from the sample centroid. The use of deviation scores also is a matter of convenience—both raw data and component scores are centered, i.e., have a mean of zero. Although this convention will be maintained, later it will be shown that centering might be unwise. Therefore, in the present context, the mean of all $\mathbf{y}$ is

$$\overline{\mathbf{y}} = \mathbf{0}$$

and the dispersion of $\mathbf{y}$ is

$$\mathbf{S}_y^2 = \mathbf{A}'\mathbf{S}^2\mathbf{A} = \mathbf{L},$$

where $\mathbf{S}^2$ is the raw data dispersion matrix. The centroid of $\mathbf{y}$ is a null vector since the effect of uniform subtraction of a constant from each individual ($-\overline{\mathbf{X}}$) is to subtract that constant from the mean. Also, recall from univariate statistics that subtraction of a constant does not change the variance. However, the dispersion of $\mathbf{y}$, $\mathbf{S}_y^2$ is $\mathbf{A}'\mathbf{S}^2\mathbf{A}$ since the effect of uniform multiplication by a constant ($\mathbf{A}'$) on the dispersion $\mathbf{S}^2$ is to multiply the dispersion by the square of the constant. Finally, as shown previously, $\mathbf{S}_y^2 = \mathbf{L}$ is a diagonal matrix of eigenvalues.

For standardized variation, the model is

$$\mathbf{y} = \mathbf{V}'\mathbf{z}$$

where each $\mathbf{y}$ is an individual's vector of p component scores (but different from the $\mathbf{y}$ for absolute variation), $\mathbf{V}$ is a matrix of p eigenvectors, and each $\mathbf{z}$ is a vector of an individual's z-scores. Since $\overline{\mathbf{z}} = \mathbf{0}$ and from the consequences of multiplying by a constant

$$\overline{\mathbf{y}} = \mathbf{0} \text{ and}$$

$$\mathbf{S}_y^2 = \mathbf{V}'\mathbf{R}\mathbf{V} = \mathbf{L}.$$

The calculation of $\mathbf{S}_y^2$ involves the correlation matrix $\mathbf{R}$ which is the dispersion matrix for z-scores. Again, $\mathbf{L}$ is a diagonal matrix of eigenvalues, but not one equal to that for absolute variation.

Further characteristics of the above models come from the characteristic equation and eigenvector function. To appreciate these, consider the two models as follows:

$$\mathbf{y}_i = \mathbf{a}_i'\mathbf{x} \text{ and } \mathbf{y}_i = \mathbf{v}_i'\mathbf{z} \ (i = 1, 2, \ldots, p),$$

where all $\mathbf{y}_i$ are p component scores and $\mathbf{a}_i$ or $\mathbf{v}_i$ are p eigenvectors of the different models. Then the models imply that the first $\mathbf{y}_i$, $\mathbf{y}_1$, is chosen on the basis of having the largest possible variance which is the largest eigenvalue derived from the characteristic equation. The first eigenvector, $\mathbf{a}_1$ or $\mathbf{v}_1$, locates the y_1 component axis where the maximum variance ($=$ distance) among individuals occurs. It will be shown later that this first component axis is defined by an eigenvector. Then, the second y_i, y_2, is chosen to have the maximum possible variance with the limitation of being uncorrelated with y_1. Again, the y_2 axis is defined by an eigenvector, one defining an axis orthogonal to the first axis. The remaining y_i are extracted so each has the maximum variance possible when all eigenvectors are orthogonal and sets of component scores have zero intercorrelation. The final result is a set of orthogonal axes characterized by successive maximum variances and zero correlations between transformed variables, component scores.

In more complete form, the above models would indicate the orthogonality of component axes, i.e.,

$$\mathbf{a}_i{}'\mathbf{a}_j = 0 \text{ or } \mathbf{v}_i{}'\mathbf{v}_j = 0 \ (i \neq j), \text{ but}$$

$$\mathbf{a}_i{}'\mathbf{a}_i = 1 \text{ or } \mathbf{v}_i{}'\mathbf{v}_i = 1.$$

The values zero and unity are cosines of the angles between pairs of axes. In the case where different axes are involved ($i \neq j$), the cosine of the angle between the axis is zero indicating an angle of 90°. In the case where the same axis is involved, naturally the cosine must be equal to one to indicate an angle of 0°.

Visualization of the nature of the transformation from original data to component scores might be helpful. Consider the bivariate model for absolute variation. The first step is to move the reference axes from those of the raw data ($\mathbf{X}$) to those of deviation scores ($\mathbf{x} = \mathbf{X} - \overline{\mathbf{X}}$). This is done as a means of convenience for the interpretive stage and is shown from figures 4.1a to 4.1b. The final step, figures 4.1b to 4.1c, transforms from orthogonal deviation score axes to orthogonal component axes. Note that only axes and not individual data points were rotated.

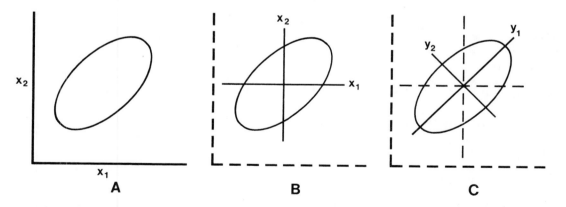

Figure 4.1 Bivariate example of transformation of raw data variable axes to component axes: A. raw data axes; B. change of original Cartesian coordinate axes to deviation axes (the origin of the x_1 and x_2 axes is at the means of x_1 and x_2); C. transformation from deviation axes to component score axis—there is zero correlation between y_1 and y_2.

Although orthogonal axes may or may not relate to correlation between variables, the lack of correlation between variables on component axes is proven by the fact that the component dispersion matrix L has zero covariances. However, it might be helpful to compare component axes with regression. The change from deviation axes to component axes is analogous to rotating x_1 to the line of regression of x_2 on x_1 while allowing x_2 to rotate and maintain orthogonality with x_1, i.e., x_1 becomes y_1 the line of regression and x_2 becomes y_2 perpendicular to the line of regression. Since y_1 is the line of regression, the slope of the line of regression of y_1 on y_2 is zero, so the correlation between variables must be zero. It must be pointed out that the relationship to regression indicates only an analogy. In actuality, no component axis is likely to correspond to a line of regression. Recall that there are $p-1$ regression and p component coefficients.

An important condition relative to method and models is that component axes approximate the principal axes of the *mnd*. This has the advantage of allowing parametric tests of hypotheses. Also, the assumption allows heuristic treatment of stimuli that explain a percentage of the total variation of the variables. In fact, without multivariate normality, components are no more than orthogonal lines which successively provide the closest fit to individuals. This means that no matter what the distribution of individuals, the first component axis passes as close as is possible to all individuals in the sample, i.e., the sum of squares of deviations of individuals from the axis is a minimum. Then, each succeeding axis satisfies the same criterion of fit, subject to the limitation that it is orthogonal to all preceding axes.

The Geometric Proof.   We have seen (Figure 4.1) the nature of the transformation to principal axes. Using the same bivariate case, we will now examine how the transformation is accomplished. Although all that follows refers to the model for absolute variation, it also holds for the model for standardized variation.

Figure 4.2 summarizes the transformation shown in the three diagrams of figure 4.1. The original raw data axes are X_1 and X_2, deviation score axes are x_1 and x_2, and component axes are y_1 and y_2. Transformation to component axes is done by developing coordinates for each individual. To project an individual P from the x-axes to y-axes requires no more than plane geometry and trigonometry. The projection is presented leaving verification to the reader.

The y_1 coordinate of P is y_{1P} which is composed of two projected parts, OA and AB. The line OA is the projection of x_{1P} (x_{1P}cosine α) and the line AB is the projection of x_{2P} (x_{2P}sine α). To generalize for bivariate raw data, any point (x_{1i}, x_{2i}) is projected to the first component axis y_1 by

$$y_{1i} = (X_{1i} - \overline{X}_1)\text{cosine } \alpha + (X_{2i} - \overline{X}_2)\text{sine } \alpha$$

By like reasoning, the y_2 coordinate is

$$y_{2i} = -(X_{1i} - \overline{X}_1)\text{sine } \alpha + (X_{2i} - \overline{X}_2)\text{cosine } \alpha.$$

Let us examine what was done. The angle α is the amount that X_1 or x_1 must be rotated to y_1, the longest or major axis of the bivariate normal distribution. Since y_2 is the last component axis it is forced to assume the only orthogonal position left, the smaller or minor axis of the bivariate normal distribution.

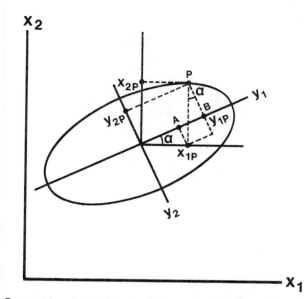

Figure 4.2 Geometric relationships in the transformation from raw score, to deviation score, and to component axes.

Note that the position of individuals would not change at all, only their reference axes were rotated. Another point, the transformation was by cosines and sines of the angle alpha. Since all sines are equal to cosines of $90° - \alpha$ and $90° - \alpha$ is the angle to the second component axis, the transformation can be expressed entirely in terms of cosines. For this reason, the component coefficients frequently are called *direction cosines*.

This proof is general—it applies to p, any number of reference axes.

The Matrix Algebra Approach. Matrix algebra applied to component vectors derived from the above bivariate example verifies that component axes are orthogonal and transformed variables, y, are uncorrelated. The basic linear equation is

$$\mathbf{y} = \mathbf{A}'(\mathbf{X} - \overline{\mathbf{X}}), \text{ where}$$

$$\mathbf{y} = \begin{bmatrix} y_1 \\ y_2 \end{bmatrix}, \quad \mathbf{X} - \overline{\mathbf{X}} = \begin{bmatrix} X_1 - \overline{X}_1 \\ X_2 - \overline{X}_2 \end{bmatrix} = \begin{bmatrix} x_1 \\ x_2 \end{bmatrix} = \mathbf{x}, \text{ and}$$

$$\mathbf{A} = \begin{bmatrix} \cos\alpha & \sin\alpha \\ -\sin\alpha & \cos\alpha \end{bmatrix}$$

The proof that the vectors of **A** are orthogonal is

$$\mathbf{AA}' = \begin{bmatrix} \cos\alpha & \sin\alpha \\ -\sin\alpha & \cos\alpha \end{bmatrix} \begin{bmatrix} \cos\alpha & -\sin\alpha \\ \sin\alpha & \cos\alpha \end{bmatrix}$$

$$= \begin{bmatrix} \cos^2\alpha + \sin^2\alpha & -\cos\alpha\,\sin\alpha + \sin\alpha\,\cos\alpha \\ -\sin\alpha\,\cos\alpha + \cos\alpha\,\sin\alpha & \sin^2\alpha + \cos^2\alpha \end{bmatrix}$$

$$= \begin{bmatrix} 1 & 0 \\ 0 & 1 \end{bmatrix}$$

Since the solution is an identity matrix,

$$\mathbf{AA}' = \mathbf{I},$$

it is possible to show a very useful property of **A**, **A**′ is equal to its inverse. Since any

$\mathbf{A}^{-1}\mathbf{A} = \mathbf{I}$, from the above

$\mathbf{A}^{-1}\mathbf{AA}' = \mathbf{A}^{-1}\mathbf{I}$, so

$\mathbf{A}' = \mathbf{A}^{-1}$.

Although trigonometry allows visualization of the bivariate case, trigonometry loses its advantages when hyperspace is involved. For this reason, further mathematical treatment is restricted to matrix algebra symbols. Such application of matrix algebra now is limited to the bivariate case, but it can be generalized to any number of variables. In algebraic symbolism

$$A = \begin{bmatrix} a_{11} & a_{12} \\ a_{21} & a_{22} \end{bmatrix}$$

and to satisfy the condition of being an orthogonal matrix,

$\mathbf{AA}' = \mathbf{I}$, where

$$= \begin{bmatrix} a_{11} & a_{12} \\ a_{21} & a_{22} \end{bmatrix} \begin{bmatrix} a_{11} & a_{21} \\ a_{12} & a_{22} \end{bmatrix}$$

$$= \begin{bmatrix} a_{11}^2 + a_{12}^2 & a_{11}a_{21} + a_{12}a_{22} \\ a_{21}a_{11} + a_{22}a_{12} & a_{21}^2 + a_{22}^2 \end{bmatrix};$$

but since $\mathbf{AA'} = \mathbf{I}$

$$a_{11}^2 + a_{12}^2 = 1$$

$$a_{21}^2 + a_{22}^2 = 1$$

$$a_{11}a_{21} + a_{12}a_{22} = 0$$

$$a_{21}a_{11} + a_{22}a_{12} = 0.$$

From the latter two equations, one of the following must be true:

$$a_{21} = -a_{12} \text{ and } a_{22} = a_{11}$$
$$\text{or}$$
$$a_{21} = a_{12} \text{ and } a_{22} = -a_{11}.$$

Orthogonality by itself does not determine which solution to use—both provide orthogonal results. The decision is made according to which solution maximizes the variance of the component. In the example, the solution is $a_{21} = a_{12}$, since we rotated through the corresponding angle, α, to cause $\mathbf{y}_1$ to have the maximum variance. If we had rotated through the angle $90° - \alpha$, $\mathbf{y}_1$ would have the smaller variance. In matrix algebra, successive variances are maximized by solving the characteristic equation.

Characteristic Equation.

Recall that the characteristic equation, or eigenvalue equation, transforms the variance-covariance matrix of $\mathbf{X}$ to $\mathbf{L}$, a diagonal matrix. Also, since all covariances in $\mathbf{L}$ are zero, all intercorrelations among transformed $\mathbf{y}$ variables are zero.

$$\mathbf{S}_y^2 = \mathbf{A'S}^2\mathbf{A} = \mathbf{L}$$

and in the bivariate case

$$\mathbf{L} = \begin{bmatrix} \lambda_1 & 0 \\ 0 & \lambda_2 \end{bmatrix}.$$

Such simplification of relationships, a symmetric variance-covariance matrix is transformed to a diagonal matrix of eigenvalues, is termed reduction to Jacobi canonical form and $\mathbf{L}$ is a Jacobi canonical matrix.

The canonical form and matrix are derived from the eigenvalues of the characteristic equation

$$\mathbf{S}^2 - \lambda_i\mathbf{I} = 0$$

which is used to define characteristic vectors, eigenvectors, in the eigenvector equation

$$(\mathbf{S}^2 - \lambda_i\mathbf{I})\mathbf{a}_i = 0$$

subject to $\mathbf{a}_i'\mathbf{a}_i = 1$. Again, although the example pertains to the bivariate case, the solutions are general, applying to any value of p. The two equations provide p eigenvalues and p sets of eigenvectors.

It is easy to prove that any λ_i is the variance of the corresponding $\mathbf{y}_i$, $\mathbf{a}_i'\mathbf{S}^2\mathbf{a}_i$. The characteristic equation is used to solve the determinant of

$$|\mathbf{S}^2 - \lambda_i\mathbf{I}| = 0.$$

If we multiply the above equation by the corresponding eigenvector $\mathbf{a}_i$ and omit the unnecessary $\mathbf{I}$ and determinant,

$$\mathbf{S}^2\mathbf{a}_i - \mathbf{a}_i\lambda_i = 0, \text{ so}$$

$$\mathbf{S}^2\mathbf{a}_i = \mathbf{a}_i\lambda_i.$$

Then, by multiplying each equation by $\mathbf{a}_i'$, the result is

$$\mathbf{a}_i'\mathbf{S}^2\mathbf{a}_i = \mathbf{a}_i'\mathbf{a}_i\lambda_i.$$

However, since

$$\mathbf{a}_i'\mathbf{a}_i = 1,$$

the equation simplifies to

$$\mathbf{a}_i'\mathbf{S}^2\mathbf{a}_i = \lambda_i$$

which proves that the variance of $\mathbf{y}_i$ is equal to λ_i.

4.2 Interpretation of Principal Components

Various mathematical and biological considerations are pertinent to drawing conclusions from any PCA. These items are interrelated and must be incorporated collectively. Perhaps somewhat arbitrarily, items are separated into mathematical considerations, biological interpretation, principal components of a correlation matrix, principal components of the covariance matrix of logarithms, size and shape components, growth, sampling and data, and derived variables.

This section is not the only one that considers interpretation of results. It does form a unit in that interpretation is closely associated with the assumptions of PCA. The same association with interpretation shall be extended in section 4.4; however, section 4.4 introduces another area of interpretation, one involving ordination and less allied to assumptions.

Mathematical Considerations.   The mathematical model is extremely important in the evaluation of principal components. For most purposes, the assumption that each data vector is composed of random variables, variates, is sufficient for meaningful conclusions. In other words, each variable assumes values of a specific set with a specific relative frequency, or probability. In general, the values of the set must be continuous rather than discontinuous over their range. This would lead to a multivariate distribution. For most purposes, other than hypothesis testing, it generally is sufficient to assume that the random variables approximate any linear multivariate distribution that can be defined by a finite centroid and a finite covariance matrix. Therefore, the assumption of random variables would appear quite reasonable.

On the other hand, the assumption of a *mnd* is simplified in terms of such statistics as centroids. Any random vector arising as a sum or average of many independently and identically distributed random vectors will approximate the *mnd*. This assumption would appear realistic for biological data, i.e., centroids derived from samples of data vectors would tend to approximate a *mnd* (Morrison, 1967). From a practical point of view this means that the assumption of a *mnd* is conservative for quantitative variables, discrete or continuous.

Interpretation of components can be limited to the aforementioned properties of the analysis:

1. The components scores of a set are uncorrelated with one another so each component can be interpreted individually.
2. The components are ordered in terms of the magnitude of their variances, the ith principal component having the ith largest variance. In conjunction with the former property, successive decreasing variation can be examined.
3. Components partition the total variance into p additive fractions that sum to the total variance, i.e., the ith component accounts for an additive portion of the variation of the original variables.
4. The first component is that linear combination which best discriminates (produces the maximum distance) between individuals of a sample. This can be applied to graphing component scores (including their ordination) and to examining eigenvector coefficients to disclose the interplay of variables involved.
5. The first through r^{th} principal component defines a subspace that provides the best r-space representation of the data. This best r-space is important in terms of extending and summarizing discrimination among individuals, by ordination; unravelling the participation of variables in discrimination; and using a few components to summarize information provided by the data.

The nature of the interplay of variables regularly is accomplished by "reading" each eigenvector in a physical sense. Positive values for coefficients are interpreted as increases in magnitude of the corresponding variables and negative coefficients as decreases. Use of the magnitude of coefficients can be extended to relate the amount of participation of each variable in a component. This can be extended to reading components in the original units of measurement. For example, if original units were millimeters, for convenience, a components coefficients can be multiplied by 100 and then the coefficients read as millimeters. Again, the sign of each coefficient would indicate positive or negative participation of the corresponding variable.

The collective signs of coefficients of any component also are important. When all signs are the same, the implication is that all variables are increasing together (positive signs) or decreasing together. Such a component is termed a general component. When signs are mixed, the eigenvector is a bipolar component.

The final mathematical consideration is a matter of definition. In the sense that a component is an interplay of variables that is related to an independent fraction of the total variance, it seems appropriate to follow Blackith's (1965) interpretation of a component as an independent pattern of variation of the variables. The latter is in accord with another mathematical property of principal component analysis. Just as component scores for each component can be derived from a data vector and the eigenvectors, a data vector's scores can be derived from the data vector's component scores and the eigenvectors.

Original Variables and Component Variables.   It is very important that the nature of the original data vectors of p measurements and the derived vectors of p component scores is known. Each element of a data vector, of course, represents a familiar single measurement. On the other hand, each element of a vector of component scores represents a fraction of each of the original measurements. In other words, the ith component score is a variable of size or of shape (a pattern of variation) and thus not a variable of a single measurement. Therefore, the identification of each size or shape factor, one from each component, can be considered as a statement of the nature of a component variable.

Biological Interpretation.   PCA is very different from methodology where biological hypotheses lead to mathematical models. In the latter, the model is constructed, data are collected, and the model is tested. In component analysis, in a sense, a single model exists, data are collected, are analyzed, and finally biological hypotheses are formulated. There is abundant evidence that biological interpretation of a PCA is possible.

For further considerations, it will be convenient to consider each component to be the response of a causal stimulus. A stimulus can be considered to be a genetic or environmental feature. The total response, the component, could be subdivided into individual variable responses that are measured by the coefficient corresponding to each variable. Such a visualization of methodology is consistent with theory (Morrison, 1967) and is discussed at length by Pearce (1965b).

In a sense biological models are easier to satisfy than components are to interpret. This can be appreciated in reference to causal stimuli and responses (components). For a single biological feature to be defined as a single component, the variables necessary to define the feature must be included and the feature must be independent of all others defined by the total set of variables. This single feature response criterion is not as simple to satisfy as it might seem—biologists regularly display the ability to separate features or processes that are not independent of one another, i.e., features that would be intermixed in the response of a single component. If this is expected, the biologist might unravel the situation. In fact, it would be useful to demonstrate the association of the features.

Unfortunately components can pertain to much more complex responses. If two processes have independent stimuli and responses, but one process also influences the other, the resulting components can be very confusing. For example, if a plant's growth stimulus is independent of that for habit, but the nature of the plant's growth influences habit, a growth-partial habit and a partial habit component might result. Since even greater complexity seems likely one should never expect to identify every component in pure biological terms. As another complication, a component can be the response of a causal stimulus of error in measurement.

Since components regularly are difficult to define in biological terms and such definitions too often amount to intelligent guesses, the conservative approach might be preferred. However, if each component is an independent response resulting from a causal stimulus, components are very attractive. In this respect, one can appreciate Tukey's (1962) recommendation for one being willing to err moderately in order that inadequate evidence might suggest the right answer. Therefore, procedures that might help unravel biological features will be developed. Naturally, one should not expect that any set of procedures will always or even usually lead to recognition of biological features. Also, even though the procedures as applied to the examples in this book eventually might seem to lead to reasonable conclusions, the conclusions are hypotheses that require further proof. In my opinion, such hypotheses are consistent with the goals of most biological research.

Components of a Correlation Matrix.   PCA frequently is applied to a correlation matrix in the sense of **R** being the dispersion of z-scores. This model is very attractive when variables involve different scales of measurement or are mixed mode. The result of the z-score model is to cause all variables to be scaled equally, in standard deviation units.

The correlation model requires examination. Principal components derived from a correlation matrix can differ a great deal from those derived from a covariance matrix. This difference is pertinent to tests of significance of eigenvalues. Tests differ in the two models. In fact, there are distinctly fewer tests of significance applicable to a component analysis of a correlation matrix. Since the tests for the latter are not very useful in practice, they are not stressed here.

The above correlation model restrictions might be surprising since many transformations of data are allowed. However, there is a distinct difference between the transformation of single variables to logarithms, square roots, arcsines, reciprocals, etc. and the transformation to z-scores. In a transformation to z-scores three variables are involved, a single variable plus its sample mean and sample standard deviation. The latter leads to an inconsistency in the z-scores of original variables. Since any z-score is dependent upon the value of a mean, a standard deviation, and an observed X, from sample to sample a given value of X generally will produce a different z-score value. Naturally this is not true for conventional transformations, e.g., a given value of X always will produce the same common logarithm value.

In spite of z-scores causing all variables to have equal standard deviations of unity, standardized variation, a plot of z-scores would generally lead to an ellipsoid rather than a spheroid of points representing individuals. Only zero correlations between all pairs of z-scores would lead to a spheroid. Also, the derived components from ellipsoids will reflect the magnitude of bivariate correlations. For example, equal coefficients in a general component would reflect equality of bivariate correlations, and might be interpreted as equality of standardized growth rates among variables.

Components of the Covariance Matrix of Logarithms.   Although principal components of raw variables that express linear dimensions are not likely to produce departures from linearity or multivariate normality that will seriously influence applying such assumptions, logarithms are useful. Both linearity and multivariate normality often are more closely approximated by logarithms than by the original variables. The convention is to use common logarithms.

There are additional advantages or features of logarithms. Since logarithms reduce differences between standard deviations, the resulting components tend to display the features of those derived from a correlation matrix, namely independence from scale and magnitude of original variables. Another feature of logarithms is that the p original variables are considered as a product, $(\Pi \mathbf{X}_i)^{1/p}$— recall the affect of a logarithmic transformation. Therefore, components of a logarithmic covariance matrix relate to possible proportional relationships between variables, the equivalence of ratio or index relationships, but without a priori assumptions as to such ratios or indices. Further utilization of this will be considered in reference to size and shape components and to growth. Finally, since logarithms of variables approximate multivariate normality, the distribution of logarithms of proportions may be expected to approximate normality.

Size and Shape Components.   Jolicoeur and Mosimann (1960) proposed that general components could be interpreted as size components and bipolar components as shape components. Mosimann

(1970) discusses this matter at some length. Although a general component can be treated as a size factor, in a more refined sense, components should not automatically be defined as size or shape factors since size might not be independent of shape. In this sense, size independent of shape is defined as *isometry* and would be indicated by equal coefficients in a general component. Size dependent upon shape would be *allometry*.

The matter is somewhat more complicated—the definition of a size variable is important. Mosimann (1970) demonstrates that from the positive variables $(X_1, X_2, \ldots, X_p)$ possible definitions of size variables are $\mathbf{X}_i$, $\sum \mathbf{X}_i$, or $(\Pi \mathbf{X}_i)^{1/p}$. In reference to a component analysis, $\sum \mathbf{X}_i$ would pertain to a covariance general component and $(\Pi \mathbf{X}_i)^{1/p}$ to a covariance of logarithms general component. Unfortunately the logarithmic general component might display equal coefficients (isometry, size independent of shape) when the covariance general component does not. In essence, when isometry is demonstrated in terms of one size variable, generally it will not be demonstrated for others. Jolicoeur's (1963b) choice of $(\Pi \mathbf{X}_i)^{1/p}$ has favorable statistical properties and is preferred in many biological situations. In applying this size variable, measurements should be linear and utilize the same scale of measurement. Since Jolicoeur's size variable can be discovered only by a logarithmic transformation of data, the aforementioned properties of such transformations are pertinent. Further treatment of size and shape leads to studies of growth.

Growth.   Huxley (1932) gave some attention to the study of relative growth in which he applied the allometry equation to two variables, X and y,

$$y = aX^b.$$

In practice, the logarithmic transformation of the allometry equation has been used since it amounts to a simple linear relationship,

$$\log y = b \, \log X + \log a,$$

the model for simple linear regression of logarithms. Jolicoeur (1963a, 1963b) has generalized the allometry equation to the multivariate case. The methodology is dependent upon the size component calculated from the covariance matrix of logarithms of original data, $\mathbf{X}$. The model of multivariate allometry is

$$\left(\frac{X_i}{G_i} \right)^{\frac{1}{\cos \theta_i}} = \left(\frac{X_j}{G_j} \right)^{\frac{1}{\cos \theta_j}},$$

where X_i and X_j are arithmetic values for the ith and j^{th} values of the p variates, G_i and G_j are the geometric means of X_i and X_j (antilogs of the mean values for $\log X_i$ and $\log X_j$ respectively), and $\cos \theta_i$ and $\cos \theta_j$ are the direction cosines of the ith and j^{th} coefficients of the size component derived from the covariance matrix of logarithms. This leads to a generalization of the allometry equation

$$X_i = \frac{G_i}{G_j^{\alpha_{ij}}} X_j^{\alpha_{ij}},$$

where $a_{ij} = \cos \theta_i / \cos \theta_j$ is the allometry exponent of the ith variate relative to the jth variate.

This is all that one needs to consider relationships between the proportions of the ith and j^{th} variates. If the proportion is constant, i.e., relative growth of the two dimensions is *isometric*, $\alpha_{ij} = 1$, so

$$cos\ \theta_i = cos\ \theta_j.$$

If the ith variate increases at a greater rate than the jth, i.e., *positive allometry* occurs in the ith relative to the jth variate, $\alpha_{ij} > 1$, so

$$cos\ \theta_i > cos\ \theta_j.$$

If the ith variate increases at a lesser rate than the jth, *negative allometry*, $\alpha_{ij} < 1$, so

$$cos\ \theta_i < cos\ \theta_j. \ .$$

The test of isometry can be extended to all p variables. One can test if the proportion of all variables remains constant as size increases. In this situation all possible allometry exponents, α_{ij}, should have unit value, i.e., all direction cosines of the size component from logarithms must be equal, the single value being

$$1/\sqrt{p}.$$

A test that can be applied to this hypothesis will soon be presented.

In addition to testing the relative growth between two dimensions by comparing their direction cosines, one can consider the hypothesis that the ith variate per se exhibits isometric growth by

$$cos\ \theta_i \approx 1/\sqrt{p}.$$

In a like sense, positive allometry would be expressed by larger direction cosine values and negative allometry by smaller values.

Sampling and Data.
The nature of sampling and data are discussed here since either can lead to erroneous conclusions.

Technically, each sample must be a random sample from a clearly defined population. Biological samples do not and most likely cannot conform to the definition of a random sample. Frequently the best sampling possible is accomplished by collecting individuals as encountered or trapped and by displaying no prejudice in what is taken. The only basis for the rejection of individuals must result from an a priori definition of the population being sampled. In no sense can one assume that every individual in the population being sampled had an equal opportunity of being represented in the sample. However, since biological studies consistently tend to be repeatable, it is assumed that conventional collecting techniques do produce samples not significantly different from random. This has frequently been shown to be the case and has become the basis for justifying such collecting. In fact, biologists must rely on such justification. Truly random sampling generally is not possible in nature.

The purpose of any morphometric study is to determine the form of a phenomenon. Both the number and nature of the measurements chosen to do this must be a primary consideration. Naturally the solution is dependent upon the particular study.

Another consideration is the nature of the measurements. Preferable are single dimensional, straight-line, continuous measurements. If multidimensional measurements are used, the data should be transformed to be sure that the linear model of PCA is approximated. For example, weights, circumferences, areas, and volumes are multidimensional variables. Most measurements of the latter type are made linear if the data are transformed to logarithms. In many instances even the former measures contribute to nonlinear relationships and thus might require transformation to logarithms.

If there is any doubt about one or more variables being linear, one can use multiple regression to test their linearity.

Derived Variables.   The main kinds of derived variables are indices and ratios. An index is any derived variable but usually consists of the average of two measurements, a weighted average of two measurements, or one measurement divided by another. The latter index, generally expressed as a percent, is a ratio.

Derived variables have various disadvantages (Pearce, 1965a; Atchley et al, 1976). Since errors of measurement are compounded, indices are less precise than single measurement variables. Also, indices' distributions often are unusual, perhaps far from normal. Most important is the fact that indices can obscure relationships between the component variables.

The purpose of derived variables often is to reduce variation owing to size differences or to provide an indication of some inherent ability. For example, the proportion of leg length/body length might be used as an index of running ability. Such a ratio is subject to various errors. If body length increases 1 percent, body weight increases by about 3 percent. Even proposing an index to correct for this does not solve potential problems. If different sized animals have different shapes, the different body weights do not increase uniformly three times as fast as length. In addition, since body specific gravity likely will change during growth, body weight has another source of variation. Also, since body size likely will influence running ability, any ratio adjusted for the previous problems, still will not accomplish the purpose of measuring running ability.

The point is not that individual measurement variables subject to PCA will automatically disclose such things as running ability. They well might not. However, any multivariate analysis provides a concise, objective disclosure of relationships among variables. When single measurements are the variables, conclusions are independent of any preconceived status of the variables. The analysis, in disclosing how variables are interrelated, can lead to disclosure of stimuli or effects underlying the relationships.

Another appraisal of the situation is provided by Corruccini (1977) in a rebuttal of Atchley et al (1976). Atchley et al utilized computer generated data to demonstrate the poor performance of pairs of variables, x and y, as ratios, x/y. Corruccini modified the ratio into what technically would be an index based upon regression. Since Corruccini's index is not the ratio criticized by Atchley et al, the poor performance of ratios is not discounted.

Corruccini's index is based upon Mosimann's (1970) implication of the actual linear relationship between two variables, the simple linear regression function $y = a + bx + e$, where e is an error term pertaining to a given y. This function easily translates into more familiar form, since the estimated point on the line of regression in reference to a given x is $\hat{y} = y - e$. Ignoring e as a trivial element in the function, Corruccini derived $((y-a)/x \approx b)$, an index that approximates a constant for any x and y whose function is linear. On the other hand, the ratio $y/x \approx a/x + b$, so y/x is correlated with x.

The purpose of Corruccini's index, as was that for many ratios, is to remove the size as a factor in comparisons between variables. In Corruccini's index, x (termed a standard size variable) is implemented for size removal from y. As a measure of the effectiveness of x, correlation coefficients were calculated between each particular index and the first principal component involving the x and y. Lack of correlation would be assumed to indicate removal of size. In 17 examples, ten disclosed correlations less than .1, two were in the .2+ range, and one was .35, which might seem to imply good performance of the index. However, the correlation of .35 appears to verify potential sources of problems.

Corruccini's index cannot be expected to be a precise description of biological reality. Over their entire range, most biological variables, x and y, will follow a non-linear function and, when $x = 0$, $y = 0$, so $a = $ o. Therefore, the index would appear to capitalize upon the fact that non-linear functions regularly do not depart significantly from linearity over short ranges of values for the function. For this reason, $(y - a)/x$, even when it performs well in adjusting for size, is not likely to be doing so in meaningful biological terms. Parenthetically, since continuous variables can be expected to approximate a multinormal distribution, for different values of x, the index could be expected to have unlike variances.

Unfortunately the problems involved in eliminating size are much more complicated (Mossimann, 1970), so complicated that Corruccini's index is a somewhat crude adjustment for size. To appreciate this, size and shape must be examined in more precise terms, those barely outlined in the discussion of isometry and allometry. In the more refined sense, true ratios are shape variables, so tend to record shape differences as organisms grow. Since ratios do not correct even for the small range of values over which they normally operate, i.e., utilize a, their flaws are readily demonstrated. Although Corruccini's index also is a shape variable, the correction of a obscures the flaws. To complicate matters more, both ratios and the index are shape variables that are correlated with size. For this reason, if one uses either to eliminate a size influence, some aspects of shape differences are removed as well. Finally, the frequent lack of correlations between Corruccini's index and the first component is not surprising, the first component generally would reflect allometry (size dependent upon shape).

4.3 Principal Component Analysis, an Example

A second example of component analysis is in order. Since the interpretation of components often is an intricate process, a superficially simple example is used to demonstrate interpretive procedures.

Table 4.1 contains the mean vector, or centroid, and variance-covariance matrix of a sample of female painted turtles, *Chrysemys picta marginata* Agassiz (data from Jolicoeur and Mosimann, 1960). Parenthetically, I wish to put aside any possible misconceptions that might arise from further discussions of the turtle example. The example was chosen only partly because of its apparent simplicity. It serves well to demonstrate how the checks and balances of the methodology operate. In the latter respect outliers will be discussed in some detail. Since true outliers often cannot be distinguished from errors in measurement or assignment to groups, true outliers and errors are discussed together. The fact that true outliers are involved is shown clearly by Mosimann (1958). In no way should any discussion of the turtles be misconstrued to imply that anything but true outliers existed.

To describe the form of turtles, the measurements length, height and width were selected. One might argue that additional measurements are mandatory. However, the analysis will demonstrate that the three measurements are a valid, even if minimal, outline of form. In addition, the measurements were derived from a previous study (Mosimann, 1958) and the analysis is remarkable in that it was performed on hand calculators.

Since the painted turtles were collected as encountered from a small pond, the sample appears satisfactory. A further aspect of a suitable sample is that no female turtle was rejected unless it did not conform to the definition of the population, "only individuals for which the sex was externally discernable are included."

Since turtle components are extracted from a variance-covariance matrix (Table 4.1), absolute rates of turtle variation are to be examined. Table 4.2 contains the eigenvalues (component variances) and eigenvectors (components) extracted from the dispersion matrix.

Table 4.1 Centroid and dispersion matrix of 24 female painted turtles, *Chrysemys picta marginata*.

dimensions	length	width	height
centroid	136.00	102.58	51.96
dispersion	451.39	271.17	168.70
	271.17	171.73	103.29
	168.70	103.29	66.65

Table 4.2 Principal components of the dispersion matrix of female painted turtle dimensions.

Dimension	Components		
	1	2	3
length	.81	.55	−.21
width	.50	−.83	−.25
height	.31	−.10	.94
variance	680.40	6.50	2.86
% of total variance	98.64	.94	.41

The first eigenvalue relates the consequence of a stimulus that contributes 98.6% of the total sample variation. Since all component coefficients are positive, the response is an increase in all dimensions, a general component. In biological terms pertinent to the turtle, the component might imply simple growth. Also in a loose sense, a size component is indicated—one does not yet know if size is independent of shape. Finally, the stimulus probably is genetic or environmental but naturally cannot be defined with certainty.

Obviously "growth" is in all three dimensions, but length increases fastest (.81), width second (.50), and height slowest (.31). A further concept of relative growth rates can be appreciated if the value of each coefficient is translated into millimeters. Also, if width is used arbitrarily as a frame of reference and other components are ignored, older females would be relatively longer and lower than are immatures.

In the more specific sense of examining size and shape plus growth, the coefficients of the first component derived from the covariance matrix of logarithms (not shown) are

$$\mathbf{a_1'} = (.6235 \quad .4860 \quad .6124),$$

again the coefficients in sequence being for length, width and height. This eigenvector is 6.3° from a vector of isometry,

$$\mathbf{u} = (.5774 \quad .5774 \quad .5774).$$

Later we will develop the test of hypothesis that the first component equals the theoretical vector of isometry. If the hypothesis is rejected, size is not independent of shape so allometry exists. Also, on an individual variable basis, width displays negative allometry and both length and height, positive allometry. Since both height and length have similar coefficients, as a pair they might be isometric.

The last two components have both positive and negative coefficients so are bipolar or shape components. The second component is the consequence of a stimulus contributing .94% of the variation among females. Like any component, it is independent of all others. Here is a feature, one not readily recognizable in biological terms, amounting to shape differences independent of other components. Some variation is the result of an increase in length (.55), a decrease in width (−.83), and a lesser decrease in height (−.10). The interpretation of negative coefficients depends upon the phenomenon being studied and is up to the heuristic judgment of the biologist. The interpretation of bipolar turtle components is analogous to the interpretation for chicken bones.

The third component represents a stimulus producing approximately .41% of the variation among females. Here height (.94) increases dramatically at the expense of a decrease in length (−.21) and width (−.25). Length and width tend to retain their relationship with one another. Later it will be shown that this component applies to older females and so might explain the fact that older females actually are relatively taller than are younger females.

The three components can be named in relation to the major variation in each. The first component can be called a size component; the second, a width component; and the third, a height component.

In any heuristic analysis of components one should pay special attention to the amount of variation represented by each. Notice that here it was assumed that all components were significant and all coefficients of each component were worthy of being interpreted. Both are potential problems in any analysis and bases for judging what to do will be considered later.

4.4 Interpretive Aids for Component Analysis

So far the analysis of turtle data was very conservative. A test of isometry was considered, but the actual test was not completed. In more general terms, the first component was also recognized as one of size or growth and the other two components as shape components or factors. If one wishes to go beyond the above and accepts that further efforts will generate only tentative hypotheses, additional interpretive aids can be helpful. Many aids are presented since they act as checks and balances against impulsive conclusions. Later it will be seen that PCA can act as an aid within a much larger framework of multivariate analysis. In essence, most of the methods to be developed will be united for heuristic purposes near the end of the book.

Along with the interpretive aids for PCA, the z-score or correlation matrix model for the turtle data will be added to that for the covariance model. The raw data are shown in Table 4.3.

Table 4.3 Data matrix for 24 female painted turtles (measurements in mm).

Individual	Length	Width	Height
1	98	81	38
2	103	84	38
3	103	86	42
4	105	86	40
5	109	88	44
6	123	92	50
7	123	95	46
8	133	99	51
9	133	102	51
10	133	102	51
11	134	100	48
12	136	102	49
13	137	98	51
14	138	99	51
15	141	105	53
16	147	108	57
17	149	107	55
18	153	107	56
19	155	115	63
20	155	115	60
21	158	118	62
22	159	118	63
23	162	124	61
24	177	132	67

The Significance of Components.   In many studies it is difficult to know which components should be interpreted and which can be ignored. The same problem often applies to coefficients within a component. Although there is no absolute method for solving these problems, fortunately there are aids. Prior to considering such aids and partly to exemplify them, the components of the z-score model will be presented.

Eigenvalues and Eigenvectors.   Table 4.5 provides the eigenvalues and eigenvectors of the correlation matrix (Table 4.4) of the turtle dimensions.

The components of the correlation matrix (Table 4.5) are similar but display important differences from those of the dispersion matrix (Table 4.2). The first components of both are general or

Table 4.4 Correlation matrix of 24 female
painted turtles

dimensions	length	width	height
length	1.000	0.974	0.973
width	0.974	1.000	0.965
height	0.973	0.965	1.000

Table 4.5 Principal components of the correlation
matrix of female painted turtle
dimensions

dimension	Component		
	1	2	3
length	−0.578	0.062	0.813
width	−0.577	0.674	−0.461
height	−0.576	−0.736	−0.354
variance	2.941	0.035	0.024
% of total	98.04	1.15	0.80

growth components, but note the remarkable near equality of coefficients for the correlation matrix. The negative signs of the first component of the correlation matrix are neither a problem nor rare. Such components relate variation from larger to smaller individuals. The near equality of coefficients implies equal growth of standardized dimensions, and results from near equality of bivariate correlations. Note that both dispersion and correlation second components are bipolar or shape components. However, the coefficients of the two sets disagree in sign as well as magnitude. The dispersion second component was mostly one of width disassociating with length, whereas the correlation second component is one of width disassociating with height. This should not be surprising since absolute and standardized variation obviously are not the same. The third components also disagree. That for the dispersion is mainly a height component and the latter a length component.

It is very important to realize that components also can be read after all signs of all component coefficients are changed. In certain cases, this makes better biological sense. For example, even though the first component of the correlation matrix is one of general growth, negative signs for all coefficients might be bothersome so all can be changed to have positive signs. In fact, when general components have all negative coefficients, it implies that all signs of all coefficients of all components might be changed. When this is done for the turtle correlation components, note that the 3rd correlation component appears to correspond to the 2nd dispersion component and the 2nd correlation component to the 3rd dispersion component. This is understandable, since the discrepancies in percent variation can represent differences in magnitude of standardized and absolute variation. Also, recall that the two sets of components might not relate well at all.

Component Correlations.   There is no method for testing the significance of individual coefficients and there is no way of being sure that a small coefficient really has no biological meaning.

However, there are statistics that can be very helpful in rejecting meaningless coefficients from interpretation. One aid is the use of component correlations, also termed *component structure*.

Component correlations are the correlation coefficients between the ith original variable and the jth component scores. The correlation often is considered to be the ith (variable's) response to the jth stimulus (Morrison, 1967). It is easy to develop the correlation between the ith variable, x_i, and the jth component scores, y_j, even for the covariance model. Since $(S^2 - \lambda_j I)a_j = 0$,

$$S^2 a_j = \lambda_j a_j,$$

the covariance of the x_i's response to y_j is

$$\lambda_j a_{ij} = a_{ij} \lambda_j.$$

Then the component correlation is obtained simply by dividing the covariance by the standard deviations of the ith variable, s_i, and the jth component, $\sqrt{\lambda_j}$, so

$$r_{ij} = a_{ij} \lambda_j / s_i \sqrt{\lambda_j}$$

$$= a_{ij} \sqrt{\lambda_j} / s_i.$$

For the correlation model, component structure simplifies to

$$r_{ij} = v_{ij} \sqrt{\lambda_j}$$

since $s_i = r_{ii} = 1$. The latter is easy to prove from a direct measure of the correlations

$$r_{ij} = \frac{1}{N-1} \sum z_{ij} y_{ij}' \sqrt{\lambda_j}$$

$$= \frac{1}{N-1} \sum (z_{ij})(z_{ij}' v_j) / \sqrt{\lambda_j}$$

$$= RVL^{-1/2}$$

but $RV = VL$, so

$$r_{ij} = VLL^{-1/2} = VL^{1/2}$$

Table 4.6 provides the component correlations for the covariance matrix and Table 4.7 for the correlation matrix. The correlations for both first components are high, but those for both second and both third components are small. For this reason we could ignore the last two components, being satisfied that the only important pattern of form is general size or simple growth.

Table 4.6 Component correlations based upon dispersion matrix of female turtles

dimension	Component		
	1	2	3
length	.998	.065	−.016
width	.986	−.162	−.032
height	.980	−.032	.196

Table 4.7 Component correlations based upon correlation matrix of female turtles

dimension	Component		
	1	2	3
length	−.991	.011	.126
width	−.989	.125	−.069
height	−.988	−.136	−.054

Variances of Variables.   Another way of judging which coefficients should be interpreted is to calculate the percentage of the variance of each variable that is accounted for by each component. For the covariance matrix model, the percentage of the ith variable in the jth component is

$$100(a_{ij}^2\lambda_j)/s_{ai}^2$$

where the total variance of the ith variable is

$$s_{ai}^2 = \sum \lambda_j a_{ij}^2$$

For the correlation matrix model, the percentage is

$$100(v_{ij}^2\lambda_j)/s_{vi}^2$$

where

$$s_{vi}^2 = \sum \lambda_j v_{ij}^2 = p$$

The variances of the variables for the two models are presented in Tables 4.8 and 4.9. Again, we see sufficient evidence to ignore the last two components if they are not biologically meaningful.

Table 4.8 Percentage of the variance of each variable for the dispersion model of turtle components

dimension	Component 1	2	3	Total Variance
length	99.54	0.43	0.03	448.29
width	97.28	2.62	0.10	174.76
height	96.06	0.10	3.84	67.98

Table 4.9 Percentage of the variance of each variable for the correlation model of turtle dimensions

dimension	Component 1	2	3	Total Variance
length	98.40	0.02	1.58	1.00
width	97.92	1.57	0.51	1.00
height	97.72	1.88	0.40	1.00

Test of Equality of Eigenvalues. Some components can be ignored on the basis that they are arbitrarily determined. If we imagine a 3-dimensional ellipsoid of data approximating the shape of a football, such a case follows. The first component axis is along the length of the football through its geometric center. The positions of the final two axes must be perpendicular to the length axis, but the last two dimensions must be placed arbitrarily within a circle. There is no next-longest possible second axis—any axis would equal the diameter of the circle, so any two orthogonal axes in the circle can be the last two component axes. Such equality of axes is rare.

To reduce the number of components to be interpreted, one might follow the quick-and-dirty rule of many data analysts: Interpret components, in sequence, until 80-90% of the variation is explained; then, reject further components unless they make sense. However, rejecting the last few eigenvectors as meaningless can omit important biological features. Therefore, one might examine such eigenvectors to design future experiments that will unravel the implied relationships.

One can test the equality of the last $p-k$ eigenvalues by the so-called *test of sphericity*:

$$\chi^2 = \left[N - k - \frac{2(p-k)+7+2/(p-k)}{6} + \sum_{j=1}^{k} \frac{\bar{\lambda}}{\lambda_j - \bar{\lambda}}^2 \right] [-ln(\lambda_{k+1}\lambda_{k+2}\ldots\lambda_p) + (p-k)ln\,\bar{\lambda}]$$

where N is the sample size and $\bar{\lambda}$ is the mean of the last $p-k$ eigenvalues,

$$\bar{\lambda} = (\lambda_{k+1} + \lambda_{k+2} + \ldots + \lambda_p)/(p-k).$$

The statistic χ^2 approximates the chi-squared distribution with $(p-k-1)(p-k+2)/2$ degrees of freedom, so $p-k$ must be 2 or more.

This is called a test of sphericity since three or more component axes of equal length define a sphere or hypersphere in which component axes are arbitrarily placed.

As was stated previously, sound statistical procedures for tests of significance of eigenvalues of a correlation matrix are unsatisfactory. The above sphericity test is not as mathematically appealing as others. Unfortunately the best test, from a theoretical point of view (Anderson, 1963), requires large samples. Anderson's test when applied to parallel a test of sphericity is

$$\chi^2 = (N-g)\,ln\,(\hat{\lambda}/\overset{k}{\Pi}\lambda_j),$$

where N is the total sample size, g is the number of groups (1 unless a pooled dispersion matrix of g groups is involved), and $\hat{\lambda}$ is the mean of the last k eigenvalues being tested for equality. The test statistic approximates the chi-squared distribution with $\frac{1}{2}k(k+1)-1$ degrees of freedom.

The first test of sphericity was formulated by Bartlett and slightly modified by Lawley. It is recommended both by Seal (1964) and Cooley and Lohnes (1971). In view of its more conservative performance, Bartlett's test is used here even if it can be considered no more than a rule of thumb. Using Bartlett's test for all three eigenvalues of either the correlation or covariance matrix leads to significant results only for the first eigenvalue. In both cases we can assume that the last two eigenvalues are equal.

The test of equality of the last two eigenvalues of the correlation matrix is

$$\bar{\lambda} = \frac{.03464 + .02404}{2} = .02934$$

$$\chi^2 = 24 - 1 - \frac{2(3-1) + 2/(3-1)}{6} + \frac{.02934}{2.94132 - .02934}^2 \times$$

$$-ln\,(.03464 \times .02404) + (3-1)\,(ln\,.02934)$$
$$= [21 + .00010] \times [7.09078 - 7.05761]$$

$$= .69664$$

with 2 degrees of freedom which is not significant at the 5% level. Also, the sphericity test for the last two eigenvalues of the covariance matrix leads to

$$\chi^2 = 3.4^21$$

with 2 degrees of freedom ($p = 0.1779$) which is not statistically significant. Therefore, the implication is that one cannot interpret either the last two correlation or the last two covariance components. However, since the sample size is small and measurements were accurate, one might question the lack of significance, so we will proceed.

Theoretical Vectors.   A useful test is one that considers the possiblity of a vector expected on biological grounds to be equal to a calculated eigenvector. The previous consideration of such a test involved comparison of a theoretical vector (one expected if growth of all turtle dimensions was isometric) with the size component derived from the covariance matrix of common logarithms. In all such cases, the theoretical vector, $\mathbf{u}$, would have coefficients all equal to $1/\sqrt{p}$ to satisfy

$$\mathbf{u}'\mathbf{u} = 1,$$

i.e., the coefficients of the theoretical would be direction cosines as are the coefficients of the eigenvectors. When direction cosines are involved, the angle θ between any theoretical vector, $\mathbf{u}$, and any eigenvector, $\mathbf{a}_i$, can be determined from

$$cos\ \theta = \mathbf{u}'\mathbf{a}_i.$$

In the case of the vector of isometry and the size component from logarithms for turtles, the value of the cosine lead to an angle of $6.3°$.

One might consider $6.3°$ too small to be of consequence; but in the size component for logarithms,

$$\mathbf{a}_1' = (.622 \quad .484 \quad .615),$$

note that the second element (width) is somewhat smaller than the first (length) and the last (height). Therefore, Anderson's (1963) statistical test of an observed vector being equal to a theoretical vector is in order, i.e.,

$$\chi^2 = N(\lambda_i\mathbf{u}'\mathbf{S}^{-2}\mathbf{u} + \lambda_i^{-1}\mathbf{u}'\mathbf{S}^2\mathbf{u} - 2)$$

which approximates the chi-squared distribution with $p-1$ degrees of freedom. In the formula, N is the sample size, $\mathbf{u}$ is the theoretical vector, λ_i is the calculated eigenvalue ($i = 1$ in the example), $\mathbf{S}^2$ is the covariance matrix (in the example, of logarithms from which λ_1 was extracted), and p is the number of variables (3). For the example, $\chi^2 = 34.212$ with 2 degrees of freedom ($p < .001$). Therefore, one might assume that length and height are isometric to one another but positively allometric to width.

Holland (1969) provides biological examples of two methods that extend the utility of theoretical vectors. First, a theoretical vector also can be examined from the point of view of how closely it corresponds to an observed component space. If the p cosines of the angles between the theoretical vector and each component axis are calculated, $cos\ \theta_u$, then the square root of the sum of squares of these cosines is

$$cos\ \alpha = \sqrt{\sum cos^2\theta_u},$$

where $cos\ \alpha$ is the cosine of the angle between the theoretical vector $\mathbf{u}$ and a vector in component space that would most closely correspond to the theoretical vector. Cosine α is termed the cosine of coincidence and α, the angle of coincidence. Naturally, a $cos\ \alpha = 1.0$ ($\alpha = 0°$) would indicate perfect coincidence.

In the second method, again an extension of methodology, one or more theoretical vectors can be involved. This method is possible only if the theoretical vectors are mutually orthogonal and if each theoretical vector has an angle of coincidence approximating zero degrees with the component space. For example, theoretical vectors $\mathbf{u}_1$ and $\mathbf{u}_2$ must have the following properties:

$$\mathbf{u}_1'\mathbf{u}_2 = 0 \text{ and } \alpha_1 \approx \alpha_2 \approx 0°.$$

The latter implies near perfect coincidence for both theoretical vectors.

Holland's second extension amounts to using the theoretical vectors as if they were eigenvectors. The amount of the dispersion involved in each of the theoretical vectors is extracted (factoring the dispersion matrix) and the eigenvalues and eigenvectors of the residual are obtained. Factoring a dispersion matrix is explained in some detail in the chapter on factor analysis. Now Holland's method can be summarized as follows:

1. Calculate the variance of $\mathbf{u}_1$ ($\hat{\lambda}_1$) and $\mathbf{u}_2$ ($\hat{\lambda}_2$):
   $\hat{\lambda}_1 = \mathbf{u}_1'\mathbf{S}^2\mathbf{u}_1$ and $\hat{\lambda}_2 = \mathbf{u}_2'\mathbf{S}^2\mathbf{u}_2$.
2. Define the amount of dispersion due to $\mathbf{u}_1$ ($\mathbf{S}_1^2$) and to $\mathbf{u}_2$ ($\mathbf{S}_2^2$):
   $\mathbf{S}_1^2 = \hat{\lambda}_1\mathbf{u}_1\mathbf{u}_1'$ and $\mathbf{S}_2^2 = \hat{\lambda}_2\mathbf{u}_2\mathbf{u}_2'$.
3. Calculate the residual dispersion, $\widetilde{\mathbf{S}}^2$:
   $\widetilde{\mathbf{S}}^2 = \mathbf{S}^2 - \mathbf{S}_1^2 - \mathbf{S}_2^2$.
4. Obtain the eigenvalues and eigenvectors of the residual:
   $(\widetilde{\mathbf{S}}^2 - \lambda\mathbf{I}) = 0$ and
   $(\widetilde{\mathbf{S}}^2 - \lambda_i\mathbf{I})\mathbf{a}_i = 0$.
5. Order $\hat{\lambda}_1$ and $\hat{\lambda}^2$ with the λ_i ($i = 1, 2, \ldots, p\text{-}2$) from largest to smallest. Use the same sequence for ordering $\mathbf{u}_1$, $\mathbf{u}_2$ and the $\mathbf{a}_i$, and examine the results as components.

Anova of Component Scores.   In many cases, samples are taken in such a way that individuals can be identified in reference to the method of sampling. For example, a species might be ''sampled at random'' along a transect and individuals can be characterized by a sequential number along the transect, by the habitat of their collection, etc. Also, in samples collected through time, a temporal sequence is definable.

Moore (1965) provides numerical biological examples of how component scores of such individuals can become univariate raw data for an anova. More specifically, each individual (based upon temporal, spatial or other sampling peculiarities) is identified with a group in the sense of an anova. Then, any one of the p sets of component scores becomes data for an univariate analysis of variances. Up to p such anovas are possible. Such anovas might be helpful in identifying the biological implication of a component. Also, these anovas can serve to determine how a biologically identified component performs between groups.

Heuristic Analysis of Each Component.   Theoretical considerations allow somewhat crude biological interpretation of components as being size or shape factors. Also, where appropriate, allometry and isometry stem from a proper mathematical base. Theoretical vectors that test biological features likewise are acceptable in the same framework. Additional bases for sound interpretation exist but are so subject specific as to be beyond the scope of this book.

Biological interpretation within the limits of mathematical theory should not be minimized. For example, it is far from trivial to be able to say that an independent facet of shape in female turtles is a length vs. width factor which accounts for .94% of the variation. However, the biologist might wish to draw further conclusions from the data. Again such conclusions cannot be considered more than intelligent guesses. In view of what was said about the problems of biological interpretation, interpretive aids seem necessary. In essence, the aids can act to circumscribe the area of guessing. Such aids as tests of sphericity, component correlations, variances of variables, and theoretical vectors can prove helpful in determining which components to interpret. Other aids will soon be added. However, even more important is that interpretation of a component makes biological sense. From this point of view only, even if all preliminary aids stress the importance of a component and many of its coefficients, the component is useless if it is biologically meaningless.

In heuristic interpretation, the first component usually presents no problem. In studies of individual species, the first component almost always is a general component, a growth component. Even in ecological studies where variables are ecological factors and the purpose is to study the patterns of variation created by factors, the first component might imply a well-known relationship. In fact, the first component from many kinds of studies tends to describe familiar features. However, components beyond the first typically are progressively more difficult to interpret. Even when a biologist is quite familiar with a phenomenon, one or more components may be extremely confusing. The biologist may be at a loss to explain the variation being described let alone the stimulus producing the variation. In such cases, a battery of statistics pertaining to each individual organism is helpful. These statistics are the total variance of the individual, the distance of the individual from the origin, the percentage of the variance of the individual accounted for by each component, and the component scores of the individual.

Variance and Distance of Individuals.   The total variance of the kth individual in algebraic form is

$$s_{xk}^2 = \sum \, \sum y_{x_{ijk}}^{\;2} \text{ or } s_{zk}^2 = \sum \, \sum y_{z_{ijk}}^{\;2}$$

and in matrix algebra

$$s_{xk}^2 = \sum y_{xk}^{\;2} \text{ or } s_{zk}^2 = \sum y_{zk}^{\;2}$$

where $i = 1, 2, \ldots, p$ components, $j = 1, 2, \ldots, p$ variables, and $k = 1, 2, \ldots, N$ individuals. For interpretive purposes, the greater an individual's variance, the more the individual contributes to the components. This is the case because the variance is the square of the distance of an individual from the origin. The distance

$$d_{xk} = \sqrt{s_{xk}^2} \text{ or } d_{zk} = \sqrt{s_{zk}^2}$$

is measured in standard deviation units.

The percentage of the variance of an individual accounted for by each component is

$$100(a_{ij}^{\;2}x_{jk}^{\;2})/s_{xk}^2 \text{ or } 100(v_{ij}^{\;2}z_{jk}^{\;2})/s^2{}_{zk}.$$

This discloses the relative contribution of each individual to each component. An important use of this statistic is to find out how many components account for most of the variance of all individuals. If one or only a few individuals contribute strongly to components that otherwise seem unimportant, the components might actually be important or the individuals might be outliers, or represent sampling or data errors. Table 4.10 present the distances and variances of individuals.

Component Scores.   The ith component score of the jth individual, y_{ij}, provides the distance of individual j along the ith component axis. The scores of all individuals can be graphed for two or three components at a time. More important, each set of component scores contains unique information about how each individual behaved in reference to each independent stimulus. In reference to a single component, if one pays special attention to which individuals have high positive and which have high negative scores, some insight as to the nature of the underlying phenomenon and its stimulus can be determined. Of course one should keep in mind which coefficients of a component are "most important" and the individuals' values for the corresponding variables.

Table 4.10 Covariance matrix distances from the origin, percentages of the variance accounted for by each component, and component scores of 24 female painted turtles.

Case	Distance	Percentage of Variance			Component Score		
		1	2	3	1	2	3
1	45.88	99.91	0.09	0.00	−45.86	−1.35	−0.03
2	40.36	99.72	0.08	2.20	−40.31	−1.13	−1.81
3	38.25	99.15	0.70	0.15	−38.09	−3.20	1.48
4	37.14	99.69	0.26	0.05	−37.08	−1.90	−0.82
5	31.70	99.41	0.32	0.27	−31.61	−1.19	1.65
6	16.88	94.53	1.29	4.18	−16.41	1.91	3.45
7	16.19	99.54	0.02	0.45	−16.15	−0.18	−1.08
8	4.77	89.27	9.14	1.59	− 4.51	1.44	0.60
9	3.20	88.96	10.83	0.21	− 3.20	−1.05	−0.15
10	3.20	88.96	10.83	0.21	− 3.20	−1.05	−0.15
11	5.13	64.42	8.07	27.51	− 4.12	1.46	−2.69
12	3.02	15.75	6.76	77.49	− 1.20	0.78	−2.65
13	4.79	13.40	86.60	0.00	− 1.75	4.46	0.03
14	4.21	1.11	97.88	1.01	− 0.44	4.17	−0.42
15	5.65	97.54	1.17	1.29	5.58	0.61	−0.64
16	13.26	98.68	0.55	0.77	13.17	0.98	1.16
17	14.06	94.71	4.89	0.40	13.67	3.11	−0.89
18	18.02	91.53	8.29	0.18	17.24	5.19	−0.77
19	25.24	97.94	0.18	1.88	24.98	−1.08	3.46
20	25.17	99.06	0.94	0.00	25.05	−2.44	0.12
21	27.19	99.46	0.06	0.48	27.11	0.66	1.89
22	29.81	99.38	0.22	0.40	29.72	−1.40	1.89
23	34.88	97.93	1.70	0.37	34.51	−4.55	−2.12
24	52.66	99.44	0.48	0.08	52.51	−3.63	−1.51

Table 4.10 summarizes covariance statistics pertinent to using individuals to analyze components. Individuals are tabulated in the same order as in Table 4.3 and the eigenvalues and eigenvectors are in Table 4.2.

The first component can be reinterpreted using the aids of Table 4.10. Since the first component has coefficients for length, width, and height of $\mathbf{a}_1' = [.81 \quad .50 \quad .31]$, and individuals were tabulated from shortest to longest, it is not surprising that magnitudes of the values of the first component scores (Table 4.10) generally range from smallest to largest individual. Since all component scores define distance and the first component accounts for 98.64% of the variance, distance provides no new information. However, the percentages of variance are very low in the first component for females 12, 13, and 14 and lower than normal in relation to distance for female 11 (Table 4.10). This is not surprising since the females are near average size, so they must have near zero component scores. For the first component, Table 4.10 and the fact that turtles grow verify the statement that approximately 98.6% of the variation in female turtles is due to growth.

The second component, $\mathbf{a}_2' = [.55 \quad -.83 \quad -.10]$ applies mostly to individuals either longer than normal for their width (positive component scores) or shorter than normal for their width (negative component scores). Note that the component scores tend to be positive for middle sized females and negative for both extremes. Also, middle size females tend to have a high percentage of their variance in the second component. The latter should be expected since so little of their variance was involved in the first component. Perhaps this component relates unique variation in the middle subset of individuals. If this is true, the nature of the stimulus, genetic or environmental is not obvious. Perhaps the middle group represents genetic adaptation via selection by environmental conditions during a critical phase of development of an age group, or purely phenotypic response to environmental conditions. It is also possible that such a component could represent error in measurement or outliers. One of the alternatives seems likely. Such specimens should be reexamined for error in measurement. Also, a canonical correlation between environmental and dimension variables over a period of contrasting environmental conditions could determine whether this is a meaningful component, and if so, which environmental features are related to morphological variation.

The third component $\mathbf{a}_3' = [-.21 \quad -.25 \quad .94]$ according to individuals' percentage variance emphasized females 11 and 12 (Table 4.10). This is surprising. The large proportions of variance indicate that these females are outliers but also could imply error, perhaps again in measurement—such errors would not have to be large. These females are lower in height than they should be. Check the raw data. Low height is the primary source of the negative scores for these females. If component three represents a meaningful pattern of variation, the high proportion of variation found in 11 and 12 is not readily explainable. On the other hand, note that 11 and 12 have negative scores, denying the component. In view of all component scores, although 11 and 12 are unusual, the component cannot be attributed solely to the unusual females. Perhaps the component does represent a genetic pattern of variation, or perhaps an almost pure phenotypic response. The component scores do not indicate to me the nature of the pattern or stimulus. The important thing is that a pattern for further exploration is exposed.

At this point the heuristic approach was hardly worth the effort. More can be done and will be when further methodology is developed.

Graphing Component Scores.   Plotting one set of component scores against another often aids in the interpretation of components. The plot of the first vs. second component scores of each female turtle is shown in Figure 4.3. On the first component axis one can see the gradation from small to

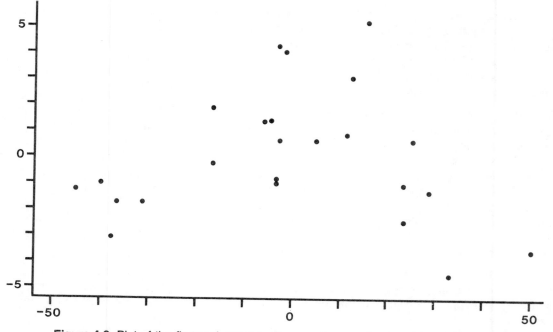

Figure 4.3 Plot of the first and second component scores for female painted turtles.

large individuals. The second axis discloses the rise in values of component scores for intermediate-sized females. No readily recognizable group is distinctly clustered away from its neighbors. Any vaguely apparent clusters can be readily explained on the basis of the small sample size. An examination of this type amounts to a crude ordination of the individual turtles.

4.5 Critique of Component Analysis—Summary

Component analysis is a powerful statistical tool. However, it is far from being an objective method that leads to direct biological conclusions. The reasons for this can be appreciated in terms of the mathematical model, the data, and decisions by the biologist. This critique shall be limited to classical applications in which a *mnd* and related assumptions are made.

Methodology. Each component of a set can be considered to be an independent response to a single stimulus and the set of coefficients to indicate the interplay of variables in the response. Unfortunately this does not translate into a one to one correspondence between each component and each biological feature. Again, two or more biological features can be intertwined in a single component. Perhaps other hypotheses, hence models, would be more appropriate. However, it would seem that one would have to know all important biological features beforehand for meaningful models to be constructed.

For surveying poorly known phenomena, no other statistical method corresponds more closely to current biological thought. Biologists appear to accept the evidence that variation can be subdivided into independent patterns of variation, some genetically controlled, others environmentally. In fact, biologists might become disturbed by a component analysis that discloses two or more biological features that are associated. Therefore, principal component analysis could well result in better understanding, perhaps even re-evaluation of biological phenomena.

The model is not the only source of difficulty. Component analysis can be considered arbitrary owing to decisions regarding variables, transformations, variation studied, vector choice, heuristic approach, sample size, and multigroup comparisons.

Variables.   The aim of component analysis can be to obtain a set of eigenvectors that describe independent actions in an overall biological phenomenon. To fulfill this aim, no further variables are needed and no included variable can warp the vectors into meaningless dimensions. The latter can occur whenever an extraneous variable is associated with those that truly describe the phenomenon. Since all parts of an organism typically display high intercorrelations, the latter is a common problem.

The goal is difficult to accomplish; however, it is more likely to be approximated when a logical set of variables is selected rather than when anything measurable is included. In short, one must know enough about a biological phenomenon to use variables to describe its shape. Usually this is easier to do in systematic problems where the goal is to compare populations. In such cases, the entire shape of each organism sampled is to be approximated or outlined by variables.

Transformations.   Random sampling and a linear relationship among variables need approximation if component models are to have meaning. It was mentioned that the latter can be tested by multiple regression. If the regression function is not linear, usually it can be made so by transforming percentages or proportions to inverse sines, counts to square roots, and continuous variates to logarithms.

Although departures from linearity are not likely to be serious enough to invalidate a component analysis, there is another reason for considering logarithmic transformation of data. The transformation tends to have the same advantages of the z-score model. Logarithms tend to provide components that are independent of scales of measurement and order of magnitude of variances. The logarithms have the additional advantage of being conceptually more closely related to the original data and of providing a component to evaluate growth (allometry and isometry).

Variation Studied.   A third arbitrary decision is whether to examine absolute variation via the dispersion matrix or to examine standardized variation via the correlation matrix. In this respect it might be added that the *SSCP* matrix provides the same eigenvectors as the dispersion matrix and that the eigenvalues of the *SSCP* matrix, if divided by the degrees of freedom, equal those of the dispersion matrix.

The choice of examining absolute or standardized variation depends upon several things. If scales of measurement differ, e.g., lengths, weights, and counts, their variances and covariances are not comparable. Then, the correlation matrix probably is the better choice. However, if one is most interested in absolute variation, perhaps the variables can be transformed to a linear scale, and the covariance matrix can be examined.

If measurements are of the same scale, the dispersion matrix may or may not be the better choice. Here again, the decision must be on a biological basis in reference to the consequences of centering and standardizing (see chapter 8). Raw data are more meaningful only if absolute rates of reactions to stimuli make better sense than do z-scores and standardized rates.

The selection of rates is exemplified by the female turtle study. Since length, width and height are linear measurements, the raw data component model is likely to be satisfactory. In using this model we learned that growth brings about different absolute rates of increase in dimensions. On the other hand, the correlation model disclosed that growth results in essentially equal standardized rates of dimension increase. Which is more important biologically? Is there really an answer? In most cases, both analyses can and will contribute to interpretation.

Vector Choice.   The choice of which vectors to interpret and which to ignore as random variation is a problem. The statistical tests, interpretive aids, and arbitrary rules are not definitive.

A significant eigenvalue can pertain to nothing more than significant experimental error, e.g., error in measurement. On the other hand, lack of significance might ignore an extremely important biological relationship. This is true because many biological phenomena are essentially all-or-none reactions. In such cases, the percentage of the variance of the component will be extremely small and not significant. For this reason, one might attempt to interpret any eigenvector. Then, all devices available should be directed towards unravelling the situation. One might well consider designing one or more experiments to diagnose the implications of such eigenvectors. If this is not easy to do, one might at least repeat the work and obtain new data to see if the same component is derived.

Heuristic Approach.   Any component analysis is subject to the criticism of lacking objectivity. Biologists familiar with more conventional biometrical techniques might consider component analysis simply, "fishing for results." In component analysis there are relatively few tests of specific null hypotheses.

Satisfying all such criticism is not easy. However, since first components often are size components, one could "verify" such components by grouping those individuals with positive and those with negative component scores and showing that the first group was significantly larger than the second. Beyond this, where really needed, verification is not simple. It might satisfy some readers to state that many component analyses have discovered important biological relationships. In fact, the field of numerical taxonomy appears to be using component analysis more and more because it makes good biological sense. However, since this is a circular argument, it cannot imply that all component analyses should stand on their own weight. Biologists sufficiently familiar with a phenomenon should be able to design a more conventional experiment that would support or deny their heuristic conclusions.

Sample Size.   Since tests of significance are not applied to individual component coefficients, the sample used should be sufficiently large to provide some reliability of the coefficients. It appears that a minimal sample might consist of as few as 25 individuals, but larger samples more definitely are preferred. In any case, the number of individuals should exceed the number of measurements so all dimensions are defined.

Recall that the assumptions of random sampling and multivariate normality, required for tests of significance in components analysis, regularly cause no problem, e.g., see Andersion, 1958; Harris, 1975; Kendall, 1957; and Seal, 1964.

Useless Analyses.   One of the purposes of component analysis is to unravel conflicting, often meaningless, correlations. If off-diagonal covariances are equal, or near equal, a condition approached in the correlation model of the turtle example, all or almost all eigenvalues might be equal. This might explain the last two eigenvalues derived from the correlation matrix.

If covariances are zero or approximate zero, again component analysis is a dubious approach. Naturally, this implies that the original dispersion matrix approximates a diagonal matrix. However, if the original dispersion is a correlation matrix, one should avoid evaluating correlation coefficients to accept the hypothesis of a diagonal matrix. Even if all tests of the null hypothesis that each product moment correlation coefficient is zero are accepted, one or more coefficients might be estimating a meaningful but small population correlation coefficient. Unfortunately, there is not a priori rule for predetermining whether or not a given dispersion matrix will provide meaningful components. All one can do is perform a component analysis and judge the results.

Correlations.   All uses of correlation coefficients in component analysis must avoid any implication of cause and effect. All proposals for their application refer to their appropriateness as a measure of goodness of fit. Also, they never are absolute criteria for judgment—biological evidence consistently allows correlation fit to be ignored.

Importance of Components.   No component is assumed to be more important, real or meaningful than any other. Just because growth often explains a great deal of variation does not imply that it is of primary biological importance. Essentially, every pattern of variation must be examined on its own merits. Perhaps the best approach is not to attempt to rate or rank the biological significance of components.

Comment.   Most of what was said applies to heuristics. Anyone attempting to extract biological meaning from components should realize the problems. Where pertinent, other points should be considered in reference to a conservative analysis. The reader should now appreciate why the conservative approach is commonplace.

The Next Step.   A purpose in this and other chapters is to expose applications of methodology in some detail. To satisfy this purpose, the PCA example seems appropriate. However, I would agree with anyone that believes more complicated examples are needed. In fact, as a next step, the reader would benefit both from further examples and from further discussion of methodology. This could serve another purpose, developing the ability to utilize the literature. Those whose knowlege of PCA is based upon this chapter might find the literature difficult.

In the sense of reviewing methodology, obtaining more examples, and entering the literature, a certain sequence of papers might prove helpful. As a first step, two works by Pearce are recommended. Pearce (1965a) presents a general account of the superiority of PCA to more conventional methods for like purposes. Also, Pearce (1965b) contains a nontechnical chapter that examines methodology, limitations, and interpretation of different biological examples.

As a second step, study of four papers that outline most of the content of this chapter and provide examples has proven beneficial to students. Jeffers (1967) reviews methodology and shows how to interpret two kinds of biological studies. One study involves ordination. The other relates components to a variable that usually would not or could not be measured. In the latter case, each of the $i = 1, 2, \ldots, p$, portions (one pertaining to each coefficient) of the ith component score is calculated from $y_{ij} = v_{ij}z_j$. First, each portion is transformed to a z-score for $k = 1, 2, \ldots, N$ individuals, $z^*_{ijk} = y_{ijk}/\sqrt{\lambda_i}$. Then, for each of the i components, compression strength is predicted by the regression function $\hat{z}_{ik} = b_{ij}z^*_{ijk}$. Finally, for all components, compression strength is predicted by $\hat{z}_k = bz^*_k$. Moore (1965), in discussing methodology, provides helpful diagrams showing variation extracted by each component. His example applies correlation and anova to component scores. Holland (1969) includes a discussion of the problems of biological interpretation and the application of theoretical vectors to a biological example. Jolicoeur (1963a) provides a biological example in the study of allometry and isometry.

The above and even a larger list of introductory papers does not solve the problems of different symbols and ways of discussing methodology. Also, many workers utilize procedures beyond PCA as aids to interpretation. Although difficulties occur, students can accomplish the transition.

For those primarily interested in ecology, a basic paper is that of Gittins. (1969). For ecological studies, PCA generally is applied to stand, species, and/or ecological factor ordination. Such applications are severely criticized by Beals (1973), whose references serve as an excellent further introduction to the literature.

The final recommendation is for the reader to select a moderately difficult, well understood problem, collect the data, and analyze the results by every device possible. The PCA part of the analysis should encompass every legitimate aid and approach that can be imagined. In my opinion, there is no better way to understand PCA or any other multivariate analysis.

Chapter 5
Multigroup Principal Component Analysis

Originally, component analysis was intended solely for the examination of a sample from a single population. However, under the proper circumstances, extension to the multigroup case is possible. Naturally, the purpose remains to examine variation, but both within each and between all populations. For within populations, all the methods and aids of a single group analysis are applied to each group. For contrasting groups, the components of each group are compared with a single set of reference components.

Potential problems of multigroup component analysis exist in possible confusion with other group comparison methods that involve component analysis, in the particular dispersion matrix used to derive reference components, and in the validity of the proposed procedures.

Other Multigroup Component Approaches. Performing a component analysis on each group and contrasting each such set of components with a reference set is not generally seen in the literature. Rather, groups tend to be compared in a single component analysis. Some of these single principal component analysis procedures are invalid. A dubious single approach exists in one of the methods of numerical taxonomy. The potential problems of such a PCA are treated in Chapter 8.

In numerical taxonomy, each group, or OTU, is represented by a single individual or centroid. By a Q-technique, a group x group similarity matrix is formed, e.g., a correlation matrix representing associations between groups. Then a principal component analysis is performed on the group x group similarity matrix. Finally, the first two or three OTU scores are plotted for analysis of OTU similarities and differences. The goal of such studies is ordination. Obviously the morphometric approach as defined here is hardly involved.

Problems of Multigroup Comparisons. Jolicoeur and Mosimann (1960) compared the three components of two groups, male and female painted turtles, directly. In their method, the component scores of males and females are calculated first in reference to the components for one sex and then in reference to the components of the other sex. Their method could be extended to angular comparisons between male and female eigenvectors and to using the components of one sex as theoretical vectors to be tested in reference to those of the other sex. Unfortunately, the computations for such comparisons become very numerous when many groups and many components are involved.

The number of possible group comparisons is $p^2(g^2 - g)/2$, where g is the number of groups and p is the number of components derived from the p variables measured for each group. Even in a modest study of five groups and five variables, 250 comparisons are involved. For this reason, the theoretically preferred Jolicoeur and Mosimann method might become impractical.

Since studies of many groups and many variables might be contemplated, a procedure that is likely to be effective in screening components seems in order. For example, if each group's components can be compared with a single set of reference components, comparisons are reduced to p^2g. In the case of five groups and five variables, reduction is from 250 to 125 comparisons; but, if the number of groups is 20, reduction if from 4750 to 500 comparisons. For the latter reason, a device that is suspect theoretically but often performs well in practice might be justified. This latter approach will be developed in this chapter. Naturally, whenever possible, the Jolicoeur and Mosimann method should be applied. In fact, the proposed method serves best when it is used as a guide to which comparisons in the sense of Jolicoeur and Mosimann should be made.

In turning to the proposed quick-and-dirty method, first we will consider which reference matrix to use and then briefly mention that the method should be related to other methodology.

Which Reference Dispersion Matrix?

As a reference for among group comparisons three matrices are available: the pooled or within group, the between group, and the total covariance matrices. These matrices are equivalent, respectively, to univariate analysis of variance residual (error or within), between group (among group or treatment) or total mean squares. The pooled within group matrix, usually termed the dispersion matrix, is preferred.

The dispersion matrix is preferred whether group dispersions are determined statistically to be equal or unequal. (The test of equality of group dispersions is not presented until the chapter on the multivariate analysis of variance.) When dispersions are equal, the total analysis reduces to examining the dispersion components. All morphometric techniques of chapter four are applied, but additional emphasis is placed upon plots of centroids and individuals on component graphs. The graphs show how individuals refer to their own and other groups and how groups (centroids) are affiliated. When group dispersions are unequal, multigroup component analysis is even more valuable—the components help unravel the bases for heterogeneous dispersions.

The between group dispersion components are a poor reference in comparison with the dispersion components. Let us contrast the two. The contrast is simplest if the nature of each matrix and its components is examined. When within group dispersions are equal, the dispersion and its components are estimates of the variation common to all populations. When the various group dispersions are unequal, the dispersion is more complex. In the sense of components, the dispersion's represent both the single component that is modified among groups and the two or more components that are present in some groups but are joined into a single component in other populations. All this might be unraveled.

An analogy with the fixed model univariate analysis of variance is helpful in evaluating the between group components. When within group variances are equal, the between group mean square estimates the variance common to all populations plus the treatment effect or variance between population means. Therefore, the between group components might relate no more than size differences in patterns of variation shared by all populations. However, it is possible that one or more groups contain variation associated with, but modified from, that found in dispersion. Then, between group components would be warped and difficult to evaluate. Also, certain between group relationships might provide entirely new vectors that again would be most difficult to interpret. When group variances are unequal, the between group mean square estimates a complex of within group and between group differences. Naturally, the components of an analogous between group dispersion matrix might be impossible to evaluate.

The total dispersion components represents even more complexity between and within variation. This can be appreciated by drawing another analogy with the fixed model univariate case. This is left to the reader.

Additional Procedures.   Multigroup component analysis is statistically suspect so must be verified by other procedures. Later the methodology will be included within the general framework of discriminant analysis. As part of discriminant analysis, multigroup component analysis is justified; but, more important, the PCA's serve to satisfy possible shortcomings of discriminant analysis. Precise methods for discriminating among groups are based upon equality of group dispersions, so discrimination becomes suspect when group dispersions are unequal. Exposition of methodology will be linked with expansion of the turtle example.

5.1 Multigroup Component Methodology

Jolicoeur and Mosimann (1960) analyzed 24 male and 24 female painted turtles in their morphometric study. The data are especially helpful in disclosing potential problems of multigroup component analysis.

The steps in methodology are:

Test of Equality of Group Dispersions.   If the hypothesis of equality of group dispersions is accepted, only dispersion components are interpreted. In the example, for the test, $F = 4.004$ with 6 and 15330 degrees of freedom ($p = .0005$, indicating very little probability of group dispersions being equal). Therefore, some difference between group components can be expected, so the following procedures are implemented:

Calculation of Group Components.   Group centroids and group dispersions are presented in Table 5.1. Note that both centroid and group dispersion are of greater magnitude in females. The table emphasizes that in contrast to males, female painted turtles get larger, are more variable, and get relatively higher.

Table 5.2 contains some of the basic statistics helpful in interpreting component analysis. In addition, the chi-squared test of sphericity of the last two components for males is

$$\chi^2 = 7.2138 \text{ with 2 degrees of freedom } (p = 0.0270)$$

and for females is

$$\chi^2 = 3.4421 \text{ with 2 degrees of freedom } (p = 0.1779).$$

Table 5.1 Centroids and Covariance Matrices of 24 male and 24 female painted turtles.

	males length	males width	males height	females length	females width	females height
$\mathbf{\bar{X}'}$	[113.38	88.29	40.71]	[136.00	102.58	51.96]
$\mathbf{S^2}$	138.77	79.15	37.38	451.37	271.17	168.70
	79.15	50.04	21.65	271.17	171.73	103.29
	37.38	21.65	11.26	168.70	103.29	66.65

Table 5.2 Group Dispersion Matrix Components A, Component Correlations R and Percentage of the Variance of each Variable F for Painted Turtles.

Component	24 males			24 females		
	1	2	3	1	2	3
variance λ	195.28	3.69	1.10	680.40	6.50	2.86
% variance	97.61	1.83	0.55	98.62	0.94	0.41
length	0.840	0.488	−0.236	0.813	0.545	−0.205
width A	0.492	−0.869	−0.047	0.495	−0.832	−0.249
height	0.229	0.077	0.971	0.307	−0.101	0.946
length	0.997	0.080	−0.021	0.998	0.065	−0.016
width R	0.972	−0.236	−0.007	0.986	−0.162	−0.032
height	0.952	0.044	0.304	0.980	−0.032	0.196
length	99.32	0.64	0.04	99.54	0.43	0.03
width F	94.42	5.57	0.01	97.28	2.62	0.10
height	90.57	0.19	9.24	96.06	0.10	3.84

The former test implies that the last two male components are different and interpretable; however, the latter test indicates that the last two female components are arbitrarily placed in space and not biologically meaningful. From a statistical point of view this means that only the first component of males and females can be compared. However, we will ignore this since the test of sphericity for females may reflect small sample size.

Note that in Table 5.2 certain statistics consistently display greater magnitude in the second and third male components than in corresponding female components. Second and third male components display a greater percentage of variance, greater correlation with variables, and a greater percentage of the variance of variables. All this might be expected from the chi-squared tests. Also note that both male and female first components are simple growth or size components. Both second components are mostly contrasts of length and width, shape components. The third male component mostly contrasts length with height, but the female's contrasts length with width and height.

Calculation of Dispersion Components.   Table 5.3 contains the statistics. For the chi-squared test of sphericity of the last two components

$$\chi^2 = 5.641 \text{ with 2 degrees of freedom } (p = 0.0592)$$

which is not significant at the 5% level.

The dispersion matrix component scores of individuals are presented in Table 5.4. Distances and percentages of each individuals variance accounted for by each dispersion component are presented in Table 5.5.

Comparison of Components.   The various statistics consistently emphasize inequality of groups. Some consideration of the statistics in Tables 5.2 and 5.3 demonstrate their importance as interpretive aids and disclose some of their attributes.

Table 5.3 Grand centroid, dispersion matrix, dispersion matrix components, components correlations, and percentage of the variance of each variable.

		length	width	height
grand centroid $\overline{\overline{X}}'$		[124.69	95.44	46.33]
dispersion S^2		295.08	175.16	103.04
		175.16	110.89	62.47
		103.04	62.47	38.96
		1	2	3
length		0.819	0.534	−0.209
width	A	0.495	−0.842	−0.213
height		0.290	−0.071	0.955
length		0.997	0.070	−0.019
width	R	0.983	−0.181	−0.032
height		0.970	−0.026	0.242
length		99.47	0.49	0.04
width	F	96.63	3.27	0.10
height		94.10	0.07	5.84
variance		437.32	5.10	2.49
% variance		98.3	1.10	0.6

Eigenvalues. Although the absolute magnitudes of corresponding group and dispersion eigenvalues differ, the percentages of total variance represented by each component are remarkably similar. In the sense of inequality of group dispersions, this suggests more difference in absolute magnitude of variation in male vs female components rather than in proportion of variation. However, if looking for differences, female percent variations more closely approximate those for dispersion in the first and third components and males are more similar to dispersion in the second component.

Eigenvectors. In comparing the group and dispersion growth components, again it is convenient to think of component coefficients having meaningful units. This can be translated into millimeters of deviation in length, width, and height equal to 100 times each coefficient. For males, the deviations are approximately (+84 +49 +23); for females, (+81 +50 +31); and for dispersion, (+82 +50 +29). In this form, one can clearly appreciate the fact that females grow to be higher than males. For the second component, the millimeter deviations are approximately (+49 −87 +8) for males, (+55 −83 −10) for females, and (+53 −84 −7) for dispersion. Therefore, in growth, part of female length is dissociated with width and height, but male length is associated only with height. For the third component, the deviations are approximately (−24 −5 +97) for males, (−21 −25 +95) for females, and (−21 −21 +96) for dis-

Table 5.4 Dispersion matrix component scores

Case	Components		
	1	2	3
1	−31.423	−1.499	0.689
2	−25.841	−1.356	−0.993
3	−23.693	−3.325	2.400
4	−22.634	−2.115	0.073
5	−17.209	−1.947	2.631
6	− 2.022	1.735	4.583
7	− 1.696	−0.509	0.127
8	9.924	1.108	1.961
9	11.409	−1.420	1.323
10	11.409	−1.420	1.323
11	10.370	1.012	−1.324
12	13.288	0.324	−1.213
13	12.706	4.086	1.338
14	14.021	3.778	0.917
15	20.027	0.184	0.923
16	27.586	0.577	2.851
17	28.150	2.629	0.736
18	31.717	4.695	0.856
19	39.342	−1.473	5.419
20	39.463	−2.945	2.130
21	41.510	0.200	3.838
22	44.104	−1.864	3.945
23	48.953	−5.175	0.134
24	66.938	−4.329	1.027
25	−39.273	1.799	2.267
26	−37.053	−0.896	−0.701
27	−34.425	−1.512	−1.544
28	−27.190	−2.495	0.379
29	−26.166	−2.733	−0.977
30	−27.616	1.242	−1.310
31	−25.228	−0.051	−0.035
32	−23.589	1.017	−0.452
33	−23.554	2.465	−1.403
34	−15.414	−0.904	−2.027
35	−15.090	0.473	−2.023
36	−15.261	2.692	−1.807
37	−10.774	0.177	−0.211
38	−10.534	0.853	−2.329
39	−10.039	0.011	−2.542
40	− 7.410	−0.606	−3.385
41	− 8.860	3.369	−3.698
42	− 5.722	−0.285	−0.730
43	− 4.492	−1.294	−3.273
44	− 1.337	2.135	−0.820
45	1.787	0.856	−1.875
46	2.111	2.232	−1.871
47	4.858	3.763	−1.543
48	13.870	−3.438	−3.763

Table 5.5 Distances and percentages of the variance of each individual's variance accounted for by each dispersion component.

Case	Distance	variance per component		
		1	2	3
1	31.466	99.725	0.227	0.048
2	25.896	99.579	0.274	0.147
3	24.046	97.092	1.912	0.996
4	22.733	99.133	0.866	0.001
5	17.517	96.510	1.235	2.255
6	5.302	14.552	10.710	74.738
7	1.775	91.260	8.226	0.514
8	10.177	95.103	1.184	3.713
9	11.573	97.188	1.505	1.307
10	11.573	97.188	1.505	1.307
11	10.503	97.483	0.928	1.589
12	13.347	99.116	0.059	0.825
13	13.414	89.725	9.279	0.995
14	14.550	92.861	6.742	0.397
15	20.049	99.780	0.008	0.212
16	27.739	98.901	0.043	1.056
17	28.282	99.068	0.864	0.068
18	32.074	97.786	2.143	0.071
19	39.741	98.004	0.137	1.859
20	39.630	99.159	0.552	0.289
21	41.688	99.150	0.002	0.847
22	44.319	99.031	0.177	0.792
23	49.225	98.894	1.105	0.001
24	67.086	99.560	0.416	0.023
25	39.380	99.460	0.209	0.331
26	37.071	99.906	0.058	0.036
27	34.492	99.607	0.192	0.200
28	27.307	99.146	0.835	0.019
29	26.327	98.779	1.078	0.143
30	27.675	99.575	0.201	0.224
31	25.228	99.999	0.000	0.000
32	23.615	99.778	0.186	0.037
33	23.725	98.571	1.079	0.350
34	15.573	97.969	0.337	1.694
35	15.232	98.140	0.096	1.764
36	15.601	95.682	2.977	1.341
37	10.777	99.934	0.027	0.038
38	10.822	94.746	0.622	4.632
39	10.355	93.975	0.000	6.025
40	8.169	82.281	0.551	17.168
41	10.175	75.832	10.961	13.207
42	5.776	98.160	0.243	1.597
43	5.707	61.961	5.140	32.899
44	2.796	22.853	68.552	8.595
45	2.728	42.907	9.837	47.256
46	3.597	34.439	38.498	27.063
47	6.336	58.790	35.277	5.933
48	14.777	88.100	5.414	6.486

persion. Here we observe a strong height component for both sexes. However, note that consistently female components are more like dispersion components. When all components and the magnitude of eigenvalues are considered, a clear picture of the known differences between the sexes is forthcoming.

Component Correlations. The three sets of component correlations all show high correlations in the first component and rather low correlations in the second and third. As was the case for examination of females alone, there is no good statistical evidence for examining the last two components of any set. On the other hand, there might be good experimental reasons for doing so. It seems reasonable to measure turtle shells to the nearest mm. Such measurements should be precise, perhaps even accurate. In the example, the measurements were accurate. Also, the components are consistent with knowledge about painted turtle shape. Therefore, the last two patterns of variation should be considered.

Percentage of the Variance of Each Variable. The first component consistently explains a higher percentage of the variance of each variable. No variable's variance is sufficiently high in the last two components to lend statistical support for their interpretation. However, again very little variation might elicit a feature of biological importance.

Angular Comparison of Components. If group dispersion matrices are all estimating the same population dispersion, there should be no difference in the same component in different groups. Since testing the equality of each group's components to corresponding components of all other groups would be tedious in a multigroup problem, even with a computer, the present compromise of calculating angles between all components of each group and all components of the dispersion matrix followed by testing the equality of these contrasts is employed. Again, this method of using dispersion components as a reference, when supported with other statistics, has proven satisfactory.

The angle θ between the ith male component $\mathbf{a}_{mi}$ or female component $\mathbf{a}_{fi}$ and the jth dispersion component $\mathbf{a}_{dj}$ is obtained from

$$\text{cosine } \theta_{ij} = \mathbf{a}_{mi}'\mathbf{a}_{dj} \text{ or cosine } \theta_{ij} = \mathbf{a}_{fi}'\mathbf{a}_{dj}.$$

Then, by considering each eigenvector of a group as a theoretical vector, its equality to each dispersion eigenvector can be tested by the procedure of chapter four. Both angles and results of chi-squared tests of equality of theoretical vectors are given in Table 5.6.

Table 5.6 discloses that only the second male component can be assumed equal to a dispersion component, the second; but all female components can be assumed equal to the corresponding dispersion components. Since it follows that females and males differ in the first and third components and are the same only in second components, we see why the test of equality of group dispersions was rejected.

In further examination of Table 5.6, if one applies the quick-and-dirty rule of angles less than 20° between vectors representing fundamentally similar vectors, a new concept unfolds. Male and female painted turtles express similar patterns of variation. This is especially true in growth, but is also true for the last two components. Therefore, one might guess that the statistical inequality of dispersions was more a function of magnitude of variation rather than patterns of variation.

Tests of Theoretical Vectors. Table 5.6 further emphasizes that tests of significance of theoretical vectors are not dependent upon angular departure between theoretical (here group vectors) and calculated (here dispersion vectors). The test is more dependent upon how well each coefficient of a theoretical vector associates with the observed dispersion matrix—review the formula.

Table 5.6 Angles between group eigenvectors rows and dispersion eigenvectors, chi-squared tests of equality of group and dispersion eigenvalues, and probability of calculated chi-squareds.

	Dispersion Components		
	1	2	3
Male component 1			
angular difference	3.70	88.96	−86.45
chi-squared	16.605	2001.095	4142.270
$P(\chi^2)$	.0002	0.0	0.0
Male component 2			
angular difference	−89.53	9.01	81.00
chi-squared	2061.763	0.449	12.788
$P(\chi^2)$	0.0	0.801	0.002
Male component 3			
angular difference	86.33	−81.05	9.69
chi-squared	4090.436	20.685	17.3398
$P(\chi^2)$	0.0	0.002	0.0002
Female component 1			
angular difference	1.06	−89.70	88.99
chi-squared	1.359	2008.727	4158.156
$P(\chi^2)$	0.509	0.0	0.0
Female component 2			
angular difference	89.67	1.92	−88.11
chi-squared	2011.549	0.081	12.947
$P(\chi^2)$	0.0	0.9609	0.002
Female component 3			
angular difference	−89.00	88.10	2.15
chi-squared	4155.902	13.421	1.293
$P(\chi^2)$	0.0	0.001	0.527

Relative to angles, note that in males, the first eigenvector (3.70° from the first dispersion eigenvector) and the third eigenvector (9.69° from the third dispersion eigenvector) are both significantly different from corresponding dispersion eigenvectors. However, the second male component (9.01° from the second dispersion eigenvector) is not significantly different from the second dispersion eigenvector. From a biological point of view, the test of significance might seem much ado about nothing.

Individual Statistics.   Consider the first component scores of males and females for the dispersion matrix. (Table 5.4 and Figure 5.1). These scores are for females (1-24) in the order of individuals ranked according to length as presented in the previous chapter, and for males (25-48) ranked in the same manner. Note that component one, the growth component, produced more negative scores for

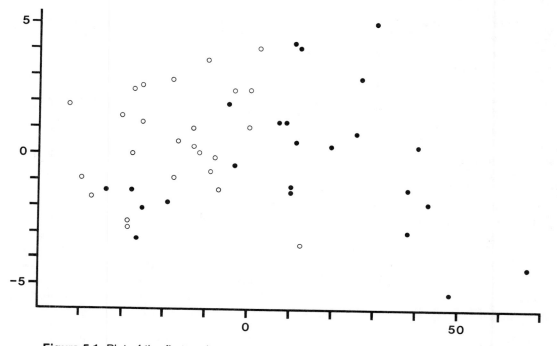

Figure 5.1 Plot of the first and second component scores for male and female painted turtles.

males than for females. This is not surprising since females become larger. In fact, examining raw data of males and females gives no indication that males and females of comparable size score differently on the component. This fact, plus the small angular departure between the first components for groups and the first component for dispersion discloses no basis for considering growth to be fundamentally different in the two sexes.

Examination of the second component is aided by reviewing angles between second components. The angular departure from the second dispersion component is $1.92°$ for females and $9.01°$ for males, but neither is significantly different from the second dispersion component. This is the case in spite of the fact that the second male component indicates a distinct increase in height being strongly disassociated with width but females resemble the dispersion in an increase in length being disassociated strongly with width and weakly with height.

Since the second dispersion component is very much like the second female component, female component scores might be more helpful as a first step towards interpretation. Recall that the scores showed that females of intermediate size respond positively and both extremes respond negatively to the component. This leads to the supposition that the component might indicate either a change in shape from small females to intermediate ones and a later change in shape to still larger females, or perhaps a response of a mid-age group. In the males, no such clear cut distinction can be made; however, reference to Table 5.4 indicates that large males having the highest positive component scores display a remarkably high percentage of their variance in this component. In spite of all this, the biological basis of the second component still is not clear.

The third component scores clearly suggests a biological interpretation of this component, sexual dimorphism. All females except 2, 11 and 12, have positive scores and all males except 25 and 27 have negative scores. With the exception of the previously mentioned outliers, numbers 11 and 12, small individuals are involved, so we might not expect well-defined sexual dimorphism in such individuals. Since this component departs little from the third male and female components, it follows that the male component was one of maleness and the female component was one of femaleness; however, sexual dimorphism represents a minor difference in angles between vectors.

A reconsideration of the analysis of females in the previous chapter is in order. The third female component is only $2.15°$ from the third dispersion component, but the previous female component scores provided no clear interpretation of femaleness. The bases for the discrepancy are three fold. First, the dispersion component, even though more like the third female component, is intermediate between those for the two sexes. This alone would not create the dispersion's dichotomy of scores, e.g., see component two. However, if the component of each group pertains to a unique feature of the group, that uniqueness would be emphasized by the dispersion component. Second, since height is emphasized most, the fact that females get larger would cause them to have more positive dispersion component scores. This is of course a function of the previous reason. Finally, 48 individuals would define a component better than would 24. Parenthetically, another reason can be added. Since females 11 and 12 are outliers, that fact would warp female components, perhaps sufficiently to cause the third component to provide poor component scores. In other studies, warping could result from incorrect assignment of individuals to groups or from data errors.

Although it follows that the third male component might be considered to represent maleness, the third male component scores (not shown) verify this no better than do the third female component scores. It also follows that sexual dimorphism in the dimensions measured both represents a minor proportion of the total variation and provides little distinction between the sexes.

Now a better attempt can be made to interpret component two. Since component one represents a size dichotomy and component three a sexual dichotomy, neither refers to the nature of individuals independent of size or sex. In the sense that the essence of a painted turtle still remains as a possible pattern of variation, this interpretation is appropriate for component two. The interpretation can be appreciated on biological grounds as well as on the grounds of solving a dilemma. It is logical to assume that independent of growth, "turtleness" might be of greater magnitude than sexual dimorphism. However, turtleness is both a poor and crude description of a pattern of variation; yes, a feeble grasping for meaning.

5.2 Overview

Jolicoeur and Mosimann (1960) considered each component of each group to be meaningful. In the present analysis, contrasting tests and the criterion of being biologically meaningful agree that all patterns are of importance. I consider this to support the quick-and-dirty approach. The tests are of special interest. Tests of sphericity indicated that all male components were different but that the last two female components were equal. However, angular comparisons and tests of theoretical vectors also separated the last two female components. In one respect this is a failure of the tests; in another it is a strength of the tests. Most definitely the factors influencing each test must be kept in mind. Then, we can appreciate why methods for accepting or rejecting hypotheses are not definitive. Always the criterion of "making good biological sense" must be a primary basis for judgment.

The lack of greater biological meaning from the component analysis might be disturbing. In reference to male or to female components alone, interpretive aids do not lead to the recognition of biological features in the second and third components. The dispersion components seem better, but the second component provides no clear biological interpretation. When such occurs, features of PCA and the study should be reevaluated.

Very pertinent is the fact that components might not provide responses recognizable within the framework of biological features. Another possibility is that poor measuring or sampling techniques result in warped components that do not define a biological feature. Also, small samples might poorly define components. However, true outliers are not likely to create problems—they did not in the example.

The foregoing analysis might appear trivial. You might wonder at all the work to find that painted turtles grow and display sexual dimorphism. Obviously the possible criticism can be countered by emphasizing what the methodology can do. Also, let us not minimize the results of the "simple example." The amount of total variation due to independent general size and shape factors was approximated as was the nature of the factors. The differences between males and females were noted. Also, the general components for males and females could be examined in terms of growth, allometry and isometry. Note that all this "trivial" information (except sexual dimorphism) is derived from a sound mathematical basis. Therefore, the biologist might consider becoming content with describing independent responses and being very conservative about hypothesizing biological implications.

Chapter 6
Factor Analysis

Factor analysis is a generic term for a class of procedures, all related in that their purpose is to analyze the intercorrelations within a set of variables. The methods almost always start from a correlation matrix, so standardized variation is examined.

PCA of a correlation matrix is a factor analysis procedure. Sometimes, true component analysis is incorrectly termed factor analysis. The other methods often are poorly defined and often are difficult to recognize due to inconsistent terminology. A consistent characteristic of factor analysis is an esoteric language that has very little carry over into other methods of multivariate analysis. This terminology will be largely ignored.

In spite of the complexity of techniques, two features generally pertain to any factor analysis. First, the correlation matrix is subdivided into fractions that add to equal the original correlation matrix. At least one fraction supposedly contains meaningful variation and another error residual. The other feature is rotation which often follows an initial solution of principal components or factors analogous to components. The avowed purpose of rotation is to provide simple solutions.

Historically, factor analysis was the concern of individuals attempting to unravel the complexities of human behavior. Even today the major practitioners are psychometricians and others limited to indirect measures of psychological and related phenomena. For such data perhaps the goals of simplification and correction to obtain order justify the application of factor analysis. In contrast, when direct measures are made, factor analysis may create more problems than it solves.

Many biometricians are very critical of factor analysis and rotation schemes. Blackith and Reyment (1971) present a detailed critique, including many references to the criticism. Their critique presents five primary bases for the criticism: First, prior to rotation, the various models of factor analysis tend to produce factors that are very similar to the components of the original correlation matrix. Therefore factors would generally lead to much the same interpretation as would components. However, the conclusions from factors cannot be comprehensive since even a slight modification from components is sufficient to negate the unique features of components.

Second, component scores are easily calculated and are based upon sound statistical theory; but factor scores often are derived only after rigorous computations that in many models involve questionable methods.

Third, there are no precise criteria either for measuring agreement of sets of factors from different methods or for selecting one set of factors over another.

Fourth, although only a few methods of rotation are in general use, the methods provide unlike results. Unfortunately there are no objective criteria for selecting one rotation scheme over another. Also, rotation is prone to reflect the pecularities that characterize any sample, and especially smaller samples.

Fifth, factor analysis consists of so many models and rotation schemes that the adept analyst can virtually control the results. Unfortunately one can become an expert user without appreciating this fact. Too often, an almost mystical faith appears to exist in methods perhaps because of their mathematical complexity, computer usage, and understandable results. Also, and perhaps most important, one or more models can be found that support the preconceived conclusions of the analyst.

For the above reasons, only an indication of methods is presented. However, in some instances, a simple form of factor analysis might help the biologist. Such an instance occurs in the predicted and residual matrices in the chapter on canonical correlation. Excellent texts on factor analysis are those of Harmon (1967) and Mulaik (1972). A brief, readable introduction is in Cooley and Lohnes (1971).

The brief treatment that follows will review PCA, barely indicate the nature of factor analysis methods, and end with rotation of factors.

6.1 Principal Component Analysis

For the sake of comparing PCA with Factor Analysis, another use of component analysis is now emphasized.

Purpose.   Like any other of the factor analysis methods, component analysis can be employed to discover the facets of and the bases for an overall phenomenon. In this sense component analysis can be applied to determine the minimum number of independent dimensions (patterns of variation) needed to explain some phenomenon. For example, many measurements may be taken to describe the form of some physiological complex. If the variables do not truly cover the phenomenon, the components would not contribute all of the independent facets of the physiology. If one were to add meaningful variables, additional significant components would result. Eventually, additional supposedly meaningful variables would not add components. In this manner, one might determine how many independent dimensions are required to diagnose the domain of the study. In adding variables, there would eventually be fewer dimensions than measurements due to high correlations between certain pairs, i.e., linear dependence would occur and the rank of the input matrix would be reduced.

Criticism.   The above exhaustive method of ascertaining dimension number is so costly as rarely to be practical. When the goal is to determine dimensions, usually a large number of variables are selected and in some cases $r < p$ dimensions are concluded to be important on the basis of some criterion. However, recall that there are no precise criteria for doing this.

There are two stated justifications for rejecting components and turning to other factor analyses. First, there is no a priori provision made for components attributable to unreliability of variables or to lack of precision in measuring techniques. I do not believe that in most biological problems such errors can be defined as a precise fraction either of a covariance matrix, or of the variables generating the matrix. If they can, another method of factor analysis could be preferred; but the analysis might be superfluous. Second, components vary under changes in scale of measurement. In the case of indirect or intangible scales of measurement, naturally the correlation matrix is preferred; but this does not necessitate eliminating components analysis.

6.2 Factor Analysis Methodology

Components analysis deals with components having zero mean and variance $\lambda_i(\mathbf{a}_i\,'\mathbf{S}^2\mathbf{a}_i$ or $\mathbf{v}_i\,'\mathbf{R}\mathbf{v}_i)$. Factor analysis regularly deals with factors having zero mean and unit variances. In many cases factor analysis often also differs by having a model which decomposes each observation into "meaningful'' and ''error'' fractions. Finally, after rotation of factors, factor analysis may or may not be involved with orthogonal axes, but most of the discussion is limited to the latter.

As a heuristic device, factor analysis might avoid some of the deficiencies of principal component analysis. However, in component analysis deficiencies do not seem to create many problems, whereas those of factor model analysis often cause even greater problems.

Principal Factor Analysis.   As considered here, principal factor analysis is exactly equivalent to a correlation model component analysis in which the variance of $\mathbf{y}$ is an identity matrix rather than $\mathbf{v}'\mathbf{R}\mathbf{v}$, and only the first r components are interpreted. Therefore, the model for principal component analysis is changed very little to become a model for principal factor analysis. Unfortunately the form of the factor model often obscures the fact that it is closely related to the component model. A purpose of this subsection is to stress the relationship.

Factoring the Correlation Matrix.   Since only the first r components are interpreted, it is of interest to return to component analysis and to examine the effects of factoring or segregating the correlation matrix into orthogonal and additional matrices. Component j extracts so called factors from the total $\mathbf{R}$ matrix as follows:

$$\mathbf{R}_j = \lambda_j \mathbf{v}_j \mathbf{v}_j'$$

and the total correlation matrix $\mathbf{R}$ is supplied by

$$\mathbf{R} = \lambda_1 \mathbf{v}_1 \mathbf{v}_1' + \lambda_2 \mathbf{v}_2 \mathbf{v}_2' + \ldots + \lambda_p \mathbf{v}_p \mathbf{v}_p',$$

$$= \mathbf{R}_1 + \mathbf{R}_2 + \ldots + \mathbf{R}_p,$$

so

$$\mathbf{R} = \sum \lambda_j \mathbf{v}_j \mathbf{v}_j' = \sum \mathbf{R}_j = \mathbf{V}\mathbf{L}\mathbf{V}'.$$

If only the first r components are interpreted, the components are derived from

$$\hat{\mathbf{R}} = \mathbf{R}_1 + \mathbf{R}_2 + \ldots + \mathbf{R}_r'$$

and eliminate

$$\tilde{\mathbf{R}} = \mathbf{R}_{r+1} + \mathbf{R}_{r+2} + \ldots + \mathbf{R}_p.$$

The above is a true model of factor analysis since the total correlation is divided by component analysis into two elements. The first, $\hat{\mathbf{R}}$, termed the *theory variance*, supposedly explains the observable variation in individuals; the second, $\tilde{\mathbf{R}}$, termed the *unreliability* or *residual variance*, supposedly measures only experimental error.

Although both $\hat{\mathbf{R}}$ and $\tilde{\mathbf{R}}$ often are also called correlation matrices, both actually are partial dispersion matrices of standardized variation. Only their sum, $\mathbf{R}$, is a true correlation matrix having all unities in the principal diagonal. Therefore, the term correlation applied to $\hat{\mathbf{R}}$ or $\tilde{\mathbf{R}}$ is descriptive rather than accurate.

Obviously nothing is new here. The above merely formalizes the situation in which the first r components are interpreted. On the other hand, there is a potential source of misunderstanding. There are methods for selecting r, methods that might seem to lend credence to the results.

The Problem, Selecting r.   The frequent inability to select r in morphometric problems was shown in the turtle example (Chapter 4). In the single component analysis of females, there was no clear definition of the nature of the second and third components. Such could also be demonstrated in a component analysis of males only. The multigroup component analysis was necessary to "define" the last two components. In spite of this, two devices for selecting r are widely applied. The first method is essentially iterative. After a guess of r, the theory variance, $\hat{\mathbf{R}}$, and residual variance, $\tilde{\mathbf{R}}$, are calculated. Then, if the residual elements of $\tilde{\mathbf{R}}$ approximate a normal distribution, a mean of zero, and an allowable magnitude of variance, r is accepted. If not, another guess of r is made and $\tilde{\mathbf{R}}$ is again evaluated. This procedure is repeated until the criteria are satisfied. Obviously the procedure is arbitrary even if fixed in a computer program. The allowable magnitude of the residual variance is purely a matter of judgment even though the judgment often is implemented by considering sample size. The second procedure is a quick-and-dirty method proposed by Kaiser (1960)—r should equal the number of eigenvalues with values greater than unity. Psychometricians have found that Kaiser's rule works well for small to moderate sized samples but many believe r should be greater for larger sample sizes (Cooley and Lohnes, 1971).

The r Component Model.   First a component analysis of the total **R** matrix is performed. The first r components

$$\mathbf{y}_j = \mathbf{v}_j'\mathbf{z},$$

for $j = 1, 2, \ldots, r$ components, define $\hat{\mathbf{R}}$; and $\tilde{\mathbf{R}} = \mathbf{R} - \hat{\mathbf{R}}$ might be rejected as meaningless error. Since component scores $\mathbf{y}_j$ have zero mean and λ_j variance, component scores can be transformed to orthogonal *factor scores* $\mathbf{f}_j$ with zero mean and unit variance by dividing $\mathbf{y}_j$; by the square root of the corresponding eigenvalue, λ_j,

$$\mathbf{f}_j = \mathbf{y}_j/\sqrt{\lambda_j}, = \mathbf{v}_j'\mathbf{z}/\sqrt{\lambda_j},$$

which for all factors is

$$\mathbf{F} = \mathbf{L}^{-1/2}\mathbf{y} = \mathbf{L}^{-1/2}\mathbf{V}'\mathbf{z}.$$

At this point the criteria of $\overline{\mathbf{f}} = 0$ and $s_{fj}^2 = \dfrac{1}{N-1}\sum \mathbf{f}_j'\mathbf{f}_j = 1$ are satisfied. If we set

$$\mathbf{B}' = \mathbf{L}^{-1/2}\mathbf{V}',$$

then

$$\mathbf{F} = \mathbf{B}'\mathbf{z} = \mathbf{z}'\mathbf{B}$$

is a model similar to a principal component analysis of **R** appropriate for examining the variation in the first r to all components of **R**, but the variance of **F** is an identity matrix

$$\mathbf{B'RB} = \mathbf{I}$$

and $\mathbf{B}$ is a matrix of *factor score coefficients* analogous to $\mathbf{V}$. Again factor analysis might stop here—the factor score coefficients being interpreted much like component coefficients.

In addition to providing a principal component analysis of $\hat{\mathbf{R}}$ rather than $\mathbf{R}$, the above model does not extract all the variance of $\mathbf{z}$. Therefore, two hypothetical vectors, $\hat{\mathbf{z}}$ and $\tilde{\mathbf{z}}$, can be defined to explain this fact. The first, $\hat{\mathbf{z}}$, is a theoretical $\mathbf{z}$ satisfying

$$\hat{\mathbf{R}} = \frac{1}{N-1} \sum \hat{\mathbf{z}}' \, \hat{\mathbf{z}}, \text{ and}$$

the second, $\tilde{\mathbf{z}}$, a residual satisfying

$$\tilde{\mathbf{R}} = \frac{1}{N-1} \sum \tilde{\mathbf{z}}' \, \tilde{\mathbf{z}},$$

so

$$\mathbf{z} = \hat{\mathbf{z}} + \tilde{\mathbf{z}} \text{ and } \mathbf{R} = \hat{\mathbf{R}} + \tilde{\mathbf{R}}.$$

Since $\hat{\mathbf{z}}$ is the proportion of $\mathbf{z}$ involved in the above formulae, $\hat{\mathbf{z}}$ can substitute for $\mathbf{z}$ in all of them. Therefore,

$$\mathbf{F} = \mathbf{B'z} = \mathbf{B'}\hat{\mathbf{z}}.$$

Factor Structure. The correlation between the original variables and the components, component correlations or component structure, also is equal to factor structure, the correlation between original variables and factors. The factor structure of $\mathbf{R}$ is

$$\mathbf{S} = \frac{1}{N-1} \sum (\mathbf{z}_i - \bar{\mathbf{z}})(\mathbf{f}_i - \bar{\mathbf{f}})'$$

but since both $\bar{\mathbf{z}}$ and $\bar{\mathbf{f}}$ equal zero

$$\mathbf{S} = \frac{1}{N-1} \sum \mathbf{z}_i \mathbf{f}_i'$$

$$= \frac{1}{N-1} \sum \mathbf{z}_i (\mathbf{L}^{-1/2} \mathbf{V}' \mathbf{z}_1)'$$

$$= \frac{1}{N-1} \sum (\mathbf{z}_i \mathbf{z}_i') \mathbf{V} \mathbf{L}^{-1/2}$$

$$= \mathbf{RVL}^{-1/2} = \mathbf{RB}$$

and since $\mathbf{RV} = \mathbf{VL}$

$$\mathbf{S} = \mathbf{VLL}^{-1/2} = \mathbf{VL}^{1/2}$$

so factor structure is equal to component correlations. Note that the first r columns of $\mathbf{S}$ will contain the factor structure of $\hat{\mathbf{R}}$.

Factor Pattern.   Further meaning of factor structure comes from the regression of predicted observations $\hat{\mathbf{z}}$ on factor scores $\mathbf{f}$,

$$\hat{\mathbf{z}}_{ji} = \mathbf{w}_j'\mathbf{f}_i$$

where $\mathbf{w}_j$ are regression weights, calculated as in multiple regression, that provide predicted observations $\hat{\mathbf{z}}$ from factor scores $\mathbf{f}$. The matrix $\mathbf{W}'$ contains slope coefficients $\mathbf{w}_j'$ as rows and is called the factor pattern. The nature of the factors is important in the sense that they are orthogonal regression predictors. Therefore, as was stated in the chapter on regression, since slope coefficients of uncorrelated variables are equal to predictor-criterion correlations, factor pattern equals factor structure, i.e., $\mathbf{W} = \mathbf{S}$ for the first r components, so

$$\hat{\mathbf{z}} = \mathbf{S}'\mathbf{f} = \mathbf{f}'\mathbf{S}.$$

Often, and in the sense of factor pattern, the elements of $\mathbf{S}$ are called *factor loadings*. A vector of factor loadings identifies a factor that is read somewhat like a component.

Theory Correlation Matrix.   In principal factor analysis, the theory matrix also relates to the structure for the first r components,

$$\hat{\mathbf{R}} = \frac{1}{N-1} \sum \hat{\mathbf{z}}_i\hat{\mathbf{z}}_i'$$

$$= \frac{1}{N-1} \sum (\mathbf{S}\mathbf{f}_i)(\mathbf{S}\mathbf{f}_i)'$$

$$= \frac{1}{N-1} \sum (\mathbf{S}\mathbf{f}_i\mathbf{f}_i'\mathbf{S}')$$

$$= \mathbf{S} \; \frac{1}{N-1} \sum (\mathbf{f}_i\mathbf{f}_i')\mathbf{S}', \text{ and since } \frac{1}{N-1} \mathbf{f}_i\mathbf{f}_i' = \mathbf{I},$$

$$= \mathbf{S}\mathbf{I}\mathbf{S}'$$

$$= \mathbf{S}\mathbf{S}'$$

Therefore, the partition of $\mathbf{R}$ is

$$\mathbf{R} = \mathbf{S}\mathbf{S}' + \tilde{\mathbf{R}}.$$

Factor Model Analysis.   The above discussion provides all material needed to provide a true factor model of principal factor analysis. This will be the first step. As a second step communalities will be applied to provide the form of the more complicated models.

Principal Factor Model. The various relationships allow presentation of the general form of the principal factor model for the first r components. Data vectors are partitioned,

$$\mathbf{z} = \hat{\mathbf{z}} + \tilde{\mathbf{z}},$$

i.e., each data vector $\mathbf{z}$ is equal to a hypothesis or theoretical fraction plus a residual or error fraction, where

$$\hat{\mathbf{z}} = \mathbf{f}'\mathbf{S}' \text{ and}$$

$$\tilde{\mathbf{z}} = \mathbf{c}'\mathbf{E}',$$

$\mathbf{f}$ being a vector of r factor scores, consecutive coefficients pertaining to successive factors (here eigenvectors), $\mathbf{S}$ here being the left $p \times r$ submatrix of the total structure matrix with the p elements for each of the r factors being factor loadings for a factor, $\mathbf{c}$ being a vector of $p - r$ elements (here the last $p - r$ of the total $\mathbf{f}$ vector), and $\mathbf{E}$ here being the right $p \times (p - r)$ submatrix of $\mathbf{S}$. Also the variance of $\hat{\mathbf{z}}$, $\hat{\mathbf{R}}$, still is termed a theory variance and that of $\tilde{\mathbf{z}}$, $\tilde{\mathbf{R}}$, a residual variance or unreliability. In practice the factor model generally is presented as

$$\mathbf{z} = \mathbf{f}'\mathbf{S}' + \mathbf{c}'\mathbf{E}', \text{ with variance}$$

$$\mathbf{R} = \hat{\mathbf{R}} + \tilde{\mathbf{R}}.$$

Principal Factor Analysis with Communalities. This model involves very little change from the previous one. Again a correlation matrix is used and subdivided in the same manner. In this example only the diagonals of $\hat{\mathbf{R}}$ will be altered. The new diagonal elements of $\hat{\mathbf{R}}$, called *communalities*, each supposedly represent the proportion of a variable's variation shared with the other variables. Communalities tend to be derived from one of three methods: 1) theoretical, computation provides values producing the minimum rank possible for the matrix; 2) iterative, a rank and initial communality is guessed or assumed, a component analysis of the resulting matrix leads to computation of new communalities for the next component analysis, and so on until the new communalities equal the old; and 3) empirical, various quick-and-dirty approximations of communalities, e.g., the multiple correlation coefficient between each variable and all others in the set, are applied.

In the sense of the principal factor model developed,

$$\hat{\mathbf{z}} = \mathbf{z}_h + \mathbf{z}_u, \text{ so}$$

$$\mathbf{z} = \mathbf{z}_h + \mathbf{z}_u + \tilde{\mathbf{z}},$$

where $\hat{\mathbf{z}}$ has been partitioned into $\mathbf{z}_h$, a so-called common factor variate representing meaningful variation shared by the variables of $\mathbf{z}$, and $\mathbf{z}_u$, a latent; unique or specific factor variable, representing variation of each variable in $\mathbf{z}$ that is uncorrelated or not shared with any other variable. Therefore, $\mathbf{R}_u$, the correlation matrix for $\mathbf{z}_u$, must be a diagonal matrix and is called the specificity or latent, specific or unique variance. Parenthetically, it must be noted that in some models no $\tilde{\mathbf{z}}$ is defined, i.e., $\mathbf{z} = \mathbf{z}_h + \mathbf{z}_u$.

A simple example of obtaining $\mathbf{R}_h$ and $\mathbf{R}_u$ from $\hat{\mathbf{R}}$ where $\hat{\mathbf{R}} = \mathbf{R}_h + \mathbf{R}_u$ is to simply supply new communalities, h^2 to the diagonal of $\hat{\mathbf{R}}$, so

$$\hat{\mathbf{R}} = \mathbf{R}_h + \mathbf{R}_u = \begin{bmatrix} h_1^2 & \hat{r}_{12} & \dots & \hat{r}_{1p} \\ \hat{r}_{21} & h_2^2 & \dots & \hat{r}_{2p} \\ \dots & \dots & \dots & \dots \\ \hat{r}_{p1} & \hat{r}_{p2} & \dots & h_p^2 \end{bmatrix} + \begin{bmatrix} d_1^2 & 0 & \dots & 0 \\ 0 & d_2^2 & \dots & 0 \\ \dots & \dots & \dots & \dots \\ 0 & 0 & \dots & d_p^2 \end{bmatrix}$$

where $d_i^2 = \hat{r}_{ii} - h_i^2$. This leads to the model

$$\mathbf{z} = \mathbf{z}_h + \mathbf{z}_u + \tilde{\mathbf{z}}$$

$$= \mathbf{f}'\mathbf{H} + \mathbf{d}'\mathbf{U} + \mathbf{c}'\mathbf{E}' \text{ with variances}$$

$$\mathbf{R}^2 = \mathbf{R}_h + \mathbf{R}_u + \mathbf{R}$$

where, since $\mathbf{H}$ and $\mathbf{U}$ are left and right submatrices of the factor structure,

$$\mathbf{S} = [\mathbf{H} \ \mathbf{U}], \text{ so}$$

$$\mathbf{R}_h = \mathbf{H}'\mathbf{H} \text{ and}$$

$$\mathbf{R}_u = \mathbf{U}'\mathbf{U}.$$

For simplicity the fact that eigenvalues and eigenvectors are derived from $\mathbf{R}_h$ rather than $\mathbf{R}$ was avoided, hence $\mathbf{f}$ and $\mathbf{S}$ are not the same as before. On the other hand, again they are not far removed from the other values.

There are many factor models and many numerical solutions. The different approaches often are based upon predetermined numerical characteristics of the factor structure (factor loadings) and/or the number of factors to be extracted. However, recall that models are not limited to those with different sets of factor loadings being orthogonal to one another. Some methods definitely create coefficients different from component coefficients, but this also is true when factors are rotated. Therefore, the concepts of unit stimuli and independent patterns of variation can become quite remote in factor analysis.

6.3 Rotation

Rotation can be from component axes to arbitrary orthogonal axes or to oblique axes. Both rotations, but especially that to oblique axes, reestablish the problems of correlations; but are justified on the basis that biological traits are in fact correlated with one another in nature. In spite of this, for some purposes even oblique factors seem to be useful. For example, selection of two shape factors to summarize maximum variation among individuals in the 2-or-3-space defined by oblique axes appears to accomplish a goal of numerical taxonomy.

A purpose of rotating factors is supposedly to obtain more meaningful positions for the estimated common factors. Unfortunately, the rationale for rotating often is "knowing" what factors are common to the p variates and which are not. In practice, different "knowledge of factors" leads to different factor models, rotation schemes, and results.

The goal of rotation often is to reduce many variables that test individuals under a wide variety of conditions to a few simple factors, each defined by a few high and many low loadings, *simple structure* that provides the essence of a phenomenon. For example, psychologists have applied different models and batteries of psychological tests to disclose the "hidden factors of the mind," the simple structure of the mind. Although some effort was made in these studies, unlike methods provided unique results and psychologists still disagree on the "factors of the mind."

Ignoring the shortcomings of simple structure, there are two biological reasons for rejecting it. First, biologists are interested in the various ramifications of a phenomenon and especially how it varies, i.e., patterns of variation rather than an estimate of static conditions. Second, biological data rarely are so encompassing as to circumscribe a phenomenon. Therefore, any simple structure probably would not be that of the total phenomenon.

Chapter 7
Canonical Correlation Analysis

The correlation between two variables can be extended to the correlation between sets of variables. Often many of the shortcomings of correlation coefficients are reduced or eliminated in the process.

Consider a system of $p + q = t$ multinormal responses. The first p variates share a common feature and so do the second q variates. Each is a logical set, a group of random variables that share an attribute, e.g., a set of ecological, morphological, physiological, or ethological variables. In fact, two definable subunits of morphological or other variable units can be contrasted. In canonical correlation, each set is on the same statistical basis, neither must be considered predictors nor criteria. In fact, in a single evaluation of the data first one set can be predictors of the other and then the opposite is possible. Of course, this differs from multiple correlation where there is always a single set of predictors and a single criterion.

This chapter starts with possible ways of studying the correlation between two sets of data. Section two introduces the basic methodology of canonical correlation. Section three summarizes limitations of the methodology. Section four develops various statistics useful as aids for interpreting the results of a canonical correlation, introduces an example, and shows how the aids are applied. The next section, five, turns to an entirely different concept of examining relationships between two data sets, an examination that involves three procedures. First, multiple regression methods segregate the part of each set predicted by the other set from the part not predicted. Second, principal component analysis provides the total, the predicted, and the residual (not predicted) patterns of variation for each set. Finally, all components are evaluated to determine the nature of the interset associations. The methodology reflects the predictive strength of multiple regression, serves as a check on canonical correlation, and leads to interpretation of independent associations (components) rather than correlation coefficients. The chapter ends with an epilogue.

7.1 Correlation Approaches to Analysis

The affinities between two data sets can be studied in many ways. Three kinds of correlation coefficients can be applied. But, first recall

$$\mathbf{R} = \begin{bmatrix} \mathbf{R}_{11} & \mathbf{R}_{12} \\ \mathbf{R}_{21} & \mathbf{R}_{22} \end{bmatrix}$$

Four submatrices are defined: $\mathbf{R}_{11}$, the submatrix of set one bivariate correlations; $\mathbf{R}_{22}$, the submatrix of set two bivariate correlations; $\mathbf{R}_{12}$, the submatrix of bivariate correlations between elements of set one and set two; and $\mathbf{R}_{21}$ ($=\mathbf{R}_{12}'$), the submatrix of bivariate correlations between elements of set two and set one. Having established the submatrices, first the bivariate correlation coefficients in $\mathbf{R}_{12}$ or $\mathbf{R}_{21}$ can be examined. Second, the bivariate correlations of all possible between set component scores or components of one set and raw variables of the other set can be evaluated. Finally, the canonical correlations between two data sets might be preferred.

$\mathbf{R}_{12}$ **Correlations.**   Although $\mathbf{R}_{12}$ or $\mathbf{R}_{21}$ bivariate correlations can be helpful in interpreting other analyses, generally they present too many problems for interpretation. All the sources of correlation coefficient errors stressed in chapter three would apply.

Correlations of Component Scores.   Correlations of interset components are vastly superior to those in $\mathbf{R}_{12}$ or $\mathbf{R}_{21}$. This is true since a correlation of interset components cannot be influenced by any other component of either set. However, spurious correlations still are possible. Neither time nor any other potential cause of meaningless between set associations is eliminated. Another potential problem is that correlation of components does not consider other possible relationships between sets. Canonical correlation can discover any association, including many very remote from components. For this reason, in a statistical sense, the correlation of component scores does not "cover" the multidimensional canonical correlation space defined by individual variables. After canonical correlation is viewed in some detail, the nature, existence, and implications of non-component associations can be evaluated.

Canonical Correlation.   This might be closest to the goal of measuring correlations between data sets. The methodology entails eigenvalue-eigenvector solutions which clarify the procedure. Eigenvalues are the squares of so-called canonical correlation coefficients. The first eigenvalue represents the maximum possible correlation between linear combinations of the two data sets and successive eigenvalues are maxima subject to relating to orthogonal axes. Two axes, defined by an eigenvector function, pertain to each canonical correlation, and one of each pair of axes is associated with one of the data sets.

The methodology defines two immediate sources of potential error. First, although correlations are eliminated within each data set, as for the above components, spurious correlations still can occur between sets. Second, each canonical correlation being a maximum, although subject to orthogonality, may or may not exist in reality.

7.2 Canonical Correlation Methodology

Consider for each individual a vector $\mathbf{X}$ of $p + q = t$ variables, where the first p elements define a logical set in a subvector $\mathbf{X}_1$ and the last q elements define another logical set in a subvector $\mathbf{X}_2$,

$$\mathbf{X} = \begin{bmatrix} \mathbf{X}_1 \\ \mathbf{X}_2 \end{bmatrix}.$$

The two sets are determined arbitrarily so $p > q$. This is important since the number of canonical correlation coefficients is the minimum of p, q. Therefore, q will always be the number of canonical correlations.

Both $\mathbf{X}_1$ and $\mathbf{X}_2$ are assumed to follow the *mnd*. The parameters of $\mathbf{X}_1$ and $\mathbf{X}_2$ exist among the partitioned population centroid vector and dispersion matrix,

$$\mu = \begin{bmatrix} \mu_1 \\ \mu_2 \end{bmatrix} \text{ and } \sum{}^2 = \begin{bmatrix} \sum_{11} & \sum_{12} \\ \sum_{21} & \sum_{22} \end{bmatrix}.$$

It is further assumed that the first q eigenvalues of

$$\Sigma_{21}\Sigma_{11}^{-1}\Sigma_{12}\Sigma_{22}^{-1}$$

are distinct, nonzero and unequal. For this to be the case requires in part that Σ_{11}^{-1} and Σ_{22}^{-1} are non-singular.

In the model most widely used, the normal score model,

$$\mathbf{z} = \begin{bmatrix} \mathbf{z}_1 \\ \mathbf{z}_2 \end{bmatrix}, \quad \mu = \begin{bmatrix} 0 \\ 0 \end{bmatrix}, \text{ and } \Sigma^2 = \mathbf{P} = \begin{bmatrix} \mathbf{P}_{11} & \mathbf{P}_{12} \\ \mathbf{P}_{21} & \mathbf{P}_{22} \end{bmatrix},$$

where $\mathbf{P}$ is the parametric matrix of bivariate correlations.

Turning to a sample, either the dispersion or correlation matrix could be applied,

$$\mathbf{S}^2 = \begin{bmatrix} \mathbf{S}_{11} & \mathbf{S}_{12} \\ \mathbf{S}_{21} & \mathbf{S}_{22} \end{bmatrix} \text{ or } \mathbf{R} = \begin{bmatrix} \mathbf{R}_{11} & \mathbf{R}_{12} \\ \mathbf{R}_{21} & \mathbf{R}_{22} \end{bmatrix}.$$

Canonical correlation coefficients can be derived from either

$$\mathbf{S}_{21}\mathbf{S}_{11}^{-1}\mathbf{S}_{12}\mathbf{S}_{22}^{-1} \text{ or } \mathbf{R}_{21}\mathbf{R}_{11}^{-1}\mathbf{R}_{12}\mathbf{R}_{22}^{-1}$$

Both expressions lead to identical canonical correlations but the latter is preferred. For the former expression, coefficients of eigenvectors reflect variable responses within sets. For this reason only the z-score model will be considered further.

Canonical correlation is canonical in the sense that it reduces the number of variates and their nonzero correlations to a minimum, q.

Linear Model.   This method is accomplished by linear functions,

$$\mathbf{x}_1 = \mathbf{c}_1'\mathbf{z}_1 \quad \mathbf{y}_1 = \mathbf{d}_1'\mathbf{z}_2$$

$$\mathbf{x}_2 = \mathbf{c}_2'\mathbf{z}_1 \quad \mathbf{y}_2 = \mathbf{d}_2'\mathbf{z}_2$$

$$\cdots \qquad \qquad \cdots$$

$$\mathbf{x}_q = \mathbf{c}_q'\mathbf{z}_1 \quad \mathbf{y}_q = \mathbf{d}_q'\mathbf{z}_2$$

with the property that the correlation between $\mathbf{x}_1$ and $\mathbf{y}_1$ is the greatest possible; that between $\mathbf{x}_2$ and $\mathbf{y}_2$ is the next greatest possible among all linear functions uncorrelated with (hence also orthogonal to) $\mathbf{x}_1$ and $\mathbf{y}_1$; and so on for all q possible pairs.

Each $\mathbf{c}_i\mathbf{d}_i$ pair are *canonical vectors* which contain p and q *canonical coefficients* respectively that define a plane. For each $\mathbf{c}_i\mathbf{d}_i$, *canonical variates* $\mathbf{x}_i$ and $\mathbf{y}_i$ are calculated for each individual and can be visualized as being plotted on the plane. The measure of association between each pair of canonical variates is a *canonical correlation coefficient*. Each pair of canonical variates indicates the amount of an individual's contribution to the canonical correlation.

The statistical model and its restrictions can be summarized as follows:

$$\mathbf{x} = \mathbf{c}'\mathbf{z}_1 \quad \mathbf{y} = \mathbf{d}'\mathbf{z}_2$$

and since $\bar{\mathbf{z}}_1$ and $\bar{\mathbf{z}}_2$ both are null vectors,

$$\bar{\mathbf{x}} = \mathbf{0} \text{ and } \bar{\mathbf{y}} = \mathbf{0}.$$

Also,

$$s_x{}^2 = 1, \quad s_y{}^2 = 1,$$

$$r_{x_i x_j} = 0, r_{y_i y_j} = 0, \text{ and}$$

$$R_c = \frac{1}{N-1}\sum \mathbf{x}_i \mathbf{y}_{i\ max}.$$

Note that the correlation between any pair of $\mathbf{x}_i$ or $\mathbf{y}_i$ is zero.

The solution becomes an eigenvalue problem. Since one wishes maximum correlations between corresponding $\mathbf{x}_i$ and $\mathbf{y}_i$ but not within either set, the characteristic equation is required to satisfy the criteria. Actually the characteristic equation operates upon the squares of interset correlations

$$\mathbf{R}_{21}\mathbf{R}_{11}^{-1}\mathbf{R}_{12}\mathbf{R}_{22}^{-1}$$

whose product is a $q \times q$ assymetric matrix.

The eigenvalues of

$$\mathbf{R}_{21}\mathbf{R}_{11}^{-1}\mathbf{R}_{12}\mathbf{R}_{22}^{-1} - \lambda\mathbf{I} = 0$$

provide q maximum squares of correlations, the scalar variable λ_i. Then, coefficients of each of the second set canonical vectors, $\mathbf{d}_i$ (each with q elements), are derived from the eigenvector function

$$(\mathbf{R}_{21}\mathbf{R}_{11}^{-1}\mathbf{R}_{12}\mathbf{R}_{22}^{-1} - \lambda_i\mathbf{I})\mathbf{d}_i = 0$$

with the restriction that

$$\mathbf{d}_i'\mathbf{R}_{22}\mathbf{d}_i = 1;$$

and coefficients of each of the first set canonical vectors, $\mathbf{c}_i$ (each with p elements) are obtained from

$$\mathbf{c}_i = \mathbf{R}_{11}^{-1}\mathbf{R}_{12}\mathbf{d}_i/\lambda_i$$

with the restriction that

$$\mathbf{c}_i'\mathbf{R}_{11}\mathbf{c}_i = 1.$$

As expected

$$\lambda_i = R_{c_i}^2,$$

the square of the ith canonical correlation between sets.

This means that there are certain similarities, but also distinct differences, between component and canonical correlation analyses. The major similarity is that in both methods original variable axes of a set are rotated to a new set of axes, selected so there are zero correlations between the resulting transformed variables. In both cases, the original Euclidian space is not warped in any way, hence there is no change in original positions, including original distances, among individuals. However, here the similarity ends. In component analysis, successive axes also are chosen to express maximum possible variance within a set of variables. In canonical correlation, axes are further selected so between set pairs have the maximum possible correlation. Most definitely the canonical axes of each set need not correspond to their component axes.

Visualization of the Transformation. Consider two vectors, $\mathbf{z}_1$ and $\mathbf{z}_2$, each representing a logical set and containing two elements of z-scores,

$$\mathbf{z}_1 = \begin{bmatrix} z_{11} \\ z_{12} \end{bmatrix} \text{ and } \mathbf{z}_2 = \begin{bmatrix} z_{21} \\ z_{22} \end{bmatrix}.$$

Next, a sample of N individuals is collected, two sets of two measurements are taken on each individual, the measurements are transformed to z-scores, and each bivariate pair of z-scores is plotted on a separate graph, z_{11} and z_{12} on one graph and z_{21} and z_{22} on another. The latter step is purely one for convenience in interpretation since the four z-scores properly define a 4-dimensional space. The next step can be imagined as relating the two graphs in that single possible 2-dimensional space in which data points will lead to the maximum possible correlation coefficient. This would define the points in terms of axes x_1 and y_1 which create the best 2-dimensional portrayal of the 4-dimensions. Best is determined in the sense of portraying a compromise between the longest possible by the narrowest possible distances between individuals, i.e., the longest by the narrowest spread of data points along a line. Finally, $\mathbf{c}_1$ in $\mathbf{c}_1'\mathbf{z}_1$ and $\mathbf{d}_1$ in $\mathbf{d}_1'\mathbf{z}_2$ provide the $x_1 y_1$ coordinates for the N individuals.

Naturally, neither $\mathbf{x}_1$ nor $\mathbf{y}_1$ need to conform to any principal component axis of either set; but it should be clear that these canonical variates are analagous to component scores. This comparison can be extended to canonical variates being analagous to size and shape variables and to canonical vectors being the basis for interpreting the nature of their corresponding canonical variates.

7.3 Limitations of Methods

Since canonical correlation is a generalization of multiple correlation most limitations pertain to that fact. The potential of nonsense and inflated correlations exists. Also, data per se and interaction between original variables require examination.

Data Limitations. As raw variables are added or subtracted from sets, canonical correlations and canonical vectors frequently are modified. This occurs because the domains of each set are functions of the original variables defining them. As variables are added or subtracted the dimensions of domains are increased or decreased. Also, variables important to the canonical correlation might be added or subtracted. In any case ipsative and related measures must be avoided. Small sample sizes should not be used. Small samples applied to repeated study of the same two data sets from the same population can lead to marked differences in results. Perhaps it is obvious that in a series of small samples one or both domains might be poorly, hence differently, defined. The minimum sample size is

$$N > p + q + 1.$$

However, this often is an inadequate sample size. Samples of 100 individuals might be considered minimal for consistent reliance in results.

Correlations and Interactions. The magnitude of correlations between variables can create another problem. If the correlation between two variables of a set is unity, for example set one, the inverse R_{11}^{-1} does not exist, i.e., R_{11} is singular. Since the characteristic equation requires R_{11}^{-1}, if it does not exist, canonical correlation is impossible. In cases where either R_{11} or R_{22} approach singularity, canonical correlations can be poorly defined.

The amount of interaction between sets is directly related to the amount of reduction of correlation between sets (Lee, 1971). If interaction is strong enough, interset correlation is zero. The contrast between high correlation and high interaction of set one z_1 and set two z_2 can be visualized by using a single vector to summarize the data of each set (Figure 7.1).

7.4 Interpretation of Canonical Vectors

Interpretation of a canonical correlation study can be more formalized, hence simpler, than that for conventional correlation analysis. In conventional correlation there are $1/2\,p(p-1)$ set one intercorrelations, $1/2\,q(q-1)$ set two intercorrelations, plus pq intercorrelations of set one and set two variables. In canonical correlation there are only q correlations. The latter have the further advantage of being independent of one another. Also, variables within any canonical vector are independent. Therefore, interpretation is more direct.

Interpretation of a canonical correlation entails examining high coefficients of corresponding pairs of canonical vectors, $c_i d_i$, for relationships between sets. The nature of an interpretation must

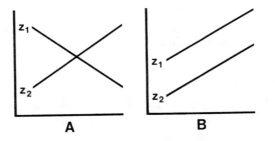

Figure 7.1 Extremes of correlations between data vectors z_1 and z_2. A. Conditions of strong interaction between sets and zero interset correlation. B. Condition of no interaction between sets and high interset correlation.

depend upon the nature of the data sets. In some cases, it is most appropriate to consider a single variable of one set as a predictor of a single variable of the other set. This one-to-one evaluation might be best when the criterion set is loosely related, e.g., when the set is composed of different species as variables, or even when the set consists of such things as weight, age group, and reproductive condition of a species. However, when all variables in a criterion set are joined so closely that scrutinizing patterns of variation is appropriate, the canonical vectors of that set can be translated as patterns of variation predicted by a single predictor of the other set. Perhaps even reference to size and shape variables is appropriate. Again, the rule of biological sense applies to interpretation.

An Example.   The example comes from data collected on the sea palm, *Postelsia palmaeformis* Ruprect, an intertidal laminarian that inhabits wave swept rocky shores along the Pacific Coast of North America. The study was made at a southernmost known locality of this brown alga (just south of Montano de Oro State Park, San Luis Obispo County, California) by Young (1971) in conjunction with other studies of this species. In part, the purpose of the field study was to determine which environmental factors might influence the annual cycle of growth through disappearance of the alga.

The plants are annuals, sporophytes start germinating in about February, and most disappear by about late December. For this study five plants were selected at random and the same five plants were measured each month from March through December, 1970. The set one variables consisted of six morphological measurements made on each plant. These were holdfast height, holdfast width, stipe length, stipe basal diameter, stipe apical diameter, and blade length, variables selected to outline the form of the alga. The set two variables consisted of six temperature records, maximum air, minimum air, average air, average sea water, maximum sea water, and minimum sea water. The temperatures were all monthly averages, hence a single record of temperatures pertained to the same five plants each month. The temperature records were from the closest recording station, Morro Bay some four miles north of the study area. It is assumed that some differences exist between the study area temperatures and those of Morro Bay. However, the temperature data are no more remote from the *Postelsia* study area than they would be for many other biological studies. Moreover, the environmental data appear to be satisfactory for the present study.

The temperature data are correlated with the plant measurements (Table 7.1). Note that the highest morphological correlations are with minimum temperatures. This table is presented only as a point of comparison with the results of canonical correlation.

Table 7.2 presents the six pairs of canonical vectors (six since $q = 6$) and the canonical correlation coefficients for each pair of vectors. Although not discussed until later, the probabilities of accepting tests of hypotheses of equality of canonical correlations are appended in the form of a probability for acceptance of each hypotheses. Evaluation of the statistics in Table 7.2 will be delayed until more interpretive aids are developed and presented.

Canonical Correlation Coefficient, R_c, and R_c^2.   A canonical correlation coefficient is a measure of the goodness of fit of one set of canonical variates to another and a square of a canonical correlation coefficient measures shared variance between a given pair of canonical variates, **x** and **y**. Since both R_c and R_c^2 are based upon transformed variables, **x** and **y**, rather than the original vector random variables, z_1 and z_2, R_c and R_c^2 are very dependent upon the success of the transformation, i.e., they may or may not be important in terms of z_1 and z_2. In essence both canonical correlation statistics are subject to the problems of correlation statistics.

Table 7.1 Intercorrelations between environmental variables, monthly average temperatures, and six measurements on the sea palm, *Postelsia palmaeformis* Ruprect.

sea palm dimensions	monthly average temperatures					
	air temperature			water temperature		
	max.	min.	avg.	avg.	max.	min.
holdfast height	.319	.780	.638	.882	.843	.883
holdfast width	.305	.752	.623	.898	.842	.915
stipe length	.261	.688	.539	.762	.702	.786
stipe basal diameter	.404	.796	.701	.904	.849	.914
stipe apical diameter	.440	.816	.733	.904	.845	.922
blade length	.211	.681	.505	.717	.653	.740

Table 7.2 Normalized canonical vectors, canonical correlations, and probabilities of canonical correlations being zero between environmental and morphological sets of data for *Postelsia palmaeformis* Ruprect.

	canonical vectors					
	1	2	3	4	5	6
holdfast height	−.210	.104	.288	−.611	.287	−.072
holdfast width	−.809	−.541	.511	.594	.111	−.155
stipe length	.276	.368	.223	−.243	−.767	.021
stipe base	−.314	.148	−.001	.181	−.248	.796
stipe apex	−.291	−.504	−.697	−.261	.114	−.580
blade length	.300	.560	−.346	.338	.493	−.009
max. air	.174	.351	.083	−.233	−.505	.065
min. air	.137	.514	.045	−.235	−.221	.004
avg. air	−.245	−.764	−.270	.357	.745	−.085
avg. water	−.822	.127	−.717	.676	.204	.816
max. water	.464	−.105	.546	−.535	−.052	−.338
min. water	.027	−.035	.324	−.138	−.310	−.457
R_c	.980	.676	.524	.338	.188	.146
probabilities	.000	.005	.170	.576	.655	.339

Canonical Variates. Both scores of each individual on each canonical correlation are a measure of that individual's participation in an association between data sets. These pairs of canonical variates are evaluated just like component scores but are simpler to apply since corresponding morphological and environmental canonical variate scores must associate if the individual is a valid participant in the relationship. The association of canonical variates must be consistent as to sign and high absolute magnitude of scores. These scores are not shown for the sea palm data.

Variable Percentage Contribution. As was the case for component analysis, the magnitude of coefficients of canonical vectors does not determine whether a given variable is important in a correlation. However, the percentage contribution of each variable to each canonical variate can be helpful. It indicates the role of each variate in each set of the canonical vectors. In algebraic form,

$$100(c_{ij}^2 \lambda_i)/ \sum_{i=1}^{q} \sum_{j=1}^{p} c_{ij}^2 \lambda_i$$

is the percentage contribution of the jth variable of the first set to the ith canonical variate, where $i = 1$, 2, ..., q sets of canonical vectors. Also,

$$100(d_{ik}^2 \lambda_i)/ \sum_{i=1}^{q} \sum_{j=1}^{p} d_{ik}^2 \lambda_i$$

is the percentage contribution of the kth variable of the second set to the ith term of the eigenvalue λ_i. These statistics are interpreted in the same way as the corresponding ones in principal component analysis. For the sea palm, the statistics are shown in Table 7.3.

Table 7.3 Percentage of the variance of each variable's contribution to each canonical variate of sea palm and environmental variables.

	canonical vectors					
	1	**2**	**3**	**4**	**5**	**6**
holdfast height	36.60	4.29	19.63	36.88	2.51	.10
holdfast width	72.88	14.03	8.31	4.67	.05	.06
stipe length	32.20	40.70	9.00	4.45	13.65	.01
stipe base	70.90	9.83	.00	3.69	2.14	13.43
stipe apex	23.53	33.53	38.48	2.24	.13	2.09
blade length	30.33	50.51	11.56	4.58	3.02	.00
max. air	28.22	54.99	1.86	6.06	8.78	.09
min. air	12.28	81.90	.38	4.27	1.17	.00
avg. air	15.15	70.52	5.28	3.83	5.17	.04
avg. water	75.00	.85	16.31	6.05	.17	1.65
max. water	62.77	1.55	24.93	9.97	.03	.75
min. water	1.71	1.42	71.77	5.46	8.46	11.17

Canonical Structure.   Examining $\mathbf{c}_i$ and $\mathbf{d}_i$ alone does not indicate which variables contribute most heavily to the maximum correlation between the sets. The measure of the latter is a set of coefficients analogous to component correlations (= component structure or component pattern). When called structure, the coefficients indicate the correlation of elements of a canonical vector with corresponding elements of data vectors. When called pattern, the coefficients are regression coefficients for the multiple regression of each element of a data vector of one set on the canonical variate of the same set. The formulae in algebraic form are

$$s_{1ij} = \frac{1}{N-1} \sum z_{1jl} x_{il} = \frac{1}{N-1} \sum z_{1jl}(c_{ij}'z_{1jl}) = \mathbf{R}_{11}\mathbf{c}_i$$

and

$$s_{2ik} = \frac{1}{N-1} \sum z_{2kl} y_{il} = \mathbf{R}_{22}\mathbf{d}_i$$

where i is the canonical variate, j the variable of set one, k the variable of set two, and l the individual. The formulae clearly show that structure is the correlation between an original variable and a canonical variate. In matrix algebra form the two formulae can simplify to

$$\mathbf{s}_1 = \mathbf{R}_{11}\mathbf{c} \qquad \mathbf{s}_2 = \mathbf{R}_{22}\mathbf{d}$$

Due to the complexity of the computation, it might not be surprising that a variable's structure can differ in sign and markedly in magnitude from the corresponding coefficient of a canonical vector. Table 7.4 displays these statistics for the sea palm.

Variance Extracted.   The proportion of the variance of the variables of each set extracted by each canonical variate, $\mathbf{x}_i$ and $\mathbf{y}_i$, discloses the magnitude of $\mathbf{x}_i$ and $\mathbf{y}_i$ contributions to all aspects of the ith dimension. This is analogous to the percentage of the variance of each variable accounted for by each principal component. For the first set, in algebraic form, these are

$$s_{c_{ij}} = \sum s_{1_{ij}} s_{1_{ij}} / p$$

and for the second set

$$s_{d_{ik}} = \sum s_{2_{ik}} s_{2_{ik}} / q$$

and in matrix algebra form

$$\mathbf{s}_c = \mathbf{s}_1'\mathbf{s}_1 / p$$

$$\mathbf{s}_d = \mathbf{s}_2'\mathbf{s}_2 / q$$

Redundancy.   Redundancy measures the actual overlap between sets in canonical space. It is useful either for individual pairs of canonical variates, or for the sum of all pairs, termed total redundancy. The overlap for individual pairs, the ith pair is measured separately for the two canonical

Table 7.4 Canonical structure (correlations between canonical variables and original variables), variance extracted and redundancy for each canonical variate of the sea palm data.

morphological structure	canonical variates					
	1	**2**	**3**	**4**	**5**	**6**
morphological structure						
holdfast height	−.913	.296	.073	−.245	.110	−.022
holdfast width	−.959	.181	.073	.146	−.033	−.138
stipe length	−.817	.483	−.017	.015	−.241	−.204
stipe base	−.952	.144	−.152	−.031	−.047	.216
stipe apex	−.954	.121	−.236	−.054	−.080	−.106
blade length	−.781	.582	−.108	.130	.048	−.140
variance extracted	.808	.121	.017	.017	.014	.023
redundancy	.775	.055	.005	.002	.000	.000
environmental structure						
max. air	−.404	−.263	−.551	−.498	−.465	−.010
min. air	−.815	.130	−.316	−.381	.243	−.121
avg. air	−.723	−.142	−.474	−.469	−.058	−.085
avg. water	−.972	−.037	.031	−.225	.014	.046
max. water	−.922	−.079	.106	−.339	.102	.081
min. water	−.981	−.017	−.015	−.136	−.105	−.086
variance extracted	.685	.019	.107	.133	.050	.006
redundancy	.657	.009	.029	.015	.002	.000

variates. The redundancy of $\mathbf{x}_i$ is equal to the proportion of the variance extracted by $\mathbf{x}_i$, $\mathbf{s}_{c_i}$, times the variance shared between $\mathbf{x}_i$ and $\mathbf{y}_i$ in set one, $R_{c_i}^2$. The redundancy of $\mathbf{y}_i$ is equal to the proportion of the variance extracted by $\mathbf{y}_i$, $\mathbf{s}_{d_i}$, times the variance shared between $\mathbf{y}_i$ and $\mathbf{x}_i$ in set two, R_c^2 ,

$$R_{dx} = (\mathbf{s}_1'\mathbf{s}_1)/p)R_c^2 = s_c R_c^2$$

$$R_{dy} = (\mathbf{s}_2'\mathbf{s}_2)/q)R_c^2 = s_d R_c^2.$$

Consider the meaning of ith canonical correlation coefficient and its extracted variance. Since $R_{c_i}^2$ is the proportion of the shared variance of $\mathbf{x}_i$ and $\mathbf{y}_i$, and $\mathbf{s}_{c_i}$ is the proportion of the shared variance extracted from set one by $\mathbf{x}_i$, R_{dx_i} is the proportion of set one, $\mathbf{x}_i$, that overlaps set two, $\mathbf{y}_i$. Also, R_{dy_i} is the proportion of set two that overlaps set one. These two redundancies often are unequal. For example, $\mathbf{x}_i$ might be a major source of variation in $\mathbf{z}_1$, perhaps close to the first principal component of $\mathbf{z}_1$; and $\mathbf{y}_i$ might be a minor source of variation in $\mathbf{z}_2$, perhaps close to the last principal component. Then, R_{dx_i} would be much larger than R_{dy_i}. Therefore, for interpretation, the larger the redundancy of sets, the greater the relationship.

The largest possible redundancy is unity. This can occur only when the canonical correlation is unity and all the variance of a set is extracted. In no case can redundancy exceed the value of the square of the canonical correlation coefficient or the variance extracted. Redundancy is more useful than the canonical correlation coefficient since it measures actual overlap of data vectors rather than canonical variates as in R_c^2 (Cooley and Lohnes, 1971). For example, a very high intercorrelation between a single pair of variables, one from each set, could produce a high canonical correlation. However, this situation would be implied by low redundancy.

There is no general rule for establishing the lower limit of redundancy for interpretive purposes. Lower limits depend upon good biological sense for any basis of judgment, and, therefore, must be left to the user. Also, redundancy is not a statistic that one accepts automatically. Since R_{dx} and R_{dy} both are functions of R_c^2 the limitations on the canonical correlation coefficient pertain to redundancy. In spite of this the influence of s_c or s_d cause redundancy to be a more conservative and useful measure than R_c.

Significance of Canonical Correlations.

One can test the null hypothesis that z_1 is independent of z_2 by

$$\chi^2 = -[N-1 - .5(p + q + 1)]\log_e \Lambda$$

where Λ is the product of all q $(1 - \lambda_i)$, and χ^2 approximates the chi-squared distribution with pq degrees of freedom. However, the actual test is one of equality of all eigenvalues which are squares of canonical correlations. This is analogous to the test of equality of principal component eigenvalues, the test of sphericity. In the present case, inequality of eigenvalues implies both that the first eigenvalue is not equal to the others and that the first canonical correlation is significantly different from zero.

If one wishes to test the null hypothesis that the last r pairs of canonical variates are independent of one another the only change in the above formula is to calculate Λ from the last r eigenvalues. The chi-squared statistic has $(p - s)(q - s)$ degrees of freedom, where s $(= p - r)$ is the number of eigenvalues deleted. This test can be applied by starting with all eigenvalues ($s = 0$) and consecutively, beginning with the first, and sequentially removing one eigenvalue at a time until there is no significant correlation between sets. Of course successive tests amount to a method for determining how many pairs of canonical vectors might be interpreted. The results of these tests have been shown in Table 7.2.

Implications of Sea Palm Study.

The pertinent results of Young's (1970) field and laboratory studies will be summarized, prior to the canonical correlation analysis. This summary is intended as a check on the analysis.

Growth of stipe and blade was most rapid from March through June, then continued slowly if at all; but definitely stopped in October (Figure 7.2). Other dimensions increased most rapidly through July but holdfast diameter grew faster than holdfast height and stipe base faster than stipe apex. The latter dimensions generally reached their maxima in July. There is no definite explanation for the decrease in both stipe diameters in August.

In 1970, plants started to appear in late February, increased rapidly in density through May, slowly increased in number through July, and maintained their maximum density through August. This was followed by a slight September decline and a rapid loss of individuals from October through December. Almost all plants were gone by February. The first fertile blades appeared in July.

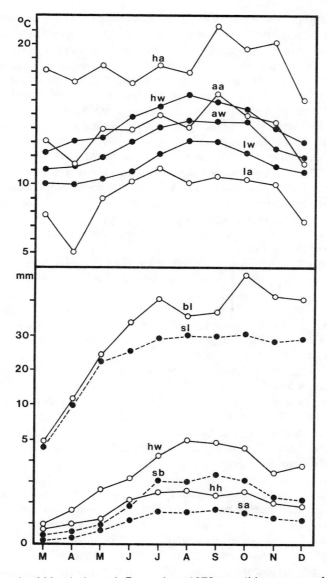

Figure 7.2 Graph of March through December, 1970, monthly averages for environmental and morphological variables for *Postelsia palmaeformis* Ruprect. Environmental temperatures are abbreviated **ha** for maximum air, **aa** for average air, **la** for minimum air, **hw** for maximum water, **aw** for average water, and **lw** for minimum water. Morphological dimensions are abbreviated **bl** for blade length, **sl** for stipe length, **hw** for holdfast width, **sb** for diameter of stipe at base, **hh** for holdfast height, and **sa** for diameter of stipe at apex (data from Young, 1971).

Three environmental factors are critical. *First* is desiccation of plants during low tides. Naturally both air temperature and wind influenced desiccation rate. Also, stipes display somewhat less water loss than do blades. At air temperatures of 10°C water loss from stipes is about 15% and from blades 18%, at 12–20°C loss from stipes is about 22% and blades 27%, and at 30°C water loss from stipes is about 31% and blades 46%. Desiccation is a major cause of plant destruction. In July and August desiccation starts being critical. Blade parts, first at the tip, become discolored and then mushy, and finally are lost by wave action. This deterioration proceeds down the blades until wave shock removes the entire head of blades. Then, deterioration proceeds to the stipe and on to the holdfast. Finally, the stipe is lost and finally the holdfast is destroyed. *Second*, wave shock is important since seemingly healthy plants were found floating in the water. *Finally*, a water temperature of 13°C approximates a critical maximum for *Postelsia*. As water temperatures rise above 8°C photosynthetic rate continuously decreases. Around 13°C respiration exceeds photosynthesis, hence plant growth ceases. The critical 13°C is reached by maximum sea water temperature in May, but not in average temperature until July. On the other hand, average water temperature dropped below 13°C in October and reached minimum in December. However, recall that by October plants are well on their way to destruction and plant density is rapidly declining.

The above results represent an interdigitation of various field observations, experimental studies, and biometrical analyses. Regularly the biometry lead the way to critical levels for experimental treatments.

Now we will judge the merits of canonical correlation for the six morphological and six environmental variables. Do not expect the raw data to enable you to appreciate the 12-dimensional hyperspace of the morphological and environmental variables. Also, do not expect data to unravel all possible maximum associations between sets.

Only the first two canonical correlations are significantly different from zero since their probabilities are the only ones less that 5% (Table 7.2), the usual magnitude selected for a Type I error. Since there is no significant association between the third through sixth pairs of canonical variates, there is no way to justify looking beyond the second pair of canonical vectors. Also, note that most of the percentages of the variances of variables occurs in the first and second pairs of canonical vectors (Table 7.3) and that the factor structure and, more important, variance extracted plus redundancy emphasize only the first pair of canonical vectors (Table 7.4).

At this point contrast the first two pairs of canonical variates as to redundancy and canonical correlation. For the first pair of canonical variates, redundancy is 77.5% for set one and 65.7% for set two, indicating considerable overlap between the first pair of canonical variates; and the canonical correlation is .980, very high. For the second pair of canonical variates, redundancy is 5.5% for set one and .9% for set two, but the canonical correlation is .676 implying association between sets. Here again is evidence of possible problems with the correlation coefficient. The second canonical correlation coefficient is significant but there is too little overlap between sets to justify considering the second pair of vectors.

The only source of interpretation, then, becomes the first pair of canonical vectors. These vectors provide the only usable maximum possible association between the data sets, an inherent property of the data. In other words, the researcher has no control over what the association will be. Hopefully it will impart something of primary interest.

The discussion that follows is based upon the "most meaningful" relationship between the first pair of canonical variates (not shown). Most meaningful is defined in terms of sign and absolute magnitude of the canonical variates. The first canonical vector for plants contrasts growth of the two longest plant parts, blade and stipe, with distinctly smaller dimensions. Canonical variates positive scores for individual plants pinpoint March through June plus November and December individuals. The magnitude of scores increase in the sequence of December, June, November, May, April and March.

In the corresponding environmental canonical vector, the largest absolute value for a coefficient is that for average water temperature ($-.822$). The negative sign implies low average water temperatures. Canonical variate scores for environment stress the same months as did scores for plants. These months make sense when the second highest coefficient (.464) for maximum water temperature, warmer months, is considered.

The implication is that morphology proceeding backward from maximum growth to the start of growth reflects the environment in the same way as does plant morphology proceeding backward from the month just prior to death to the previous month. To me the biological ramifications of this are unclear. Perhaps the relationship, in spite of considerable redundancy, is spurious. If not spurious, we must assume that a temperature rise relates to growth in the same way that a temperature decline does to loss of parts. This is an association; however, to me a meaningless one. I know of no fundamental biological process producing both growth and decay, let alone a process whose consequences are switched by the sequence of environmental events. Obviously, parts or all of two or more biological features are involved. Also, unfortunately the analysis verifies none of the fundamental biology of *Postelsia*.

If you are disappointed by the findings, good. I believe it is more important that you know about the pitfalls and frustrations of canonical correlation analysis than have you consider it a panacea for discovering any and all associations between sets of data. Also, these results emphasize the importance of looking for another approach to examining relationships between two data sets. These other methods generally will serve either to provide information when canonical correlation is not too helpful, or to add further implications to a useful canonical correlation analysis.

Do not assume that canonical correlation is always a useless device for associating data sets. It frequently is most helpful. On many occasions results will be convincing, stressing fundamental features of a phenomenon.

Digression—Canonical Structure. The problems of interpreting canonical vectors are familiar to users of methodology. Many users have found that canonical structures often lead to greater biological meaning than do canonical vectors. Cooley and Lohnes (1971) by way of example imply that structure rather than canonical vectors is the basis for interpretation.

Let us examine the performance of structure in reference to the sea palm (Table 7.4). Recall that interpretive aids emphasized only the first canonical correlation. Then, note the structure pertaining to the first pair of canonical variates. Structure is consistent with small plant size being associated with low temperatures. This might be extended to general growth closely paralleling, perhaps being limited to, lower temperatures, i.e., the first relationship shown later by the principal component method. However, structure does not imply the stress condition.

7.5 Relating Patterns of Variation by Components

The methodology of multiple regression was used to determine the portion of the criterion predicted by the regression and the portion not predicted, the residual. This can be extended to the two set case where application is to the set one submatrix R_{11} and the set two submatrix R_{22}. Since either submatrix can be designated the criterion or the predictor and designations can be changed in a single analysis, each submatrix can have a predicted and a residual portion. The extension of multiple regression methodology to provide the various submatrices will be accomplished first. Second, since the regression methods include the final information needed, the pitfalls of canonical correlation are defined for recognition in a real problem. Third, the various component analyses are presented. Fourth, components of the two submatrices, each treated as a set, are associated by an entirely new and different application of canonical correlation. Finally, all methods are applied to the sea palm example.

Multiple Regression Approach.   Most of what is developed here follows Hope (1969).

First, let us compare multiple regression with canonical correlation. In both methods the correlation matrix is partioned. The only difference is that in canonical correlation all submatrices are truly matrices containing more than one element in each row and each column—review multiple regression, only $\mathbf{R}_{11}$ is truly a matrix; $\mathbf{R}_{12}$ is a vector and $\mathbf{R}_{22}$, a scalar.

The amount of the criterion that can be predicted or accounted for by the predictors in regression is

$$\mathbf{bR}_{12} = \mathbf{R}_{21}\mathbf{R}_{11}^{-1}\mathbf{R}_{12} = R^2.$$

The square of the multiple correlation coefficient, R^2, is the variance of the criterion accounted for by the predictors. The above formula for the amount of criterion predicted, or the amount due to regression, is the same for canonical correlation; but in canonical correlation the result is a matrix, $\hat{\mathbf{R}}$, rather than the scalar coefficient of determination. In canonical correlation, the matrix of $\mathbf{c}_i$ canonical vectors is the equivalent of the vector $\mathbf{b}$ of slope coefficients; and the vector of canonical correlations R_c is equivalent to the multiple correlation coefficient R.

The deviation from regression, the part of the criterion not accounted for by the predictors, or the residual, is

$$\mathbf{R}_{22} - \mathbf{R}_{21}\mathbf{R}_{11}^{-1}\mathbf{R}_{12} = \mathbf{R}_{22} - \hat{\mathbf{R}}_{22} = \tilde{\mathbf{R}}_{22}$$

for both regression and canonical correlation. In regression, both the predicted (R^2) and residual ($1 - R^2$) parts are scalars. In canonical correlation, both predicted $\hat{\mathbf{R}}_{22}$ and residual $\tilde{\mathbf{R}}_{22}$ are matrices.

Since in canonical correlation analysis the two sets of variables can have equal logical status, the regression of the $\mathbf{z}_1$ variables on the $\mathbf{z}_2$ variables also can be calculated. The amount of $\mathbf{R}_{11}$ predicted by $\mathbf{z}_2$ is

$$\mathbf{R}_{12}\mathbf{R}_{22}^{-1}\mathbf{R}_{21} = \hat{\mathbf{R}}_{11}$$

and the residual is

$$\mathbf{R}_{11} - \hat{\mathbf{R}}_{11} = \tilde{\mathbf{R}}_{11}.$$

7.5.1 Possible Pitfalls of Canonical Correlation.

Canonical vectors and regression weights resemble principal components. In all, dimensions are chosen to maximize some value. This is not much of a problem in components or regression weights. Both involve assumptions of linear functions; however, this rarely is a problem. In regression, tests of linearity and significance (not developed in this book) can be applied to determine if departures from the assumptions exist. In component analysis, nonlinearity can lead to an extraneous component. For example, growth is extracted as a linear component but since actual absolute growth rate often is nonlinear, some of growth (a shape factor) might not be covered by a single growth component. Therefore, deviations from linear growth could possibly contribute to one or more linear fractions and components. Such components are readily recognizable as shape factors through interpretive aids. Moreover, the linear simplification of growth can be more useful, especially in morphometric comparisons of different groups—the two or more growth components separate fractions that often are more meaningful. Naturally, the above problems might be eliminated by a logarithmic model.

The situation in canonical correlation is not so simple. Four potential areas of difficulty exist. First, dimensions are chosen to maximize a correlation. If the raw association between sets is spurious, canonical correlation maximizes a meaningless association. Even when the correlation is meaningful, the condition indicated by maximization might not be approached in nature.

Second, canonical correlation does no more than propose possible relationships between sets. One cannot proceed safely beyond an assumption of how one data set fits to another. On the other hand, biologists regularly are more interested in reasons for prediction or association rather than success per se. The goal in most studies is to determine the "meaning" of predictors.

Third, reading weights of canonical vectors as one reads components must be done with care. Two conditions can lead to erroneous weights, variables of a set being inadequate and/or illogical and canonical correlations being due to invalid (residual) portions of the variation rather than to valid (predicted) portions. Since the latter often is possible, methodology should display how valid and invalid parts behave. Although such methods usually are not employed, they are emphasized here. When residual variation is involved in canonical correlation, the cause frequently is nonlinear functions. This is the case in the *Postelsia* study. The period of study was an annual cycle of rise and fall of environmental and morphological measures, a curve. Of course this should be and was recognized. More important, in many instances recognition might not be possible, so the methodology becomes very important.

Fourth, a most serious difficulty exists when two canonical correlations are equal. This is equivalent to sphericity of principal components. Again, vectors are not uniquely determined. Such canonical vectors and their canonical variates assume any possible orthogonal positions in their hyperplanes, so no biological importance can be assigned to coefficients of the canonical vectors. This is true no matter how high the values of the canonical correlation or the redundancy coefficients.

7.5.2 Principal Component Approach

Most of the above pitfalls can be recognized and overcome by adding eight component analyses of correlation matrices to the canonical correlation analysis. In the sense of the principal component approach, these component analyses are of the total R, R_{11}, $\hat{R}_{11}$, $\tilde{R}_{11}$, R_{22}, $\hat{R}_{22}$ and $\tilde{R}_{22}$ matrices. Also involved will be the correlation matrix developed from $\mathbf{R}_{11}$ and $\mathbf{R}_{22}$ component scores. In all cases, the complete battery of tests and statistics of conventional component analysis are applied. The eight analyses are helpful even though the assumptions of some component analyses might be suspect.

The above principal component analyses of z-score dispersions, or standardized variation, might require further clarification. Recall that the p set one variables, $\mathbf{z}_1$, and the q set two variables, $\mathbf{z}_2$, also represent subvectors of $\mathbf{z}$ of $p+q=t$ elements,

$$\mathbf{z} = \begin{bmatrix} \mathbf{z}_1 \\ \mathbf{z}_2 \end{bmatrix}.$$

In addition, the total correlation (standardized dispersion) matrix contains submatrices of standardized variances and covariances of elements of $\mathbf{z}_1$ ($\mathbf{R}_{11}$) and of $\mathbf{z}_2$ ($\mathbf{R}_{22}$). Also, in the multiple regression approach, predicted and residual fractions of data vectors were implied:

$$\mathbf{z}_1 = \hat{\mathbf{z}}_1 + \tilde{\mathbf{z}}_1 \text{ and } \mathbf{z}_2 = \hat{\mathbf{z}}_2 + \tilde{\mathbf{z}}_2$$

with corresponding dispersions

$$\mathbf{R}_{11} = \hat{\mathbf{R}}_{11} + \tilde{\mathbf{R}}_{11} \text{ and } \mathbf{R}_{22} = \hat{\mathbf{R}}_{22} + \tilde{\mathbf{R}}_{22}.$$

The foregoing leads to seven component analyses involving seven different dispersion matrices in characteristic equations and eigenvector functions:

1) $\mathbf{y} = \mathbf{V}'\mathbf{z}$ involving $\mathbf{R}$,

2) $\mathbf{y}_1 = \mathbf{V}_1'\mathbf{z}_1$ involving $\mathbf{R}_{11}$,

3) $\hat{\mathbf{y}}_1 = \hat{\mathbf{V}}_1'\hat{\mathbf{z}}_1$ involving $\hat{\mathbf{R}}_{11}$,

4) $\tilde{\mathbf{y}}_1 = \tilde{\mathbf{V}}_1'\tilde{\mathbf{z}}_1$ involving $\tilde{\mathbf{R}}_{11}$,

5) $\mathbf{y}_2 = \mathbf{V}_2'\mathbf{z}_2$ involving $\mathbf{R}_{22}$,

6) $\hat{\mathbf{y}}_2 = \hat{\mathbf{V}}_2'\hat{\mathbf{z}}_2$ involving $\hat{\mathbf{R}}_{22}$, and

7) $\tilde{\mathbf{y}}_2 = \tilde{\mathbf{V}}_2'\tilde{\mathbf{v}}_2$ involving $\tilde{\mathbf{R}}_{22}$.

The eighth principal component analysis, that of $\mathbf{R}_{11} - \mathbf{R}_{22}$ component scores involves a somewhat peculiar dispersion matrix. The original "data vectors" used to calculate the dispersion matrix also are unusual. Each data vector for each individual contains $p + q = t$ elements, the first p being the $\mathbf{y}_1$ component scores (derived from 2 above) and the last q being the $\mathbf{y}_2$ component scores (derived from 5 above). Thus, the t-element "data vector", $\mathbf{y}^*$, is

$$\mathbf{y}^* = \begin{bmatrix} \mathbf{y}_1 \\ \mathbf{y}_2 \end{bmatrix}.$$

These data indeed lead to an unusual dispersion matrix,

$$\mathbf{S}^{*2} = \begin{bmatrix} \mathbf{S}_{11}^* & \mathbf{S}_{12}^* \\ \mathbf{S}_{21}^* & \mathbf{S}_{22}^* \end{bmatrix} = \begin{bmatrix} \mathbf{L}_1 & \mathbf{S}_{12}^* \\ \mathbf{S}_{21}^* & \mathbf{L}_2 \end{bmatrix},$$

where $\mathbf{S}_{11}^*$ is the dispersion submatrix of $\mathbf{y}_1$, $\mathbf{S}_{22}^*$ of $\mathbf{y}_2$, and $\mathbf{S}_{12}^* = \mathbf{S}_{21}^{*}{}'$ contains covariances between elements of $\mathbf{y}_1$ and $\mathbf{y}_2$. However, since both dispersions are of component scores of a set, each submatrix is a diagonal submatrix of eigenvalues, $\mathbf{L}_1$ (from 2 above) and $\mathbf{L}_2$ (from 5 above).

Since the first seven component analyses all were z-score models, that model is retained for the eighth analysis. All elements of $\mathbf{y}^*$ are transformed to z-scores, $\mathbf{z}^*$, with dispersion

$$\mathbf{R}^* = \begin{bmatrix} \mathbf{R}_{11}^* & \mathbf{R}_{12}^* \\ \mathbf{R}_{21}^* & \mathbf{R}_{22}^* \end{bmatrix} = \begin{bmatrix} \mathbf{I} & \mathbf{R}_{12}^* \\ \mathbf{R}_{21}^* & \mathbf{I} \end{bmatrix}.$$

Note that the effect is to transform $\mathbf{S}_{11}^*$ and $\mathbf{S}_{22}^*$ to identity matrices, $\mathbf{R}_{11}^*$ and $\mathbf{R}_{22}^*$, and to cause $\mathbf{R}_{12}^* = \mathbf{R}_{21}^{*}{}'$ to contain correlations between components of the two sets. Therefore, the eighth model for principal component analysis is

8) $\mathbf{y} = \mathbf{V}^{*}{}'\mathbf{z}^*$, involving $\mathbf{R}^*$,

but the component scores, $\mathbf{y}$, are entirely different from those in principal component analysis from #1 above.

At the end of this section, a canonical correlation of component scores will be added. The model for this canonical correlation will involve the characteristic equation

$$\mathbf{R}_{21}^* \mathbf{R}_{11}^{*-1} \mathbf{R}_{12}^* \mathbf{R}_{22}^* - \lambda \mathbf{I} = 0$$

which, since both $\mathbf{R}_{11}^*$ and $\mathbf{R}_{22}^*$ are identity matrices, reduces to

$$\mathbf{R}_{21}^* \mathbf{R}_{12}^* - \lambda \mathbf{I} = 0.$$

Components of Total R.   In the components of total $\mathbf{R}$, especially, there is justification for studying standardized variation, total $\mathbf{R}$, rather than absolute variation, total $\mathbf{S}^2$. This is the case since the two sets often will have contrasting scales of measurement. For this reason, correlation matrices are the dispersion for all component analyses.

The consequences of subdividing a $\mathbf{R}_{11}$ or $\mathbf{R}_{22}$ into predicted and residual fractions should be mentioned. Any predicted or residual matrix has lost one of its dimensions; however, not one identical with a principal component. For the latter reason, any total $\mathbf{R}_{11}$ or $\mathbf{R}_{22}$ plus its predicted and residual fractions have the same number of component axes. Naturally this allows all possible comparisons. In practice, I have found this to be true. However, there is the possibility that a predicted or residual matrix is singular. When this occurs, unfortunately some of the analyses are not possible.

The first component analysis, that of total $\mathbf{R}$, encompasses the $p + q = t$ variables as a single set of variables. By itself, this analysis is questionable since two logical sets might not elucidate biologically meaningful patterns of variation. However, the analysis clarifies how two sets of variables may or may not work together. In displaying linkage and/or contrasts of data sets in components, insight on further analyses is gained.

Total, Predicted and Residual Components of Submatrices.

The total, predicted and residual components of R_{11} and R_{22} are compared with canonical vectors and among themselves. The yardstick for comparison is the angle between any pair of vectors. In practice angular contrasts are decisive.

Total, predicted, or residual components of a set regularly include a component that agrees with a canonical vector of that set. Biologically, most interest might be in canonical vectors resembling predicted components. Since predicted components pertain to fit to the other data set, the canonical correlation would be verified. However, linking a canonical vector to a residual component is not a useless outcome. Rather, the linking proposes a spurious canonical correlation.

The total, predicted and residual fractions can be compared completely apart from canonical correlation. This is most useful when canonical correlation analysis is unsatisfactory. The predicted components still translate variation associated with the other set and the residual components, variation that is independent of the other set. This procedure resembles factor model analysis but here the components of all three matrices, not just the predicted or its subunits, are evaluated.

Another important aid in comparing total, predicted, and residual components is the communality, or proportion of the variance of each variable, achieved in the predicted vs. residual components. The total communality for any variable is its correlation with itself, unity as expressed in the diagonal of the total matrix. Since the sum of predicted and residual equals total, the communalities for each variable in predicted and residual matrices sum to unity. As is the case for the total matrix, communalities for both predicted and residual variables are found in the principal diagonal. Since no predicted or residual variable is likely to have a communality of unity, elements of their principal diagonal usually are not unity. Communalities are direct measures of each variable's involvement in predicted and residual. Also, the predicted communalities are the simplest measure of how much variables relate to the other data set.

Principal Components of R_{11}-R_{22} Component Scores.

The last component analysis, that of R_{11}-R_{22} component scores, can aid in finalizing interpretation. Component analysis is performed in the same manner as before. The only difference is the data. The data for each individual consists of a vector of the 1st through p^{th} component scores of set one from R_{11}, plus the 1st through q^{th} component scores of set two from R_{22}. Therefore, each individual has $p + q = t$ component scores and these provide a $t \times t$ correlation matrix, R^*. The submatrices for set one and for set two, since each consists of components with zero intercorrelations, naturally will be identity submatrices; but there should be nonzero intercorrelations between component scores of set one and set two. The fact that each component score within a set is independent of all others of its own set causes the interset component correlations to be extremely useful. Interset correlations of components can be examined and the implied associations noted. It will not be done in the example, but it would add further verification of the results.

The component analysis provides an interesting way of evaluating both sets. The components within a set already are uncorrelated. Therefore, new components calculated from original component scores of both sets are not overly influenced by within set variables. However, this does not mean that all variables (original components) but one within a set will have zero new component coefficients. In other words, if two or more original components of one set are actually associated with two or more original components of the other set, a new component will emphasize this fact. Therefore, a potential danger exists. One should limit interpretation to if and how independent patterns of variation of set one relate to independent patterns of variation of set two. The battery of tests and statistics of any comprehensive component analysis apply to that interpretation.

Canonical Correlation of R_{11} ***vs.*** R_{22} ***Component Scores.*** This canonical correlation analysis could be derived from S^{*2} or R^*. Although the canonical correlations based upon S^{*2} are the same as those from R^* the elements of canonical vectors from S^{*2} are in units proportional to those of the respective responses, variances, of the original components. This may or may not prove useful. When R^* is used, original component variances are equalized. Therefore, R^* does not allow the amount of original variance due to components to stress major patterns of variation. In other words, in a sense all patterns of variation of either set will be on an equal footing. This has both advantages and disadvantages. The disadvantages are mostly corrected by the battery of other analyses and awareness on the part of the investigator. R^* will be used here.

Whether S^{*2} or R^* is selected, the canonical correlation can combine only interset components. Since correlation of components within each set are zero, methodology does not allow the conditions found in canonical correlation of sea palm raw data. However, the current method still has the possible shortcoming of maximizing correlations. This potential shortcoming is rectified by comparing the canonical correlation of components findings with those of principal component analysis of the component scores of both sets. Actually each method serves as a check upon the other. When both agree, there is a strong basis for interpreting the analysis.

7.5.3 Return to the Example.

The discussion will proceed in the following steps: First, the components of total R will be examined. Second, angles between the total, predicted or residual R_{11} morphological and R_{22} environmental components, six sets, and the corresponding morphological or environmental first canonical vector are compared. Third, final judgment is made of the meaning and/or value of the canonical correlation analysis. Fourth, the morphological and environmental total, predicted and residual components are examined. This is facilitated by step five, canonical correlation of component scores and by the final step, principal component analysis of component scores.

In the various component analyses and the new canonical correlation analysis the importance of each step will be stressed. However, since emphasis is upon relationships between sets, unrelated components are generally ignored.

Principal Components of **R.** Step one, the principal components of total R, is a "quick-and-dirty" device for setting the scene for what is to follow. The first two components (Table 7.5) do this very well. The first component is a general one that summarizes 79.1% of the variation. Obviously it does not imply that the plant and environment are growing together. Overall the first component with its interpretive aids (not shown) demonstrates the well known fact that *Postelsia* growth occurs over a period of increasing temperatures (March through June). Later comparisons will establish if this is pertinent or not. Although such a total R component is not necessary for important relationships to exist between sets, it is encouraging when the first total R component is of this nature. The second component, 11.5% of the variation, contrasts the morphological variables with the environmental variables. In the sense of two sets of variables, this shows that part of the variation of each set is independent of the other set. Component scores (not shown) stress all March, July, August, and September plus some April and October individuals. Output of other components (not shown) verifies

Table 7.5 First two principal
components of sea palm
morphological and
environmental variables

variables	components	
	1	2
holdfast height	−.302	−.178
holdfast width	−.305	−.214
stipe length	−.283	−.296
stipe base	−.306	−.086
stipe apex	−.313	−.082
blade length	−.274	−.333
max. air	−.168	.663
min. air	−.290	.189
avg. air	−.265	.472
avg. water	−.313	.053
max. water	−.300	.076
min. water	−.314	.050
variance	9.489	1.380
% variance	79.1	11.5

that the early time is one when plants are growing slower than the rate of general growth. The later months are when environmental temperatures are above 13°C, respiration exceeds photosynthesis, and plants are being sustained on stored nutrients. Since the plants are surviving upon stored nutrients, in a sense they might be independent of temperatures. However, later we shall see that July through August and perhaps September plants do, in part, reflect the environment.

Still part of step one is to observe if and how the total $\mathbf{R}$ components might join with canonical vectors. We see no such connection, but it is not critical here. Total components might not associate well with canonical vectors. For best alliance two conditions would have to exist. First, corresponding 1st through q^{th} components of $\mathbf{R}_{11}$ and $\mathbf{R}_{22}$ would be indicated in the first q total components. Second, the same q pairs of $\mathbf{R}_{11}$ and $\mathbf{R}_{22}$ components would correspond to the appropriate vector of the 1st through q^{th} pair of canonical vectors. This situation neither seems likely nor does it assure significant canonical correlations let alone redundancy. In studies like that of the sea palm, there is no reason to expect or require this. Fundamental morphological patterns of variation need not reflect like variation of the environment. For example, a limiting environmental factor need not be an environmental component and the response of an organism might be independent of a component.

Angular Comparisons.   Step two is angular comparison between total, predicted or residual $\mathbf{R}_{11}$ or $\mathbf{R}_{22}$ and the appropriate canonical vector. In the example, there is no close correspondence between a first canonical vector and any total or predicted component. Although the angles are not tabulated, the components (Tables 7.6 and 7.8) and their communalities (Tables 7.7 and 7.9) are presented.

Table 7.6 Principal components of sea palm R_{11}, predicted sea palm $\hat{R}_{11}$ and residual $\tilde{R}_{11}$ correlation matrices

| | **R_{11} components** | | | | | |
	1	**2**	**3**	**4**	**5**	**6**
holdfast height	−0.407	0.189	0.849	0.198	−0.174	−0.091
holdfast width	−0.415	0.018	−0.254	0.701	0.517	0.058
stipe length	−0.407	−0.494	−0.173	0.090	−0.570	0.477
stipe base	−0.404	0.543	−0.107	−0.482	0.200	0.508
stipe apex	−0.414	0.309	−0.404	−0.055	−0.395	−0.641
blade length	−0.401	−0.574	0.101	−0.476	0.426	−0.303
variance	5.422	.281	.117	.081	.061	.037
% variance	90.4	4.7	2.0	1.4	1.0	.6
probability	.00	.00	.02	.09	.10	1.00

| | **$\hat{R}_{11}$ components** | | | | | |
holdfast height	−0.414	0.011	−0.600	0.644	−0.166	0.161
holdfast width	−0.426	0.225	−0.439	−0.740	−0.060	0.154
stipe length	−0.385	−0.451	−0.044	−0.013	0.744	−0.303
stipe base	−0.422	0.361	0.255	0.071	−0.283	−0.736
stipe apex	−0.422	0.411	0.521	0.147	0.280	0.531
blade length	−0.378	−0.668	0.330	−0.102	−0.506	0.188
variance	4.886	.105	.022	.010	.003	.001
% variance	97.2	2.1	.4	.2	.1	.0
probability	.00	.00	.00	.00	.00	1.00

| | **R_{11} components** | | | | | |
holdfast height	−0.298	0.576	0.711	−0.248	−0.007	−0.110
holdfast width	−0.313	−0.295	0.020	0.124	0.247	−0.859
stipe length	−0.585	−0.296	−0.125	−0.511	0.409	0.356
stipe base	−0.174	0.659	−0.501	0.294	0.445	−0.004
stipe apex	−0.300	0.201	−0.458	−0.368	−0.690	−0.222
blade length	−0.592	−0.138	0.134	0.664	−0.313	0.272
variance	.622	.138	.091	.058	.032	.031
% variance	64.0	14.2	9.4	6.0	3.3	3.1
probability	.00	.00	.00	.00	.00	1.00

Table 7.7 Communalities and square of multiple correlation coefficients for morphological R_{11}, predicted morphological $\hat{R}_{11}$, and residual morphological $\tilde{R}_{11}$ variables of the sea palm

variable	R_{11} communality	mult R^2	$\hat{R}_{11}$ communality	mult R^2	$\tilde{R}_{11}$ communality	mult R^2
holdfast height	1.000	0.867	0.849	0.987	0.151	0.307
holdfast width	1.000	0.910	0.901	0.986	0.099	0.628
stipe length	1.000	0.921	0.749	0.996	0.251	0.737
stipe base	1.000	0.914	0.887	0.998	0.113	0.349
stipe apex	1.000	0.935	0.895	0.996	0.105	0.557
blade length	1.000	0.895	0.746	0.991	0.254	0.717

Table 7.8 Principal components of environmental R_{22}, predicted environmental $\hat{R}_{22}$ and residual environmental $\tilde{R}_{22}$ correlation matrices

	R_{22} components					
	1	2	3	4	5	6
max. air	−0.312	0.784	−0.336	−0.066	0.410	0.045
min. air	−0.420	−0.062	0.773	0.132	0.449	0.065
avg. air	−0.423	0.357	0.268	−0.044	−0.781	−0.101
avg. water	−0.431	−0.299	−0.252	−0.044	0.117	−0.804
max. water	−0.423	−0.328	−0.174	−0.699	0.002	0.442
min. water	−0.428	−0.239	−0.352	0.697	−0.087	0.378
variance	4.905	.807	.218	.062	.005	.002
% variance	81.8	13.5	3.6	1.0	.1	.0
probability	.00	.00	.00	.00	.00	1.0

	$\hat{R}_{22}$ components					
max. air	−0.222	0.750	−0.359	0.078	0.502	−0.039
min. air	−0.406	0.080	0.836	−0.158	0.324	0.013
avg. air	−0.371	0.466	0.068	−0.078	−0.795	0.044
avg. water	−0.471	−0.269	−0.173	0.083	−0.006	−0.818
max. water	−0.448	−0.283	−0.360	−0.671	0.098	0.359
min. water	−0.475	−0.249	−0.088	0.711	0.023	0.446
variance	4.047	.183	.034	.006	.001	.000
% variance	94.8	4.3	.8	.1	.0	.0
probability	.00	.00	.00	.00	.00	1.0

Table 7.8 *Continued.*

			$\tilde{R}_{22}$ components			
	1	2	3	4	5	6
max. air	−0.684	−0.586	−0.130	−0.109	0.387	0.099
min. air	−0.356	0.598	0.528	0.115	0.461	0.104
avg. air	−0.560	0.078	0.252	−0.077	−0.761	−0.182
avg. water	−0.170	0.272	−0.447	0.078	0.190	−0.809
max. water	−0.190	0.463	−0.585	−0.436	−0.072	0.461
min. water	−0.167	0.061	−0.314	0.879	−0.132	0.282
variance	1.241	.318	.132	.033	.003	.001
% variance	71.8	18.4	7.6	1.9	.2	.1
probability	.00	.00	.00	.00	.00	1.0

Table 7.9 Communalities and square of multiple correlation coefficients for environmental R_{22}, predicted environmental $\hat{R}_{22}$, and residual $\tilde{R}_{22}$ variables.

variable	R_{22} communality	mult R^2	$\hat{R}_{22}$ communality	mult R^2	$\tilde{R}_{22}$ communality	mult R^2
max. air	1.000	0.971	0.307	0.992	0.693	0.974
min. air	1.000	0.977	0.691	0.992	0.309	0.958
avg. air	0.999	0.992	0.598	0.998	0.401	0.988
avg. water	1.000	0.997	0.913	1.000	0.087	0.977
max. water	1.000	0.991	0.835	0.999	0.165	0.965
min. water	1.000	0.988	0.926	1.000	0.074	0.850

No total morphological component is closer than 54° and none of the predicted morphological components is closer than 45° from the first morphological canonical vector. Also, no total environmental components is closer than 23° and none of the predicted environmental components is closer than 33° from the first environmental canonical vector. Since correspondence generally should be within 20° for close relationship Blackith (1965), except for the 23°, close correspondence can be dismissed. However, the 23° involves the last total environmental component, so the relationship must be suspect. Recall that any last component occurs in the last dimension in component space. Since the last component axis must be orthogonal to all others, there is only one place that it can be. For this reason, a last component might be, but need not be, meaningless.

Residual components are especially interesting. The last residual morphological component is only 19° from the first morphological canonical vector and the last residual environmental component is only 15° from the first environmental canonical axis. Since both components are the last, they both become suspect. However, if these components can be justified, the canonical correlation can be ignored as spurious. Recall that a meaningless canonical correlation can arise in two ways. First, if data sets do not cover their phenomena, the canonical correlation might be spurious. In the present case it would mean that the measurements do not define the organism, the environment, or their possible association. Of course this is possible, but known biology of *Postelsia* makes this unlikely. Second, actual relationships, if they exist, might be too complex to be unraveled by canonical correlation. This appears likely for the sea palm since stored nutrients enable the plant to complete its life cycle through unfavorable conditions of high temperature and desiccation. It still might seem surprising that plant growth and increasing temperatures were not associated by canonical correlation analysis. However, this is understandable because some of the same temperatures late in the year are associated with progressive destruction of the plants.

Evaluation of the Canonical Correlation.   At this point final judgment on the canonical correlation analysis, step three, is possible. Generally this is done best in reference to three major pitfalls in using canonical correlation. First, is the phenomenon described overemphasized? In the sea palm, is the association of an increase in blade and stipe with a low average water temperature overemphasized? This can be suspected since the correlation applies to conflicting phases of the life cycle. Second, is the meaning of the predictors clear? The environmental variables are the predictors and their meaning is unclear. Fortunately, the sources of poor meaning in predictors are available to clarify the situation. Again, the sources are correlations involving illogical sets and/or residual variation. Since the sets are logical ones, the fact that the last residual components represent the closest association between canonical vectors and components must be reevaluated. The only alternative left is that canonical vectors resemble residual components, so canonical vectors are biologically meaningless.

The final pitfall of canonical correlation is passe at this time. However, to review methodology, is the first canonical correlation equal to any other? Of course the analysis would not have proceeded this far if the first canonical correlation was equal to any other one.

At this point a definite conflict in results exists. Canonical correlation analysis per se implies no relationship between environmental and morphological variables. However, component analysis of total **R** indicates just the opposite. This conflict provides a strong justification for further component analyses.

Principal Component Analysis of Submatrices.   The fourth step is comparing the component analyses of the three morphological and three environmental matrices. This approach can produce conclusions drastically different from those of canonical correlation since the methodology of regression rather than correlation is involved. Contrasting total, predicted and residual components is accomplished by angular comparisons (angles not tabulated here). Closest agreement is between morphological total and predicted components in corresponding first (2°) and second (21°) components. Closest agreement is between morphological total and residual components in corresponding third (29°), in fifth total and fourth residual (28°), and in sixth total and fifth residual (13°). Both predicted and residual components are more than 40° from fourth total. As expected, in no case is a residual component more like a predicted component than is a total component.

Now the first and second total components can be assumed related to the environment and three through six can be assumed unrelated to the environment. Since the first two total components summarize 95.1% of the alga's variation and the first two predicted components 99.3% of the alga's variation related to environmental temperature, not surprisingly the life cycle of *Postelsia* is shown to be "adapted" to the environment. Note that the latter statement is further justified by the communalities for predicted variables.

Next the independent patterns of plant variation that are predicted by the environment can be interpreted. In doing this both total and predicted coefficients should be compared (Table 7.6). The plant's first total and first predicted components are much alike—the vectors are only 2.3° apart. Both components indicate general growth of the plant. Component scores (not shown) for both vectors indicate growth ends in July. This interpretation is vastly superior to examining raw data (Figure 7.2) because component analysis indicates that growth stops at the time when biological stress starts. The second total and predicted components, 20.7° apart, do not correspond closely in coefficient magnitude. In view of the nature of the variation this is understandable. Both second components disclose stipe and blade lengths lessening and other dimensions increasing. The ''lessening'' and ''increasing'' must be judged on the basis of time of occurrence and raw data for that time. Both total and predicted component scores indicate March, April, July, August, and September plants. The March and April scores can be dismissed on the basis of allometric growth since stipe and blade growth rates increase later. The July, August, and September plants are of special interest. This is the period in which water temperature stress occurs and the plant becomes dependent upon stored nutrients, and in which air temperature and wind cause desiccation leading to loss of parts. Note that the blade (the part lost first) has the largest absolute, but a negative value, in both total and predicted second components. The negative coefficient for stipe might imply shrinkage, perhaps owing to desiccation of that structure, if not the loss of parts that actually occurs. Perhaps the other dimensions merely maintained their magnitude. The raw data are of minimum help here since raw scores represent a complex of all patterns of variation.

Analysis has pinpointed the July-September period of stress, the period immediately followed by marked destruction of parts and loss of individuals. The question remains, can this period be related to the environment? As a first attempt, step five, canonical correlation of component scores can be used to force maximum association between components (Table 7.10). Applying the procedures for determining which pairs of canonical vectors to interpret, the first two pairs have significant canonical correlations, the redundancy for both canonical variates of the first pair is equal, about 16%, and the redundancy for both of the second pair is about 9% (Table 7.11). Low redundancies should be accepted here. Recall that this canonical correlation of component scores is based upon correlation matrices rather than covariance matrices, thereby causing all original component variances to be equalized to unity. This would reduce redundancies. Also, there are biological reasons for assuming that an organism's pattern of variation need not relate closely to an environmental component to be pertinent.

The first canonical correlation pairs the first morphological and first environmental component. Growth of the sea palm reflects seasonal increase in temperatures. Most definitely it is not surprising that growth of a plant is adapted to the environmental conditions during growth. However, the fact that both growth and the environment are cyclic and the cycles correspond in the sense of all variables increasing together could mean the association is spurious. This possibility will be negated in the principal component analysis of component scores.

The second canonical correlation discloses maximum association between the second morphological and fifth environmental components. Component scores of the fifth environmental component emphasize April, June, July, August, and November. Recall that the second morphological components emphasized March, April, July, August, and September. Therefore, plants and environment correspond only during April, July, and August. In discussing the second morphological component, April was dismissed on the basis of lagging growth in stipe and blade. The fifth environmental component, as a predictor, pertains to moderately high maximum and minimum air tempera-

Table 7.10 Canonical vectors, canonical correlations and probability of canonical correlations being zero between sea palm morphological R_{11} component scores and environmental R_{22} component scores

	canonical vectors					
	1	**2**	**3**	**4**	**5**	**6**
R_{11} components						
1	0.941	−0.318	0.048	0.027	−0.075	0.074
2	−0.292	−0.729	−0.139	−0.247	−0.469	0.288
3	0.063	0.352	0.454	−0.587	−0.066	0.563
4	−0.141	−0.449	0.749	0.028	0.405	−0.231
5	−0.078	0.004	0.158	0.739	0.006	0.651
6	0.006	0.207	0.432	0.217	−0.778	−0.344
R_{22} components						
1	0.902	0.056	0.161	0.393	−0.007	0.052
2	0.330	−0.248	−0.663	−0.391	−0.111	−0.474
3	0.136	0.359	−0.433	−0.260	0.462	0.620
4	−0.184	0.331	−0.291	0.566	0.477	−0.473
5	0.057	0.834	0.108	−0.269	−0.403	−0.233
6	0.146	−0.024	0.500	−0.483	0.620	−0.333
R_c	.983	.713	.540	.347	.212	.181
probability	.000	.001	.101	.456	.496	.229

Table 7.11 Variance extracted and redundancy of morphological R_{11} components and ecological R_{22} components

component	R_{11} variance extracted	redundancy	component	R_{22} variance extracted	redundancy
1	0.167	0.161	1	0.167	0.161
2	0.167	0.085	2	0.167	0.085
3	0.167	0.049	3	0.167	0.049
4	0.167	0.020	4	0.167	0.020
5	0.167	0.008	5	0.167	0.008
6	0.167	0.005	6	0.167	0.005

tures and to very low average temperature. Naturally, the very low average air temperature relates best to April, the month having the recorded lowest average temperature. The high minimum and maximum air temperatures logically pertain to the July-August period. Since, high air temperature promotes desiccation and both lab and field studies confirm the importance of such air temperatures plus wind in desiccation, the results of this canonical correlation analysis are most attractive. If verified, some plant features both during growth, March through June, and during stress, at least July and August, reflect the environment.

Principal Component Analysis of Component Scores.

The support for the canonical correlation of component scores is left to principal component analysis of both sets of component scores (Table 7.12). The first component clearly joins variable one (morphological component one) and variable seven (environmental component one). Component two unites variable two (morphological component two) and variable eleven (environmental component five). Although the latter relationship is not as strong as the first, it does not depart from what might be expected if a minor environmental pattern of variation approximates the conditions influencing the plant's form over a small period of time. Also, since the time of plant stress was clearly defined by a component predicted by the environment, a total environmental component might be expected to indicate the environmental source of stress. In addition, even if a single limiting environmental factor is believed to explain plant stress, the factor might be expected in a component since stress occurs over three months. Here, the fact that biological knowledge justifies the interpretation should be secondary to the analysis justifying the interpretation. I believe the analysis does.

In conclusion, the various methods beyond canonical correlation of raw variables have joined 90.4% of plant variation (growth) with 81.8% of environmental variation plus 4.7% of plant variation (stress response) to .1% of environmental variation. The life cycle of *Postelisa* is adapted to the environment in March through June growth and in July through August stress, during which spores are finally produced. After sporulation, largely independent of environmental factors, the plants die and disappear.

Table 7.12 Principal component analysis of sea palm morphological R_{11} (scores 1-6) and environmental R_{22} (scores 7-12) component scores (only components 1 through 6 are shown).

scores	components					
	1	2	3	4	5	6
1	−0.665	−0.225	−0.034	−0.019	0.053	−0.052
2	0.207	−0.515	0.099	0.174	0.332	−0.204
3	−0.045	0.249	−0.321	0.415	0.047	−0.398
4	0.099	−0.317	−0.530	−0.020	−0.286	0.163
5	0.055	0.003	−0.111	−0.522	−0.004	−0.460
6	−0.005	0.146	−0.305	−0.153	0.550	0.243
7	−0.638	0.039	−0.114	−0.278	0.005	−0.037
8	−0.234	−0.176	0.469	0.276	0.078	0.335
9	−0.096	0.254	0.307	0.184	−0.326	−0.438
10	0.130	0.234	0.206	−0.400	−0.338	0.335
11	−0.040	0.590	−0.076	0.190	0.285	0.165
12	−0.103	−0.017	−0.354	−0.341	−0.438	0.235
variance	1.983	1.713	1.540	1.347	1.212	1.181
% variance	16.5	14.3	12.8	11.2	10.1	9.8
probability	.00	.00	.00	.00	.00	.00

7.6 Epilogue

This entire chapter was written and the example chosen from the point of view that morphometric analysis of a species was involved. The entire approach is applicable to other situations, but interpretation must reflect the particular problem.

The example was chosen because it shows how factors causing a canonical correlation analysis to fail are not likely to negate the component analysis approach. The March through December rise and fall of all variables virtually destroyed the correlation approach but not that of regression coupled with component analysis.

Other Examples. Although biologists frequently examine relationships between two data sets, rarely has canonical correlation been employed. Usual approaches to such data regularly are linked to multiple regression. Often data are categorized into a predictor and a criterion set. Then, each of the criterion variables might be involved in a single, unique regression on the predictors. Frequently, one or both sets of original variables are transformed to component scores. In such cases, predictor components might be allied sequentially to single, raw criterion variables. Again, a single regression would pertain to a unique criterion. In other instances, individual criterion components might be regressed upon the set of predictor components. Still other approaches examine the between set bivariate correlations of raw variables or component scores.

The total methodology of this chapter has not been applied to single biological problems. However, Cassie's (1972) method is similar to the PCA approach and Calhoun and Jameson (1970) used canonical correlation to associate weather variables with tree frog morphology.

Chapter 8
Ordination and Cluster Analysis

In studying a group of organisms, primary interest often is in the relationship between the individuals rather than in variation between the variables. In such cases, the goal is to arrange the individuals in a manner that discloses their fundamental relationships. In a correct sense, this amounts to classification; but it will be referred to by the two methods involved, ordination and cluster analysis. This is done since "classification" as a statistical term has been long preempted for procedures of discriminant analysis. True classification pertains to known groups and to the assignment of individuals to the group of closest resemblance. In both ordination and cluster analysis, no a priori groups are presumed. This is the case even if one has strong evidence to the contrary.

There are two sources of agreement between ordination and cluster analysis. The first is in application, both methods pertain to situations where a priori groups are undesirable. The other is that both methods can involve identical starting points, a matrix containing coefficients of resemblance (similarity or dissimilarity) between pairs of individuals. A possible resemblance matrix is one composed of the product moment correlation coefficients between pairs of individuals.

The two methods differ mathematically, thus so do their goals. Ordination is concerned with reducing p-dimensional space to a smaller r-dimensional space. A true ordination amounts to examining the graph of component scores on the first few, generally two or three, principal component axes. Cluster analysis is concerned with assigning individuals to discrete sets, clusters.

This chapter contains two parts. Part I, ordination, follows the first section on measurements of resemblance; and, in turn, is followed by part II, cluster analysis. Each part includes an introduction to the sections included.

8.1 Measurements of Resemblance

Resemblance coefficients are of two types, similarity and dissimilarity. However, the two types are sufficiently related for the distinction to be of minor importance. In general, one type can be easily transformed to the other type. A more important difference exists in reference to the nature of the original data. Qualitative data coefficients are used for presence-absence data and most are derived from three simple counts of these data. Quantitative data coefficients are less alike mathematically. Three of the latter will be mentioned, the correlation coefficient, the Euclidean distance, and the cos θ coefficient. In addition, attempts to handle mixed qualitative-quantitative data exist; one involves data transformation and another, a so-called general similarity coefficient.

Measures of resemblance utilize $k = 1, 2, \ldots, p$ variables to provide a matrix of coefficients between all ij pairs of N individuals. For general purposes, the resemblance matrix is symbolized $\mathbf{A} = (a_{ij})$; but different symbols are applied to the different coefficients. Although such $N \times N$ matrices involving a Q-technique are the rule, the matrices on occasion are $p \times p$ and an R-technique is involved.

In biology, resemblance coefficients, ordination, and cluster analysis are most commonly applied to ecology and numerical taxonomy. In both areas, the nature of an individual and the p-variables can be confusing. In ecology, an individual regularly is a sampling site, perhaps a quadrat; and a variable, a species. The p-variables would be the p-species encountered in a given ecological analysis. In numerical taxonomy, an individual often is a single organism or a centroid derived from a sample of organisms; and a variable pertains to a qualitative or quantitative measure of a character state, i.e., a conventional measurement of some feature of the organism. These distinctions should be kept in mind in further discussion.

Most resemblance coefficients are measures of similarity; the Euclidean distance is a measure of dissimilarity. Similarity coefficients are typified by a fixed range of values, often 0 to 1 (especially when recording the absence or presence of a variable), with the upper limit indicating maximum possible similarity. The Euclidean distance has a fixed lower limit, 0, and a theoretical upper limit of infinity. The larger the value of this distance, the greater the dissimilarity.

Presence—Absence Coefficients.   Three counts allow computation of any of the five more commonly utilized presence-absence similarity coefficients. The counts are the number of the p characteristics found in samples from two populations, n_i and n_j (where $n_i \leq n_j$), and the number of characteristics common to both samples, c_{ij}. Such counts are common in ecology and are of the number of species in each sample. In spite of conceptual differences between the coefficients, Hazel (1970) found little difference in their performances. The coefficients are those of Dice,

$$D_{ij} = 2c_{ij}/(n_i + n_j);$$

of Fager,

$$F_{ij} = c_{ij}/(n_i n_j - 1/2n_j)^{1/2};$$

of Jaccard,

$$J_{ij} = c_{ij}/(n_i + n_j - c_{ij});$$

of Otsuka,

$$O_{ij} = c_{ij}/(n_i n_j)^{1/2};$$

and of Simpson,

$$S_{ij} = c_{ij}/n_i.$$

D gives matches twice the weight of mismatches, and J (the most widely used) emphasizes differences and ignores mismatches. O is the presence-absence counterpart of cos θ (see below) and F is O modified by half the characteristics present in the larger sample. S emphasizes similarities.

Correlation Coefficient.   This is the familiar product moment correlation coefficient; but in an unfamiliar form, the correlation is between individuals as defined above rather than between variables. Although

$$r_{ij} = \frac{s_{ij}}{s_i s_j},$$

the covariance, s_{ij}, is between individuals i and j and the standard deviations, s_i and s_j, are for individuals i and j. For example, s_i is based upon the deviations from their mean value of the p measurements for individual i. In some cases, the mean is the average of p conventional measurements, e.g., length, width, and height. Aside from the peculiarity of such a mean, the standard deviation of an individual is not a meaningful concept (Eades, 1965). From a statistical point of view, such standard deviations often are invalid because they are based upon dependence of errors. This would be the case with conventional measurements on a single organism because such measures would tend to be highly correlated.

In spite of the statistical invalidity of applying this correlation coefficient, it remains among the most popular of the similarity coefficients. The frequent justification for this is better performance than from other measures of resemblance.

Euclidean Distance.   The Euclidean distance between individuals i and j based upon p variables is

$$d_{ij} = \left(\sum_{k=1}^{p} (X_{ik} - X_{jk})^2 \right)^{1/2}.$$

In the case of two variables, this formula is that for solving the length of the hypotenuse of a right-angle triangle, given the lengths of the two sides. Since d_{ij} increases as p increases, d_{ij} commonly is normalized by division by p; but normalization tends to decrease d_{ij} as p increases.

Two other difficulties are of primary importance. Distance is influenced by scales of measurement. One solution is to transform each variable to z-scores prior to calculating distances. A more serious problem occurs because of correlation between variables. Distances are underestimated or overestimated in reference to the sign and magnitude of correlations between the variables involved. The relationship can be appreciated in the distance formula modified to account for the correlation between variables X_k and X_m,

$$d_{ij} = \left(\sum_{k=1} \sum_{k=1} (X_{ik} - X_{jk})(X_{im} - X_{jm})r_{km} \right)^{1/2}.$$

Another possible correction is the substitution of component scores for variables.

Angular Separation or Cos θ.   This similarity coefficient is obtained for $k = 1, 2, \ldots, p$ variables from

$$\cos\theta_{ij} = \sum X_{ik} X_{jk} / (\sum X_{ik}^2 \sum X_{jk}^2)^{1/2}.$$

Cos θ has certain advantages over d. Cos θ ranges in values from $0 - 1$ ($90° - 0°$); d from $0 - \infty$. Since data often are nonlinear, cos θ in implying a great circle distance can be more appropriate than d. However, cos θ should not be applied to standardized variables; then, cos $\theta_{ij} = r_{ij}$.

Mixed Coefficients.   The theoretical problems of providing a similarity coefficient based upon mixed quantitative-qualitative variables are not completely satisfied. Parks (1969) suggests a transformation of the original data. Quantitative variables are scaled to a $0 - 1$ range by

$$y_k = (X_k - x_{\min})/(\text{range of } X);$$

and discrete variables are coded to conform to the same range. For example, dichotomous variables are coded 0 or 1; three-state variables to 0, ½, and 1; etc. Unfortunately both presence-absence and multistate variables can suffer from the arbitrary assignment of values and from the inequality of scales of measurement. The latter may alter the positions of individuals from their original Euclidean distribution. After transformation, the Euclidean distances can be computed.

Gower (1971) suggests a remarkable statistic, the general similarity coefficient,

$$G_{ij} = \frac{\sum w_{ijk} a_{ijk}}{\sum w_{ijk}},$$

where a_{ijk} is a measure of similarity between individuals i and j, and w_{ijk} is the weight of the kth variable. For quantitative data,

$$a_{ijk} = 1 - X_{ik} - X_{jk} / (\text{range of } X_k)$$

and ranges from 0 to 1. For presence-absence data, a_{ijk} is 0 for a mismatch and 1 for a match. The weight, w_{ijk}, is 0 when comparison between individuals i and j is not possible (mismatch, missing observations, etc.) and 1 when possible. The weight can be set to 0 or 1 in the case of negative matches (joint absences, etc.). The nature of the data and the choice of weighting for negative matches lead to extreme flexibility in this coefficient. In its various forms, the general similarity coefficient can resemble most of the other similarity coefficients.

I: Ordination

Ordination pertains to techniques for placing individuals into a theoretically continuous sequence that reflects some fundamental property of the individuals. The techniques may or may not involve multivariate analysis. When they do, ordination amounts to examining reduced dimensional space, $r < p$ dimensions to determine relationships between individuals. PCA was said to be so used.

Prior discussion of PCA was allied to theoretical limitations of the methodology. As already indicated, these restrictions are not necessary. Neither the assumption of multivariate normality, nor even that of random vector variables is mandatory for a PCA ordination. Without these assumptions, PCA still provides views of data determined by the eigenvectors of the particular dispersion or resemblance matrix. The component axes are orthogonal and successively provide the maximum possible distance between individuals. Methodology culminates in a higher dimensional plot of individuals, an ordination. Within this framework, discrete and even qualitative character states have produced satisfactory results. Unfortunately, a conventional PCA of such data cannot be relied upon to provide the biological basis for ordination that one might desire. In many cases, a PCA ordination might be virtually meaningless.

The interpretation of any kind of ordination should follow from and be dependent upon the particular study. However, in most methods pattern of individuals on a graph will approximate one of three general types—serial axes, nodal axes, and polynomial axes. On serial axes, individuals are plotted from one extreme to the other without any recognizable breaks (Figure 8.1a). Such axes indicate an underlying trend, a gradient, in the individuals. On nodal axes (Figure 8.1b) individuals are grouped into two or more clusters, the clusters being poorly to clearly defined. Such axes imply distinct groups of individuals. On polynomial axes, individuals conform to a curve. Normally when a polynomial axis occurs, it will be the second axis (Figures 4.3 and 8.1c). The individuals may be continuous or clustered along the curve. Again, continuous plots indicate gradients and discontinuous ones, clusters. The curve results from the second axis not being clearly independent of the first, a condition reflecting both nonlinearity and nonnormality of the data.

Although ecological ordination will be the major emphasis of further discussion, the nature of applications in numerical taxonomy requires at least a brief comment. In numerical taxonomy a frequent goal is to determine the relationship within an assemblage of organisms, an assemblage that cannot be separated into definable groups or populations. The assemblage may or may not conform to the concept of a single sample from a single site, potentially a single population. Jeffers (1967) provides an example of such a ''single population'' study. Aphids were collected from a single habitat and ordination disclosed four distinct clusters. More frequently in numerical taxonomy, ordination regularly pertains to a known heterogeneous assemblage of organisms. Individuals are presumed to be of unknown taxonomic rank (Operational Taxonomic Units or OTU's) but usually the taxonomist knows that the study is of diverse taxa. For example, individuals collectively might represent a genus or higher taxonomic category (Sneath and Sokal, 1973).

The formal techniques of ordination originated in ecology where recognition of the potentially continuous nature of vegetation lead to arranging sites or samples (quadrats, etc. from a community) into a continuous order that presumably reflects ecological information. Ordination of sites involves

Figure 8.1 Two-dimensional patterns of ordination of individuals: A, serial; B, nodal; and C, polynomial.

four general approaches, each of which possesses certain limitations. First is ordination in reference to environmental gradients. The major weakness of environmental ordination of sites is that one must presume which environmental factors are important in determining vegetation patterns. Such ordination is useful in examining specific environmental factors and their responses in terms of vegetation but need not provide the key to overall vegetation patterns. Second is ordination by weighted species values based upon a prior sequencing or ordering of the species, a type of species ordination. In this approach the final ordering of sites can suffer from the subjectivity in the sequencing of the species. Third is ordination by measures of interstand distances based on vegetation, another type of species ordination. Much of this distance analysis has relied upon a PCA solution. Related techniques for the same purpose include Polar Ordination and Nonmetric Multidimensional Scaling. Finally, Correspondence Analysis utilizes the presence of species in samples to place species and site samples on corresponding gardients. In general, ecological ordinations pertain to an evaluation of community structure, sites; but it often pertains to an evaluation of species as well.

No single method of ecological ordination can be presented because no one method is so robust as to be best for all studies. The reasons for this are multifaceted, but naturally reflect the nature of species distributions in nature. Nonnormality and nonlinearity might cause any method to provide poor ordinations in a given situation. This topic is expanded in the next section, a section that examines the features and the problems of ecological ordination.

The above is reflected in the literature by the active evaluation of new and old ordination techniques. Sections 8.3-8.6 introduce four techniques, Principal Coordinate Analysis, Correspondence Analysis, Polar Ordination, and Nonmetric Multidimensional Scaling. Principal Coordinate Analysis is mostly a technique of numerical taxonomists and the other three of ecologists. Both Principal Coordinate Analysis and Correspondence Analysis are closely related mathematically to PCA. Polar Ordination is mathematically quite simple and Nonmetric Multidimensional Scaling is a nonparametric method applicable to nonlinear data.

The section on Correspondence Analysis is concluded with an outline of Indicator Species Analysis. The latter method utilizes Correspondence Analysis and other devices to classify vegetation sites into groups. In a sense, Indicator Species Analysis completes Correspondence Analysis by placing the results or ordination within a hierarchical classification scheme.

The last three methods are especially useful for ecological ordination and with PCA are evaluated in the final section (8.7). The most popular ecological ordination techniques are PCA and Polar Ordination (Bray and Curtis, 1957). Gauch et al (1977) performed Monte Carlo studies comparing PCA, Polar Ordination (PO), and Reciprocal Averaging (RA, the most useful algorithm for Correspondence Analysis). Fasham (1977) conducted like studies of PCA, Nonmetric Multidimensional Scaling (MDS), and RA. Active discussion of these and other methods exist in current issues of *Ecology, Journal of Ecology* and *Vegetatio*.

8.2 The Problems of Ecological Ordination

The ordination of sites or environmental factors from species data presents unique analytical difficulties. Any kind of ecological analysis is subject to biases in its results. These biases stem from different analytical perspectives, numerical methods of analysis, kinds of data and transformations, and properties of ecological distributions. Each of these biases will be considered in sequence.

Analytical Perspectives.   From the analytical point of view of the ecologist, methodology can be dichotomized into a zonal or a gradient analysis approach. In the zonal approach, distinct communities are recognized and classified. Classification frequently pertains to some overall system or scheme in reference to the structure of communities. Frequently, the structural basis pertains to important organisms, usually the dominants. For a zonal analysis, each sample pertains to a homogeneous stand. Therefore, the approach performs well when stands are distinct, dominants are clearly recognizable, and the "correct" dominants are present in sampling sites. These conditions often prevail when environmental gradients are sharp and/or topographic discontinuities occur. On the other hand, the zonal approach becomes progressively more difficult as gradients become more gradual and uniform. In essence, sampling becomes virtually impossible. The real danger of such sampling is the potential bias of defining sampling sites on the basis of a preconceived notion about a classification scheme and its structural basis.

Numerical Methods.   These methods more often pertain to gradient analysis and so presume both that vegetation is potentially continuous and that sites can be placed into a continuous order reflecting ecological information. Methodology utilizes ordination based upon species data to place sites along a single or multidimensional gradient (often environmental) and is based upon such statistics as frequencies, similarity coefficients, distances, etc.

Sampling can be essentially as for zonal analysis, i.e., represent attempts to select homogenous stands, and share the sampling problems of zonal analysis. Statistically more appealing is objective sampling, efforts at randomization that result in heterogeneous sites being selected. The resulting data generally cannot be placed into a zonal classification scheme. For this reason, numerical analysis must be applied.

Numerical methods are not a panacea for the problems of ecological analysis. However, certain methods perform well and are theoretically better than others. In the latter sense, numerical methods are monothetic or polythetic. In monothetic methods, at each step of the analysis, a single criterion or attribute is the basis for decision; in polythetic methods, many attributes are utilized at each step. Also in the same sense, methods are agglomerative or divisive. Agglomerative techniques gradually build up clusters, typically start by creating all possible pairs of "most similar" sites and proceed by pairing pairs, etc. until all samples are united in a hierarchy. This approach suffers from higher levels of the hierarchy (late pairing) being strongly dependent upon the lower levels (early pairing). Divisive methods successively split the whole set of sites.

Theoretically, the best methods are both divisive and polythetic. Divisive polythetic methods are preferred since they utilize all available information at the initial step and are less likely to be lead by chance to poor ordinations. However, most divisive methods have polythetic tendencies and most agglomerative, monothetic. Perhaps obviously, monothetic methods tend to misclassify and polythetic do not. All methods to be discussed are divisive, polythetic techniques.

Hill et al (1975) summarize many problems of ecological analysis, but especially those of numerical analysis, within the framework of four criteria and two requirements of any good method:

1. The method should be able to classify even visually heterogeneous data so as to provide a convenient framework with which to summarize the range of variation in the vegetation. As far as possible, the resulting groups should have a straightforward ecological interpretation.

2. The method should be open-ended. That is to say, new sites should be capable of being assigned to their correct place in the resulting classification without necessitating further laborious calculations.
3. The method should be computationally efficient. To be of practical utility it should be applicable to full-scale surveys.
4. The method should not be capable of making serious errors of misclassification.
5. The methods of analysis should be applicable to the total site data as well as to the individual samples.
6. An ordination should be used to extract the major trends in the data, thereby presenting an overall picture of the vegetational variation and providing a basis for correlation with environmental factors.

Agglomerative methods suffer from (2), (3), and (4) and some divisive methods from (4). However various methods of analysis either tend to reflect, or could be made to reflect, the above criteria. In all of the methods, the nature of the data is very important. In the case of PCA, there are various ramifications of good data; however, all considerations reflect the fact that the precision of scores for ordination becomes critical.

Data and Its Transformation.   Part of the problem in obtaining appropriate ordinations from component scores pertains to the nature of the ecological measures. The species data can be quantitative (counts, densities, frequencies, cover, etc.) or qualitative (generally a simple indication of the presence or absence of each species at each site). Therefore, data regularly do not even approximate normality. Perhaps even more important for ordination, data often assume marked nonlinear functions.

There is an additional data problem. The ecological relationships of both the species and the collecting sites are of importance to the ecologist, but data often pertains only to species. The ecologist often wishes to ordinate species in terms of sites and sites in terms of species. For such studies, either the quantitative measure (count, etc.) of each species, or the presence of each species, is determined for each sample. Two possible kinds of dispersion (more often, similarity) matrices can be calculated, an R-type species $\times$ species matrix, or a Q-type site $\times$ site matrix. In the former case, a PCA pertains to species as variables, to a so-called species-space, and to component scores pertaining to sites. In the latter case, sites are variables and they define a so-called site-space with component scores pertaining to species. The purpose of such studies is to recognize vegetation types and/or species assemblages.

Noy-Meir (1973a) and Noy-Meir et al (1975) demonstrated that part of the success of an ecological ordination pertains to the nature of the data; but also to how the data are transformed, specifically centered and/or standardized. Centering was involved in all previously considered PCA models—all mean component scores were made equal to zero by transforming raw variable scores to deviations from their mean, a fairly common practice. Standardization is accomplished by measures of "size" (the mean, the maximum value, or the sum of all values) or by measures of "dispersion" (the standard deviation or the range). For example, z-scores are both centered (have a mean of zero) and standardized (by the standard deviation). Also, z-scores are generally accepted as the best standardized measures, both for quantitative and presence data. All of these transformations are important because they influence the measures of similarity between sites and species (Table 8.1).

Table 8.1 The influence of raw and transformed data on the contributions of species and sites to ordination.

data	species 2 present	2 absent	favored species	favored sites
raw	equally	none	abundant	rich in species
species centering	equally	equally	abundant; high intersite variance	rich?
site centering	equally	none?	abundant?	rich; high variance between species
species standardization	equally	none	rare; common species absent in some sites	rich in rare species
site standardization	equally	none	dominant or only species in many sites	poor
simultaneous	equally	none	from poor sites; rare species	poor; having unique species

To appreciate the above, the nature of the results from untransformed data is important. For presence-absence data, the presence of two species at a site contribute equally to their association and their joint absence at another site does not contribute. In like manner, for quantitative data the weighting of species is by their abundance and of sites by their species richness. Both kinds of data favor abundant species or rich sites, so are useful.

Centering by species, or sites, causes the contribution of each species, or site, to be proportional to its variance—variance among sites for each species and variance among species for each site. Centering by species differs from non-centering in that joint absences contribute positively and unmatched occurrences contribute negatively to their association. Also with presence data, species with 50% frequency contribute most. With quantitative data, the variance tends to increase with the mean, so abundant species are stressed. For these reasons, simply centering data can be a poor practice.

Standardization is more complicated in the sense that it can be by species, by sites, or by both. For both qualitative and quantitative data, z-score standardization of species equalized species contributions independent of the nature of the sites; but site weights are proportional to species rarity. The consequence of standardization might be surprising when first read. The z-scores make rare and absent common species more important than abundant species and emphasize sites rich in rare species. This can be appreciated by calculating z-scores for any pair, triplet, etc. of hypothetical presences in a set. A single presence in a set of a given size will receive the largest possible z-score for that set. Generally this is not what the ecologist is seeking.

In site standardization, joint occurrence of any two species in a species-poor site contributes much more than in a species-rich site. The z-score transformation of quantitative site data also equalizes site weights. Therefore, species weights are dependent both upon the abundance of the species and the poorness of the sites in which the species occur. The largest species weights go to the dominant or to the only species in the greatest number of sites. This sounds similar to no transformation at all and it is. The difference is that without transformation, rich or dense sites are emphasized most. It would seem that z-scores of sites would appeal to many ecologists.

In double standardization, joint presences of species are most important in poor sites and in rare species. Quantitative data are affected in a like manner. However, standardization generally cannot produce true z-scores for both sites and species. For example, if first site data are transformed to z-scores and then species data, the latter will modify the site z-scores, causing the site scores to only approximate z-scores. The influence of such stepwise standardization depends upon which was standardized first and the nature of the data. Simultaneous double standardization avoids the above since the standardization is symmetric for both species and sites. In symmetric standardization, one can expect the highest weights for rare species in poor sites. Two or more sites containing the only occurrence of a few species will produce a unique cluster, the intermediates tending to be ignored. In essence, disjunct clusters are dependent upon the uniqueness of their species or sites, not upon the number and abundance of the species or the number and richness of the sites. This also is useful ecologically.

Utilization of a particular transformation should be decided in reference to the goals of a particular study. If the purpose of a study is to evaluate species abundance and site richness (a diversity oriented study), data should be neither centered nor standardized. To evaluate species independent of their abundance (a species oriented study), centering and standardization of species is appropriate. Such species oriented studies are inappropriate for site evaluation, since sites rich in rare species are emphasized. To evaluate sites independent of their richness (a site oriented study), z-scores are applied to site data. These site oriented studies are appealing, since species are weighted in reference to their abundance and occurrence in species-poor sites. Finally, double standardization further balances species and sites. The uniqueness of sites, especially species-poor sites containing rare species, is stressed. This standardization is least understood from a theoretical point of view. In spite of its emphasis of rare species confined to species poor sites, simultaneous double standardization would seem very useful.

The considerations here are general. They pertain to any ordination. Also, they serve to provide some further appreciation of the nature of the correlation model of PCA.

Properties of Ecological Distributions.   Successful ordinations must reflect the distribution of individual species and sampling sites (Gauch et al, 1977; Gauch, 1973). It appears safe to assume that the distribution of a single species approximates a normal curve relationship between species success and the environment. Also, this relationship should be reflected in a set of samples. Furthermore, physiological and ecological studies imply a distinct tendency for the total species of a community to be distributed along a 1-dimensional (1-D) gradient, a coenocline. It also appears likely that species can assume a 2-D community pattern, a coenoplane. Still higher order gradients might be possible, but methods have not been evaluated in this respect. Coenoplanes possibly can assume a variety of patterns but likely reflect a 2-D gradient. The 1-D response of all species to a coenocline generally is nonlinear so extends to a 2nd-dimension in a pattern that is somewhat bell-shaped, the previously mentioned polynomial second axis. Such bell-shaped distributions are termed Gaussian, since they resemble a normal curve, or the horseshoe effect, because ends of the curve often are involuted. The polynomial axis represents the inability of a linear model to summarize a nonlinear gradient in a single axis; more technically, a quadratic dependence of the 2nd axis on the 1st (Hill, 1973, 1974).

Various features of samples can create ordination problems. First, sampling errors can create sequence reversals or scattering of samples into the 2nd axis. Second, on the basis of sample estimates outliers can be more centrally located than they should be and for this reason compress other samples into a tight cluster. However, even when this occurs, the true coenocline might appear on the 2nd or 3rd ordination axis. Third, subsets of samples containing somewhat unique species (disjunctions), especially when leading to clusters unique in species composition, can distort the total ordination, even creating erroneous subsets. Fourth, samples of like species composition tend to be more centrally located. Fifth, sample clusters warp PCA ordinations since eigenvectors are strongly directed to the clusters. Finally, secondary gradients as found in coenoplanes can round the 2-D distribution of samples and involute or upturn the edges of the distribution. Also, the 2nd ordination axis often is a polynomial extension of the true 1st axis and the 3rd ordination axis contains the true 2nd dimension.

Species ordination have certain unique features. On a graph, dominants tend to be more distally located; species of wide distribution, more centrally located; and species of narrow distribution, more distally located.

Especially pertinent and related to the former is beta diversity, the degree of species changes among samples along a gradient. No beta diversity exists in a set of samples having identical species composition; and maximum beta diversity, in a set where each sample contains a unique set of species. As beta diversity increases, some distortion of the ordination is likely. Distal samples might involute causing sample reversals and the horseshoe effect becomes more pronounced. Also of importance is within-sample variation, alpha diversity, especially if it is surpassed by sampling error. Then, the consequences of sampling error are more likely to become most serious.

Certain difficulties are more statistical in nature. These can be summarized in reference to models, similarity coefficients, and monotonicity. The horseshoe effect is an artifact of linear models being applied to a nonlinear gradient. An ideal model would be nonlinear and would, for a coenocline, produce a single axis that precisely displayed the gradient. Commonly applied similarity coefficients are nonlinear functions of the ordination of samples along a gradient. This is true even in Nonmetric Multidimensional Scaling, the only nonlinear model currently examined for performance (Fasham, 1977). This means that the ideal of a perfect single axis ordination of a 1-D coenocline is not yet possible. The horseshoe effect must be accepted now as a necessary "evil" of ordination. In a monotonic function, the ranked sequence of 1st and 2nd axis scores must agree. If one set of axis scores are ranked from minimum to maximum values, the ranking must also cause the corresponding other axis scores to be ranked from minimum to maximum. The only possible exception to this is that successive other axis scores can remain the same. Most definitely, after ranking of one set, if any successive corresponding scores of the other axis decreases in magnitude, monotonicity does not occur. Actually for errorless ordination, after ranking the corresponding axis scores should always increase, be positively monotonic. When monotonicity is not preserved, as the dissimilarity between samples increases their distances need not increase.

8.3 Principal Coordinate Analysis

Conventional PCA appears to perform best within the framework of each variable being a continuous variate. Unfortunately, such variables might not exist as measures of certain features. For many traits discrete variables or qualitative appraisal of character states are the only measures possible. Also, certain features might be unavailable for measurement on certain individuals. Especially in the latter case, a PCA might be unsatisfactory.

The above problems are commonplace in numerical taxonomy. Both missing data and qualitative character states are frequently encountered. Also, the numerical taxonomist often is more interested in examining variation among individuals rather than variation among measurements. All this has caused some shift from PCA to Principal Coordinate Analysis, PCORD. Whichever method is employed, the purpose of such studies is ordination.

PCORD, since it examines variation, resembles PCA. Both methods create a new set of orthogonal axes in Euclidean space. There are two primary differences between the two methods. First, PCA extracts eigenvalues and eigenvectors from a dispersion matrix of variables or individuals, but PCORD only from a matrix of OTU's (individuals). The latter matrix usually is an association matrix; and, owing to frequent mixing of quantitative and qualitative variables, a matrix of general similarity coefficients is appropriate. As is often the case in numerical taxonomy, the coefficients and matrix of similarity will be called association coefficients or association matrices.

The second difference follows from the principal coordinate eigenvalues and eigenvectors being derived from an $N \times N$ symmetric association matrix having measures of association between individuals rather than a $p \times p$ symmetric dispersion matrix of measurements. Since ordination still pertains to individuals, coefficients of eigenvectors rather than principal coordinate scores are plotted.

Gower (1966) provides a mathematical foundation for PCORD. Blackith and Reyment (1971) present a source listing of a computer program for the analysis. Sneath and Sokal (1973) include PCORD in their exhaustive treatment of numerical taxonomy.

Methodology.   The methodology of PCORD is as follows:

1. Measurements on p-variates are taken on a sample of N individuals. Each individual can actually be a centroid from a prior sample of a single OTU. Original variables can be coded character states, dichotomous measures and/or quantitative measurements. In general p should be, but need not be, greater than N.
2. Coefficients of association between all i and j individuals, a_{ij}, are calculated to form a symmetric $N \times N$ association matrix, $\mathbf{A}$.
3. The characteristic equation is employed to obtain up to N possible values of λ,

$$|\mathbf{A} - \lambda \mathbf{I}| = 0.$$

4. The eigenvector function is next applied to obtain the same number of eigenvectors, $\mathbf{b}_k$,

$$(\mathbf{A} - \lambda_k \mathbf{I}) \mathbf{b}_k = 0,$$

   subject to standardization so $\mathbf{b}_k' \mathbf{b}_k = \lambda_k$.
5. As is the case in PCA, the first two or three eigenvalues often summarize most of the variation in $\mathbf{A}$. Therefore, a principal coordinate graph of the first two or three coordinate axes can be expected to summarize much of $\mathbf{A}$. Such graphs contain plots of corresponding coefficients of pairs or triplets of eigenvectors, not scores. The coefficients of eigenvectors pertain to individuals whereas the scores pertain to variables. This is the crux of this Q-technique.
   In the case of a 3-D graph the plot of b_{ik} is for $k = 1, 2, 3$ axes and for $i = 1, 2, \ldots, N$ individuals.

6. As a check on the performance of the principal coordinate graph, the total distance between the ith and j^{th} individual, d_{ij}, can be obtained from

$$d_{ij}^2 + \sum_{k-1}^{N} b_{ik}^2 + \sum_{k-1}^{N} b_{jk}^2 - 2 \sum_{k-1}^{N} b_{ik} b_{jk}.$$

Although not part of a conventional PCORD, most of the interpretive aids for PCA would be helpful. For example, the percentage of the variance of a variable (individual in PCORD) in each principal coordinate would be an excellent basis for judging how many coordinate axes should be interpreted.

Application. The above methodology has been applied mostly to ordination of OTU's. Blackith and Reyment (1971) provide a biological example for the method. Unlike ordination in PCA, PCORD is not designed to include the possible recognition of patterns of variation and variable bases for the ordination.

Unfortunately the performance of PCORD is not well known. It should perform better than PCA; however, it would seem that both data and nonlinearity could bias the results.

8.4 Correspondence Analysis

A frequent goal in biology is to arrange values of a variable in some meaningful manner. Such ordinations are difficult, especially when the variable consists of many qualitative character states, e.g., an ecologist might wish to arrange sites along a natural gradient that is meaningful in reflecting environmental differences. Gradient analysis can become even more difficult when two such variables are involved. For purposes of exposition, an ecological situation involving two variables, collecting sites and species, will be a constant frame of reference for further discussion.

Various means of analysis, including PCA, are in current practice. In such studies, sites and species data are either analyzed separately or in a single analysis as a single set of data. From the chapter on canonical correlation, it should be obvious that either separate or joint analysis of the two sets might be ineffective. The problems of any analysis can be appreciated in reference to those of PCA. Even when data are collected and transformed properly, there simply is no reason to expect that any single pattern of variation must coincide with the desired gradient. The gradient could be a facet of a single component or perhaps an amalgam of two or more. One might say that conventional analyses are prone to failure since sites and species might be related in a manner that cannot be detected by the analysis. For this reason, conventional approaches are most dangerous when there is no clear appreciation of the nature of the gradient. Then, one might consider an actually erroneous gradient as reasonable.

Beals (1973) reviews such ordination procedures and strongly criticizes PCA as a device for gradient analysis. The basis for the criticism is that PCA does not relate to the nonlinear features of ecological data. Beals summarizes the situation as follows:

Principal component analysis, as a method of ordination to detect environmental influences, makes many unreal assumptions about ecological data. It does not take into account the normal-curve relationship between species success and environment, nor the ecological ambiguity of species absence in a stand. Furthermore, it uses an ecologically nonsensical centroid, and presumes a species-dimensional space. The latter is shown not to relate in any Euclidean way to environmental space. Each plant species in a pair of stands responds to the total environmental difference of those two stands, not to factors independent of those to which other species are responding.

A model preferable to species-dimensional space is one which is defined by changes in vegetation from point to point (Δ-vegetation space).

Beals' criticism emphasizes the Q type analysis, ordination of site samples in a species-dimensional space. Since centering and standardization of species data generally does not produce what the ecologist desires, it is not surprising that evidence of shortcomings in resulting ordinations can be found. In fact, Fasham (1977) and Gauch et al (1977) demonstrate that even a PCA based upon the best transformation of data might perform poorly. Since they also show that Correspondence Analysis does perform very well, the latter bears consideration.

The Nature of Correspondence Analysis.
Although many ecologists might disagree with the criticisms of PCA on the basis of having found meaningful results, the potential hazards in gradient analysis via PCA should not be ignored. Fortunately there is a method, Correspondence Analysis, that should have minimal theoretical problems.

Correspondence Analysis was developed in 1935, was long neglected, and was recently reexamined in some detail by Hill (1973, 1974). The papers by Hill and the text of Lefebvre (1976) are the primary bases for this presentation. Hill (1974) demonstrates that the analysis is equivalent to Fisher's contingency table analysis, is a generalization of Whittaker's gradient analysis, and is a refinement of Kendall's serration. Also, Correspondence Analysis is related to Canonical Correlation Analysis and to PCA. In the sense of a broad definition, Correspondence Analysis is a kind of PCA.

A primary advantage of Correspondence Analysis is that a single application results in ordination of both samples and species in terms of the best fit of one to the other. Other advantages are that it can be expected to perform better than many other procedures and that it requires only presence-absence data for each species at each site. The method is one involving simultaneous double standardization.

Correspondence Analysis tends to resemble gradient analysis more than does a conventional PCA. In gradient analysis, where possible, the species and/or the sites are scaled along a moisture, altitudinal, or other ecological gradient by means of direct observation. Unfortunately, a conventional PCA for such purposes often is difficult or impossible to interpret. In many instances, gradient analysis also is unsatisfactory. The physical criteria for ordination of species or samples frequently are so subtle that only an approximation of a sequence of species is possible.

To apply Correspondence Analysis, one must assume that a maximum association between species and sites is ecologically most meaningful. This would seem acceptable, but the approach of mathematical solution might appear circular. To paraphrase Hill (1973) solution involves a circularity that is not vicious. Rather, the solution is consistent with those in much of scientific research. A potentially more serious problem exists in that Correspondence Analysis involves a linear model. In the final section on ordination, it will be shown that even the cruder modified PCA solution of Correspondence Analysis performs well in spite of this.

Mathematical Solutions.
Two solutions for Correspondence Analysis are available. Hill (1973, 1974) applies an inerative procedure termed Reciprocal Averaging (RA) and Lefebvre (1976) applies a modified PCA. If the modified PCA is of standardized data, both models will produce similar

results. Although important differences in results are unlikely, Hill's method more closely approximates the avowed goals of Correspondence Analysis. For RA, only presence-absence data are collected for each species in reference to each sample. For the modified PCA, either the number of individuals of each species are counted for each stand, or presence-absence data are obtained.

Applications.   There are many possible applications of Correspondence Analysis. From an interpretive point of view, the applications fall into three general categories. First is the relationship of two variables along a single gradient. For example, a variable of species and another of sites can lead to a sequence of species that corresponds to a sequence of sites. Second is the "equivalence" of various states of a single variable. For example, like correspondence scores for two or more species, or sites, implies their equivalence. Third is the equivalence of various states of one variable with various states of the other variable. For example, two or more equivalent species corresponding to two or more equivalent sites can be interpreted as different species performing similar roles in similar habitats.

From a discipline point of view, most fields of biology should find applications. The applications to ecology are stressed here. Naturally such studies can be of a local, regional, or even of a global nature. The applications to biogeography should be self evident. Perhaps an application to numerical taxonomy might indicate the broad possibilities of Correspondence Analysis. A consideration of OTU's $\times$ character states (features measured) could lead to a greater appreciation of how characters relate to OTU's.

Reciprocal Averaging.   This iterative procedure starts from a very crude vector of species scores. Each species is characterized by one of two contrasting ecological features. For the first trial species vector, most definitely no definitive attempt must be made to approximate a precise ecological sequence for the species. For example, the main criterion for a gradient might be assumed to be moisture, so the contrasting features would be "wet" vs. "dry." Since the final scores for ordination are an inherent property of the data, even if the assumption of the importance of moisture is wrong the only consequence is to make the RA computations more lengthy.

The preliminary steps are simple and indicate the nature of RA. From field observations that can be recorded while collecting the presence-absence data for each species in each site, species are guessed as being "wet" or "dry" species. On the other hand, since neither the moisture criterion for classification, nor an incorrect assignment of wet or dry will alter the end results, even a trial vector with elements chosen at random is suitable.

Once all data are collected, they are arranged into a two-way table, a species by sites matrix containing unity for each presence of a species in a site and zero for each absence. This table becomes a reference for the RA of species and site scores. The first trial vector is one for species scores, 0's for "wet" species and 100's for "dry" species. Zero and 100 also will be the extreme values for the final two vectors of scores. Then, for each site, the average of the species scores, 0's and 100's, but only for the species present in the site, becomes that site's trial score in the first trial vector of site scores. After all site scores are obtained, this first sites vector is scaled so the range of scores is 0 to 100. At this point there should be some intermediate site scores, i.e., some scores other than 0's and 100's. The scaled site scores then are used to derive new species scores, each a mean of the scores for all sites in which the species is present. Rescaling the new species vector, 0 to 100, a new set of site scores are calculated and scaled as before. The iteration proceeds until neither a new species vector, nor a new sites vector changes from the previously iterated vector. One set of scores must stabilize immediately after the other. Since species scores are averages of site scores and site scores are averages of species scores, the method is called reciprocal averaging. A numerical example is provided later.

One can see that RA involves non-centering and simultaneous double standardization, so it has advantages of both features. Although both species and site scores are made to conform to a range of 0 to 100, 50 is not a mean or other direct measure of central tendency. However, each species score is standardized by the total number of sites in which the species occurs and each site score by the total number of species at the site. The symmetric nature of the standardization should be evident.

RA might seem very remote from PCA, but this is not the case. Later, it will be shown that both RA and conventional PCA of a specific "dispersion" matrix lead to similar results. In fact, a true RA algorithm for large data matrices might be called an interactive PCA approach (Hill, 1973).

Since RA is analogous to PCA, more than one species or one samples axis must exist. A second set of species and site scores is derived from the same procedure, but scores are computed differently. Each species score remains as the average sample score for all sites in which the species occurs, and each sample score remains as the average species score for all the species in that sample. The only difference is that a correction is applied to remove the first axis influence. Again, the rescaling from 0 to 100 applies in the computations as before. Several axes are possible, but two or three often summarize the pertinent information for correspondence analysis.

The vectors of species and samples scores for each axis can be expected to display specific properties. In coenoclines, the first axis scores tend to be serial and to stress the desired gradient. The second axis scores often portray the consequences of a polynomial axis. A curve is expected on the second axis since a linear model is being applied to a nonlinear function. In coenoplanes, the second axis might display the second axis of the coenoplane; but a third axis might be required to portray the true second axis.

Data.   Data are summarized in an $m \times n$ data matrix

$$\mathbf{A} = (a_{ij}) \quad (i = 1, 2, \ldots, m; j = 1, 2, \ldots, n)$$

whose only possible values, 0 or 1, are determined by a bivariate random variable, $\mathbf{K} = (I,J)$, I of m species and J of n species. If the ith species occurs in the jth site, $\mathbf{K} = (i,j)$ and a_{ij} is set to unity. If the ith species does not occur in the jth site, neither is there a value of $\mathbf{K}$, $\mathbf{K} \neq (i,j)$, and a_{ij} is defined as zero. Therefore, $\mathbf{A}$ is simply a matrix containing 1's for occurrences of species in sites and 0's for absences of species in sites.

The m character states of I or n character states of J can be derived from discrete variables, dichotomous variables, qualitative character states, or even continuous variables. In the case of continuous variables, their ranges must be divided into a number of parts, each part becoming a separate variable. The only criterion is that I and J must consist of discrete states of a single variable. This means that an environmental variable is possible. For example, "discrete states" of climatic, edaphic, topographic, etc. measures can be associated with species.

Approach.   From $\mathbf{A}$, correspondence analysis provides functions, f and g, on the ranges of I and J such that the correlation of $f(I)$ and $g(J)$ is a maximum. This amounts to obtaining scores, $\mathbf{x}' = (x_1, x_2, \ldots, x_m)$ and $\mathbf{y}' = (y_1, y_2, \ldots, y_n)$, which are defined by $f(i) = x_i$ and $g(j) = y_j$ for up to m or n dimensions. In practice, only the first two or three dimensional sets of scores of $\mathbf{x}$ and $\mathbf{y}$ are plotted on separate graphs or on a single graph, either graphing approach portraying the maximum correspondence in the sequence of m-species and n-sites.

Since I and J are maximally correlated, the solution of x_i must be in terms of y_j. For this reason computations are iterative, the scores x_i and y_j are constantly being refined in reference to one another, reciprocal averaging.

Computations.   Consider the data matrix

$$\mathbf{A} = (a_{ij})$$

of 0's and 1's, where rows represent species and columns, sites. the sum of the occurrences of each of the m species in the n sites, the sum for each row, is

$$r_i = \sum_{j=1}^{n} a_{ij}$$

and the sum of the species occurring in each site, the sum of each column is

$$c_j = \sum_{i=1}^{m} a_{ij}.$$

Then, the algebraic mathematical formulation for RA is

$$x_i = \sum_{j=1}^{n} a_{ij} y_j / r_i$$

$$y_j = \sum_{i=1}^{m} a_{ij} x_i / c_j$$

which provides m x_i's and n y_j's, where each element of x_i corresponds to a species and each of y_i to a site. In matrix algebra notation, the latter become

$$\mathbf{x} = \mathbf{R}^{-1}\mathbf{A}\mathbf{y}$$

$$\mathbf{y} = \mathbf{C}^{-1}\mathbf{A}'\mathbf{x},$$

where $\mathbf{R}$ is a $m \times m$ diagonal matrix containing the r_i and $\mathbf{C}$ is a $n \times n$ diagonal matrix containing the c_j.

Since both x and y are expressed in terms of one another, the solution must be an iterative procedure.

Example.   Table 8.2 contains hypothetical presence-absence data for five species (rows) and seven sites (columns). For this example, consider two sets of possible scores

$$x_{ki} \text{ and } y_{kj} \ (i = 1, 2, \ldots, m; j = 1, 2, \ldots, n; k = 1, 2, \ldots, t)$$

Table 8.2 Hypothetical presence-absence data for an example of reciprocal averaging computations.

species	samples						
	1	2	3	4	5	6	7
1	0	0	1	1	0	0	1
2	1	1	0	0	1	0	1
3	1	0	0	0	1	1	0
4	1	0	1	0	0	0	1
5	0	0	0	1	0	0	0

where i pertains to species ($m = 5$), j to sites ($n = 7$), and k to iterations; t is the total number of iterations necessary for the solution.

The following steps are involved:

1. "Guess" a first trial species vector, x_1, with elements of either 0's or 100's based upon some assumed dichotomy of species response. In the example, $x_1' = (100, 0, 100, 0, 100)$.

2. Calculate $y_{1j} = \sum\limits_{i=1}^{m} a_{ij} x_{1i}/r_i$. This amounts to the sum of the 100's and 0's in x_{1i} for all a_{ij} equal to

   unity divided by the number of a_{ij} involved. In the example, $y_1' = (33, 0, 50, 100, 50, 100, 33)$. Then, after defining the range of scores in y_{1j} values as u and the minimum score in y_{1j} as v, scale the values of y_{1j} so the minimum value is 0 and maximum value is 100,

   $$y_{1j} = 100(y_{1j} - v)/u.$$

   In the example, since the range already is 0 to 100, the scaling need not be done. Note that in y_1 there are scores other than 0's and 100's.

3. Calculate $x_{2i} = \sum\limits_{j=1}^{n} a_{ij} y_{1j}/c_j$ which amounts to the sum of y_{1j} scores for all a_{ij} equal to unity divided

   by the number of such scores. In the example, $x_2' = (61, 29, 61, 39, 100)$ and after rescaling, $x_2' = (45, 0, 45, 14, 100)$.

4. At this point, $k = 2$. Following the same procedure as before, the kth y is $y_2' = (20, 0, 29, 73, 23, 45, 20)$; and after rescaling, $y_2' = (27, 0, 41, 100, 31, 62, 27)$.

5. Then, the iteration proceeds by setting $k = k + 1$, the kth x is found as per (3) and the kth y as per (2). Again, k is incremented, etc. until the scores of $k = t$ satisfy some criterion of approximation to the scores of $k = t - 1$. Here, the criterion is that 100 times the sum of absolute deviations between x_{t-1} and x_t are less than the number of elements in x,

   $$100 \left(\sum\limits_{i=1}^{m} |x_{t-1,i} - x_{ti}| \right) < m.$$

The steps (1) through (5) determine the first axis scores for x and y.

6. The last value of the range, u_t, that pertaining to $\mathbf{x}_t$ before scaling, provides the eigenvalue

$$\lambda_1 = u_t/100,$$

the 100 being involved, since the vector from which the unscaled $\mathbf{x}_t$ was derived had a range of 100.

The columns of Table 8.3 contain representations of the successively iterated values of $\mathbf{x}$ and $\mathbf{y}$. The number above each column pertains to the iteration. Notice that numbered columns are followed by a column having the same number plus an "*s*", e.g., 12 and 12s. The former column represents the unscaled and the latter the scaled scores for a particular iteration.

Since the range in values in the final unscaled set of $\mathbf{x}$ scores is 89.75, the first eigenvalue is .8975.

7. To obtain the second axis scores for $\mathbf{x}$ and $\mathbf{y}$, one must first calculate

$$r_i x_{ti}$$

$$\bar{z} = r_i x_{ti}/ \sum_{i=1}^{m} r_i$$

$$r_i \bar{z} \text{ and}$$

$$d_i = r_i x_{ti} - r_i \bar{z}$$

Table 8.3 First axis computations for reciprocal averaging of hypothetical data in Table 8.1. Computations include trial x vector (1s), representative intermediate values of x and y, and final x and y (16s).

	1s	2	2s	3	3s	12	12s	16	16s
	100.00	61.11	45.09	55.85	43.91	69.86	66.48	70.16	66.75
	0.00	29.16	0.00	21.28	0.00	24.05	15.55	24.40	15.77
x	100.00	61.11	45.09	40.09	23.89	10.06	0.00	10.24	0.00
	0.00	38.88	13.77	31.53	13.01	43.51	37.19	43.92	37.52
	100.00	100.00	100.00	100.00	100.00	100.00	100.00	100.00	100.00
	33.33	19.60	27.06	12.30	17.09	17.58	21.12	17.76	21.30
	0.00	0.00	0.00	0.00	0.00	15.55	18.68	15.77	18.91
y	50.00	29.41	40.54	28.46	39.56	51.84	62.27	52.13	62.53
	100.00	72.54	100.00	71.95	100.00	83.24	100.00	83.37	100.00
	50.00	22.54	31.08	11.94	16.60	7.77	9.34	7.88	9.45
	100.00	45.09	62.16	23.89	33.20	0.00	0.00	0.00	0.00
	33.33	19.60	27.02	18.97	26.37	39.74	47.77	40.01	47.99
λ	.8975								

8. Then, obtain x_l, a vector from the above iteration for the first axis. This vector is chosen on the basis of crudely approximating the end result. In this instance, the criterion is to obtain x when

$$\sum_{i=1}^{m} x_{li} - x_{l+1,i} < m \text{ first occurs.}$$

In the example, the scaled vector is $x_{13}' = (66.61, 15.65, 0, 37.35, 100)$.

9. Then, calculate a first trial x vector for the second axis

$$x_{1i} = x_{li} = (\sum d_i x_{li} / \sum d_i x_{ti}) x_{ti}.$$

Next, x_{1i} is scaled from 0 to 100.

The results for all computations in (7), (8), and (9) are presented in Table 8.3.

10. Then, the procedures of (2) through (5) are followed to obtain the final x_k and y_k. However, to prevent x_k and y_k from converging to 1st axis scores and to expedite the procedure, each of the k iteration values is corrected by

$$x_{ki} = x_{ki} - \sum d_i x_{ki} / \sum d_i x_{ti}) x_{ti},$$

where x_{ti} contains the first axis scores for x.

11. Again, the associated eigenvalue is estimated as before. The second axis computations are shown in Table 8.4.

12. The first and second axis scores for species and for stands are each plotted on a single graph (Figure 8.2). As expected, in both cases the second axis is a polynomial axis.

13. The above procedures could be applied in conjunction with a correction to obtain third axis scores.

Table 8.4 Second axis computations for reciprocal averaging of hypothetical data in Table 8.1. See text for explanation.

r	$\sum r x_t$	$\sum r \bar{z}$	d	x_{13}	c	1	1s	4	4s
3	200.30	101.01	98.29	66.61	66.71	−.10	24.07	24.13	24.07
4	63.12	136.01	−72.89	15.65	66.76	−.11	21.90	22.15	22.08
3	0.00	102.01	−102.01	0.00	0.00	.00	72.23	72.20	72.10
3	112.61	102.01	10.60	37.35	37.51	−.16	0.00	0.04	0.00
1	100.00	34.00	66.00	100.00	99.94	.06	100.00	100.12	100.00
14	476.03		−0.01						

$\bar{z} = 34.00$

$\sum d_i x_{12i} = 12402.62$

$\sum d_i x_{16i} = 12410.41$

$\bar{d} = .9994$

$\lambda = .7063$

	1	1s	4	4s
	31.38	32.13	31.39	32.23
	21.90	16.39	22.08	16.73
	12.03	0.00	12.03	0.00
	62.03	83.05	62.03	83.23
	47.07	58.19	47.09	58.36
	72.23	100.00	72.10	100.00
	15.32	5.46	15.38	5.57

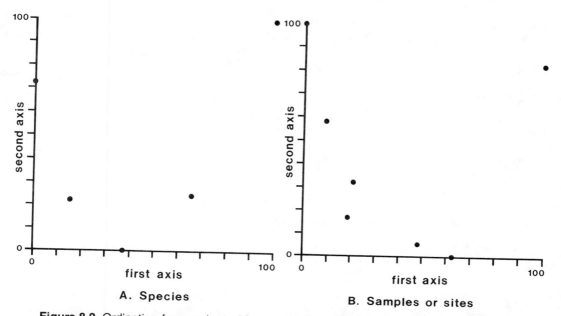

Figure 8.2 Ordination from reciprocal averaging of hypothetical data (Table 8.1): A, first and second axis ordination of species; and B, first and second axis ordination of samples.

8.4.1 Principal Component Analysis

For a more conventional Correspondence Analysis via PCA, counts might be taken and transformed. On the other hand, as mentioned previously, PCA of a specific matrix derived from presence-absence data can provide satisfactory results if species scores are first transformed to z-scores. The latter might be surprising in view of the results of such a transformation.

Lefebvre (1976) develops methodology in reference to a data matrix of centered but untransformed counts. However, Hill (1974) demonstrates that counts cannot be expected to provide better results than presence-absence data. Therefore, unless counts or some other quantitative measures are required for other evaluations of sites and/or species relationships, sampling for a given study can be just as simple as for RA.

Computations.   In reference to previous notation, two changes are required in the formulation. First, a single expression involving only one of the sets of scores is required. The formula for species, $\mathbf{x}$, can be rewritten, replacing $\mathbf{y}$ with its value in terms of $\mathbf{x}$

$$\mathbf{x}_1 = \mathbf{R}^{-1}\mathbf{A}\mathbf{C}^{-1}\mathbf{A}'\mathbf{x}.$$

In this form, $\mathbf{x}$ is used to derive $\mathbf{x}_1$. The only problem is that $\mathbf{R}^{-1}\mathbf{A}\mathbf{C}^{-1}\mathbf{A}'$ is not a symmetric matrix. Second, for simplicity in computing eigenvalues, $\mathbf{x}$ is transformed to $\mathbf{x}^*$

$\mathbf{x}^* = \mathbf{R}^{1/2}\mathbf{x}$, so

$\mathbf{x}_1^* = \mathbf{R}^{-1/2}\mathbf{A}\mathbf{C}^{-1}\mathbf{A}'\mathbf{R}^{-1/2}\mathbf{x}^*.$

Then, defining

$\mathbf{B} = \mathbf{R}^{-1/2}\mathbf{A}\mathbf{C}^{-1/2}$

$\mathbf{B}\mathbf{B}' = \mathbf{R}^{-1/2}\mathbf{A}\mathbf{C}^{-1}\mathbf{A}'\mathbf{R}^{-1/2} = \mathbf{L},$

which demonstrates that $\mathbf{L}$ is a $m \times m$ symmetric matrix. Therefore,

$\mathbf{L} - \lambda\mathbf{I} = 0$

will have m positive eigenvalues and a complete set of eigenvectors.

Lefebvre (1976) carries the computations one step farther prior to calculating eigenvalues. This extra step centers but does not standardize the data. Proceeding from above, $\mathbf{V}$ is obtained from $\mathbf{L}$

$\mathbf{V} = \mathbf{L} - \mathbf{r}^{1/2}\mathbf{r}^{1/2'}.$

Then, utilizing the characteristic equation,

$\mathbf{V} - \lambda\mathbf{I} = 0,$

and

$(\mathbf{V} - \lambda_i\mathbf{I})\mathbf{u}_i = 0$

leads to the ith set of final scores for $\mathbf{x}$ and $\mathbf{y}$

$\mathbf{x}_i = \mathbf{u}_i'\mathbf{R}^{-1}\mathbf{A}\mathbf{C}^{-1/2}$

$\mathbf{y}_i = \mathbf{C}^{-1}\mathbf{A}'\mathbf{x}_i\lambda_i$

where $\mathbf{x}_i$ consists of m species scores and $\mathbf{y}_i$ consists of n stand scores.

Changing the latter to a standardized model is very simple. all that is required is to transform each set of scores for each species to z-scores. This would amount to a standardized by species model.

8.4.2 Indicator Species Analysis

Hill et al (1975) extended the application of RA to the determination of indicator species for "vegetation types," sets of sites that can be grouped within the ordination. Six steps can be recognized in the analysis. First, RA leads to ordination of first axis site scores. The first axis scores are the only ones involved in indicator species analysis. Second, the site ordination is divided at its center of gravity to recognize a negative group below the center of gravity and a positive group above it. The terms "positive" and "negative" have no special meaning, being applied purely as designations of convenience. Third, the five "best" indicator species of the groups are chosen and used to obtain site indicator scores. Fourth, an indicator threshold (a point near the center of gravity) is selected on the basis of optimizing correct classification of sites to groups. Fifth, a zone of indifference, a narrow area containing borderline sites assigned to neither group and agreeing closely with the indicator threshold, is recognized to redefine the groups. Finally, the process is repeated, as before, on each defined group to obtain a classification hierarchy. Various levels of classification can thereby be recognized.

The purpose here is only to outline the method sufficiently to indicate how it is applied and how well it can be expected to perform. Many details will not be discussed. As is the case for RA, both source listings of the computer programs and details of procedures, especially sampling, are available from M. O. Hill.

Reciprocal Averaging.
The RA algorithm is modified to contain a weighting factor to downweight rare species. The downweighting prevents the recognition of a few species–poor sites, typified by the total distribution of a few rare species, as a distinct "vegetation type."

For purposes of weighting species, only a slight modification of the original RA algorithm is involved. If the number of sites in which a species, x_i, occurs, r_i, is greater or equal to one-fifth the total number of sites, $n/5$, the weight for species x_i is set to unity,

$$w_i = 1 \qquad\qquad \text{if} \quad r_i \geqslant n/5.$$

If not,

$$w_i = [r_i/(n/5)]^2 \qquad\qquad \text{if} \quad r_i < n/5.$$

In this sense, $n/5$ is defined as a threshold of rarity. Therefore, the only change in the original reciprocal averaging formula is to replace x_i with $w_i x_i$.

Divide Site Ordination.
To divide the site scores, y_j, at their center of gravity requires the weighted mean site score, $\overline{y}$, the natural zero point,

$$\overline{y} = \sum_{j=1}^{n} c_j y_j / \sum_{j=1}^{n} \sum_{i=1}^{m} a_{ij}.$$

Then, all site ordination scores are transformed to deviations from their weighted mean to redefine the y_j as

$$y_j = y_j - \overline{y}.$$

Next, adjusted site scores are plotted. For future needs, the minimum site score, y_{min}, and maximum site score, y_{max}, are noted. Also, a critical zone between $y_{min}/5$ and $y_{max}/5$ is recognized and divided into eight equal parts (Figure 8.3). The latter defines area 1 to the left and below the critical zone, the critical zone containing areas 2-9, and area 10 to the right and above the critical zone. Later, the

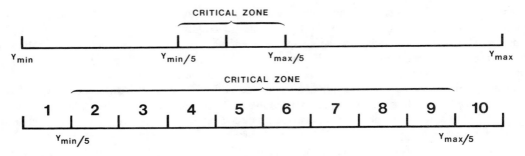

Figure 8.3 Zones and areas pertaining to indicator species analysis.

critical zone will be utilized to determine both a final indicator threshold and a final zone of indifference.

Indicator Species Selection.   To select the five best indicator species, a possible division is examined in terms of the mean deviation score, zero. Although this mean score is not the final indicator threshold (maximum value of the negative group), sites with positive site scores are said to comprise a positive group and sites with negative scores, a negative group.

The choice of five indicator species also is a matter of convenience, a single species would be no more effective than classical association analysis and more than five would prove difficult to handle. Selection is based upon how well the five species fit the positive-negative groups dichotomy. A perfect indicator species would be confined to all sites of a group, positive or negative, and a species with no indicator value would be equally distributed in both groups. Defining N_1 as the total number of negative sites, n_{1i} as the number of sites in which species x_i occurs, and N_2 and n_{2i} as positive site statistics corresponding respectively to N_1 and n_{1i}; then, the indicator value of x_i is

$$I_i = |n_{1i}/N_1 - n_{2i}/N_2|.$$

Therefore, a perfect indicator species would receive a value of unity and a species with no indicator value, zero.

The above algorithm would lead to the five species having the largest indicator values being chosen as indicator species. However, the actual application of the algorithm is not in reference to an indicator threshold value of zero. Occurrences in areas 4-7 of the critical zone are ignored, since the four areas contain borderline sites too intermediate for practical consideration. In this sense, areas 4-7 are said to define a zone of indifference. Although this is the zone of indifference utilized for the choice of indicator species, the zone is changed later for the final classification of sites.

Classification of Sites.   Classification involves the calculation of site indicator scores, a preliminary classification of sites, and a final refinement of the classification.

Site indicator scores are derived from the site distribution of indicator species, a zero preliminary indicator threshold, and a preliminary zone of indifference (areas 4-7). To compute site scores, first each species is assigned a value of +1 (if it occurs in more positive sites) or of −1 (if it occurs in more negative sites). Then, for each site, a site indicator score is obtained by summing the foregoing values for the indicator species present. This defines six groups for sites (Figure 8.4):

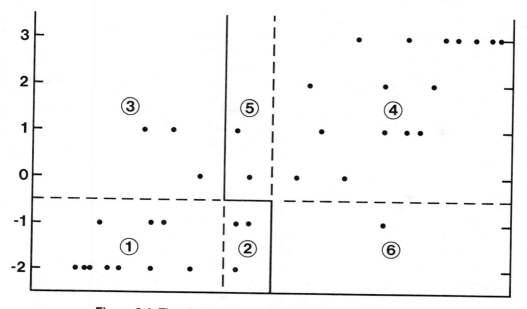

Figure 8.4 The six basic groups of indicator species analysis.

1. negative group;
2. negative-borderline group, sites within the zone of indifference;
3. misclassified negative group, sites with negative ordination scores that receive positive site indicator scores;
4. positive group;
5. positive borderline group; and
6. misclassified positive group.

Further refinements are made to reduce the number of misclassifications. First, five additional classifications, each derived from a different indicator threshold are compared. The five possible threshold values are derived from one of the five possible boundaries in areas 4-7 of the critical zone. The indicator threshold that produces the fewest misclassifications (conflicts between site indicator scores and site ordination scores) becomes the final indicator threshold.

Next, five possible combinations of adjacent areas in the critical zone are evaluated as a final zone of indifference. The zone from areas 2-5, 3-6, 4-7, 5-8, and 6-9 producing the fewest misclassifications becomes the final zone of indifference. The final indicator threshold value and zone of indifference lead to final assignment of groups.

The above procedures provide the primary level for site classification, a dichotomy between sites, the positive and negative group. The same procedures can be applied to produce a second level of four groups, two within the positive group and two within the negative group. In the case of r levels, 2^r groups are defined in a hierarchical classification. The total number of divisions that can be made, although somewhat data dependent, is determined by the ecologist.

Comments.   The use of trial values represents an attempt to obtain the best performance of indicator species and the best classification of groups. The trial values optimize correct classifications by allowing some flexibility in the choice of both a threshold value and a zone of indifference. Since the species indicator scores and the indicator threshold value define "vegetative types," new sites can be assigned to the original classification by a key to groups. In practice, Hill et al (1975) find approximately 3% misclassification per level. Unfortunately for five levels of division, this can mean 15% misclassification. For this reason, an additional refinement is included, preferential species are determined. A preferential species is one that is twice as likely to be within its group that in another. Likelihood is actually evaluated as the percentage or proportion of correct group occurrences rather than the absolute number.

8.5 Polar Ordination

One of the simplest and oldest methods of ordination is that of Bray and Curtis (1957). It is termed Polar Ordination, Wisconsin Comparitive Ordination, or Bray-Curtis Ordination. For some time, this method has been considered crude mathematically. Part of this is due to the simplicity of the polar ordination algorithm and to the lack of objectivity in selecting the endpoints for any ordination.

Polar Ordination (PO) is one of the two most popular techniques for ordination, the other being PCA. In a study of performance with PCA and correspondence analysis by PCA (Gauch et al, 1977), PO performed well, regularly outperformed PCA, and was outperformed by correspondence analysis. Although true RA remains untested, following Gauch et al (1977) and Fasham (1977), in comparing tested methods, RA will also symbolize correspondence analysis by PCA.

In some cases, PO outperformed RA. These will be considered later. In such instances, a theoretical imperfection in PO might become a biological strength. The theoretical problem is that PO requires the ecologist to select the extreme samples, other samples then automatically being placed between the extremes. If the extremes are chosen correctly, the ordination tends to be held in the proper sequence. Therefore, when RA is likely to produce distorted results, PO can serve as a check or even as an alternate procedure.

Some of the details common to all ecological ordination studies were presented previously in the discussion of RA so will not be repeated here or in the discussion of Nonmetric Multidimensional Scaling. Further details of PO will pertain to its nature and methodology.

The Nature of Polar Ordination.   Polar ordination was formalized in a classic study of Wisconsin forest communities (Bray and Curtis, 1957). In application, species data are collected for each site and, for each pair of samples, species data are used to calculate a coefficient of similarity between sites. In the original Bray-Curtis Ordination, the similarity coefficient was a measure of the number of species found in both sites in reference to the sum of the total number of species in each site. Other similarity measures were added later—we shall consider three of them in the presentation of methodology, three evaluated for performance by Gauch et al (1977). These other measures involve the "counting" of individuals in each sample and the calculation of an importance value for each species in each sample. Aside from the effort in counting each species, computations remain quite simple.

Next, similarity measures are transformed to dissimilarity coefficients, usually by subtracting each computed similarity value from the theoretically largest value possible for the particular similarity coefficient. Then, from field observation and perhaps the aid of dissimilarity coefficients, the two extreme samples are selected. The selection is statistically arbitrary and critical. The utility of any PO is dependent upon a correct selection. Finally, in reference to the two selected sites (the endpoints of the ordination) the positions of the other samples in the ordination are determined.

Methodology.   A possible interruption in the otherwise straightforward procedure of PO is the choice of the similarity coefficient to be used. The study of Gauch et al (1977) disclosed no meaningful difference in performance among the variations of PO discussed here, except for the poor performance of Euclidean distance based upon importance values. For this reason, the selection of any one of the other measures would be a matter of personal preference or perhaps a matter of importance to other aspects of a particular study. It will be seen that the original Bray-Curtis index is the simplest to obtain.

Methodology is as follows:

1. A set of n random or systematic random samples are selected from a community, each of the n sampling sites is a quadrat or similar subset of the community. Within the set of n samples, m species are found to occur.
2. For each pair of samples, i and j, a coefficient of similarity is computed. If the Euclidian distance is computed, the coefficient is one of dissimilarity.

    The original coefficient applied by Bray and Curtis (1957) was the similarity coefficient of Dice. Other measures of similarity tend to be more popular now. Many of the latter measures involve importance values for each species in each site, hence data more detailed than simple presence-absence. The importance value of species h in sample i is

    $$P_{hi} = u_{hi} + v_{hi} + w_{hi};$$

where u_{hi} is the relative density of species h in sample i, v_{hi} is the relative dominance of species h in sample i, and w_{hi} is the relative frequency of species h in sample i. The relative density of species h in stand i is 100 times the density of species h in sample i divided by the total density of all species in sample i. The relative dominance of species h in sample i is 100 times the basal area of species h divided by the total basal area of all species in sample i. The relative frequency of species h in sample i is 100 times the frequency of species h in sample i divided by the total frequency of all species in sample i. Therefore, the minimum importance value approaches zero and the maximum value is 100.

Two other popular similarity measures to be considered are the percent similarity and the coefficient of community (Sørensen variant). The percent similarity involves importance values and is

$$S_{ij} = 200 \frac{\sum\limits_{h=1}^{m} \min(P_{hi}, P_{hj})}{\sum\limits_{h=1}^{m} (P_{hi} + P_{hj})}$$

The coefficient of community (Sørensen variant) involves double standardization, but not importance values. Since the coefficient is merely 100 times the coefficient of Dice, it simply transforms the original Bray-Curtis statistic to a scale of 0-100.

Also popular, the Euclidean distance utilizing importance values, a coefficient of dissimilarity, is

$$D_{ij} = \left(\sum_{h=1}^{m} (P_{hi} - P_{hj})^2 \right)^{1/2}.$$

3. Measurements of dissimilarity are obtained for the measures of similarity by subtracting each from its maximum possible value. In both the percent similarity and coefficient of community (Sørensen variant), subtraction is from 100.
4. Two samples are chosen as endpoints. Then for each of the remaining i samples, the distances of the sample from the first endpoint, D_{1i}, and from the second endpoint, D_{2i} are calculated. The distance between endpoints is defined as L.
5. Then, the position of the other i samples, X_i, along the ordination axis is

$$X_i = \frac{L^2 + D_{1i}^2 - D_{2i}^2}{2L},$$

and the distance, E_i, of sample i from the ordination axis is

$$E_i = (D_{1i}^2 - X_i^2)^{1/2}.$$

6. Next, a 2-D graph is provided by all $X_i E_i$ pairs.
7. From the above first axis ordination of X_i it might be found that dissimilar sites on the basis of E_i might be more or less equidistant from the two ends. The most extreme pair of deviates can become the basis for a second axis ordination. Further axes can be determined in the same manner.

8.6 Nonmetric Multidimensional Scaling

Excellent ordination performance of Nonmetric Multidimensional Scaling (MDS) is indicated for ecology by Fasham (1977) and for numerical taxonomy by Rohlf (1972). For this reason, the technique should become more popular among biologists. Historically, MDS was developed for the fields of psychology and sociology and still finds its greatest application in those disciplines.

Modern usage of MDS was introduced by Shepard (1962a, b). Kruskal (1964a, b) formalized methodology and provided algorithms for computation. Shepard has termed the method as the analysis of proximities and recently (1974) summarized procedures.

The Nature of MDS.   MDS is a nonparametric technique for inferring multidimensional structure. It resembles PO in that similarity coefficients are calculated and then used to derive dissimilarity measures between pairs of samples. Beyond this MDS departs from PO. Three primary departures occur. First, although the dissimilarity measures can be metric, such quantitative measures are not utilized in further steps. Dissimilarity measures of samples are transformed to an ascending numeric rank order. The "nonmetric" in MDS pertains to this nonparametric rank ordering. Second, in addition to the rank order of samples, the distance between all pairs of samples is obtained. In a sense, these distances are an additional set of dissimilarity coefficients. Generally, the rank order measures and distances will not agree. The goal is to obtain the correct ordination from the disagreement. Third, MDS involves a nonlinear model.

In concept, MDS methodology is fairly simple. The rankings and distances become part of an iterative procedure that extracts the best fit of the two sets of measures. This is analogous to RA, but MDS is applied to two measures of samples, not measures of both samples and species. The ultimate goal in MDS is to obtain a precise $r < n$ dimensional ordination of the samples where r is the number of dimensions of the community structure and n is the number of samples.

The major strength of MDS is its nonlinear, nonparametric basis. There are three potential sources of weakness. First, MDS requires a rank order relationship between dissimilarity coefficients and distances. The ideal is a perfect fit of the two rankings, both increasing together, a monotonic relationship is sufficient. The permissable departure from monotonicity is not known. Second, MDS can be trapped in an excellent fit between dissimilarity rankings and distances in a subset of the total samples (a local minimum). In such a case, the total set of samples will fit the global minimum poorly in those portions not in the subset. Prevention of such trapping in the algorithm appears to require a good trial vector, one approximating the true ordination. Third, good performance requires that the number of dimensions of the community structure is known. In a Monte Carlo Study where the number of dimensions was known (Fasham, 1977), MDS sometimes outperformed RA.

Methodology. The methodology is as follows:

1. A set of n samples that include m species is collected at random from a community. From each of the i samples, counts of each of the h species defines X_{hi}.
2. The counts, X_{hi}, or simply the presence-absence of species are used to derive a measurement of similarity between each pair of samples, i and j. Five such coefficients (Pearson's product moment correlation coefficient, angular separation or cos θ, percent similarity, Kendall's rank correlation coefficient, and Jaccard's coefficient of community) were examined by Fasham (1977). All but Kendall's rank correlation coefficient were already defined.

    The Kendall's rank correlation coefficients are calculated from

    $$r_{ij} = \frac{1 - 4q_{ij}}{m(m-1)}.$$

    The value of each q_{ij} is determined as follows: First, the species in samples i and j are ranked separately, 1 to m on the basis of ascending number of individuals. Second, for sample i the ascending rank order of species, 1 to m, is listed in a row. Third, a second row is added. This row contains the same order of the species as in sample i but contains the ascending rank order numbers of the species in terms of sample j. Finally, for each species in sample j, row 2, the number of *smaller* rankings of species to the right are counted. The sum total of all such counts is q_{ij}.
3. Each similarity coefficient is transformed to a dissimilarity coefficient, δ_{ij}, generally by subtraction of the particular similarity coefficient from an appropriate constant.
4. The distance between each pair of samples is obtained from some value of r in

    $$d_{ij} = \left(\sum_{h=1}^{m} |X_{hi} - X_{hj}|^r \right)^{1/r}.$$

When $r = 1$, the formula provides the Manhattan or city block distance; when $r = 2$, the conventional Euclidean distance.

As presented here, the X_{hi} represent counts. They can represent other quantitative measures, but remain more difficult to obtain than presence-absence data.

5. Samples are ordered so their dissimilarities, δ_{ij}, are in ascending order. In the unlikely event that the order corresponds exactly to increasing distances, d_{ij}, the samples are monotonically increasing and calculations are completed by nonparametric regression (Kruskall, 1964b) to produce an estimated distance, $\hat{d}_{ij}$, the value on a regression curve for a given δ_{ij}.

6. In the general case of imperfect correspondence, calculate the stress (departure from monotonicity) by

$$S = \frac{\sum_{i < j} (d_{ij} - \hat{d}_{ij})^2}{\sum_{i < j} d_{ij}^2}$$

Stress is a goodness of fit measure, the smaller the value for stress the better the fit. Notice that as long as $\hat{d}_{ij}$ is within $2d_{ij}$, $S < 1$. The stress of the X_{hi} is invariant with respect to transformation of X_i in terms of location, scale, rotation and reflexion—for $b \neq 0$, $X_i = a + by_i$. Therefore, the final X_i can be ordinated in reference to any center or axis.

7. Using the method of steepest descent (Kruskall, 1964b), minimize the stress and obtain the values for ordination.

Two Computational Problems.   The major problem in MDS is the initial ordering of samples, step 5. An inappropriate ordering can cause stress to be minimized in reference to a segment of the n samples, a local minimum, rather than the n samples, a global minimum. To assure the latter, Fasham (1977) recommends that the sequence of samples provided by a prior RA be used as the initial ordering.

Another problem with MDS is determining the number of dimensions actually represented by the data. Stress is of no help, it decreases with the number of dimensions examined even if additional dimensions are meaningless. Both Kruskall (1964a) and Shepard (1974) consider this difficulty. Kendall (1971) suggests that the horseshoe effect be used as a measure of one more axis than actually exists. Fasham (1977) suggests using the minimum spanning tree (Gower and Ross, 1969). In a 2-D ordination, one dimension is implied by the 2nd axis being a polynomial axis and the 2-D ordination agreeing with the order of connections in the minimum spanning tree. In the same 2-D, 2-D data are indicated by one or more main lines that may have side branches. Many crossing side branches indicate more than two dimensions. In the case of 2 or more dimensions, the same number of axis ordinations from RA can be applied to disclose the other MDS axes.

MDS Performance.   Although details of performance of ordination methods is the subject of the next and final section of this chapter, some aspects of MDS performance require introductory remarks.

In a comparison of PCA, RA and MDS, Fasham (1977) applied a measure of the true order of the samples and a measure of sample displacement to evaluate the methods. The latter measure would disclose samples out of their exact positions even if all samples were in the proper sequence.

In the only coenocline studies, neither centered nor uncentered PCA recovered the true order of samples, all MDS similarity indices except the Jaccard index recovered the true order, and RA recovered the true order. All MDS ordinations except the rank correlation coefficient had less displacement than RA, even the Jaccard index which did not recover the true order. In a 2-D MDS ordination of the coenocline, only percent similarity and cos θ recovered the true order, but all MDS similarity coefficients had smaller displacement indices than RA. All this is in terms of fit to the true 1-D coenocline.

Displacement of the coenocline into the second dimension was also measured to account for the horseshoe effect beyond the true dimensionality. For this study, only the best MDS performer, log transformation of raw data and cos θ, was compared with RA under four conditions of low to extremely high beta diversity. All RA and all 1-D MDS ordinations obtained the correct sequence of samples. In the 2-D MDS ordinations, only the high and extremely high beta diversity conditions produced an incorrect sequence. However, every MDS measure of displacement in 1- and 2-D was less than that for RA. The "better fit" of the two 2-D incorrect MDS sequences can be explained on the basis of a lesser horseshoe effect.

In studies of three coenoplanes, performance statistics favored MDS in two cases, but both MDS and RA provided realistic ordinations. In the case of a rectangular grid of sites with unequal beta diversities, the RA ordination displayed the horseshoe effect into the lesser beta diversity and the MDS ordination a compression into the lesser beta diversity. However, RA recovered the 2-D axis on its 3rd axis.

8.7 Critique of Ordination Techniques

As of now, no method of ecological ordination can place every sample into its precise position with reference to all other samples. A perfect method of ordination does not exist. The methods examined here in some detail, RA, PO and MDS can determine the correct sequence of samples along a coenocline but all fail in some degree to maintain the precise distance between samples along the gradient. Much the same situation exists for coenoplanes. It seems that these methods usually approximate community structure very well but in varying degrees provide distortion in the details of community structure. Our purpose here is to summarize briefly the performance of PCA, RA, PO and MDS.

Although not treated here, other ordination techniques deserve like characterization. Any serious study of ordination will quickly indicate that it was barely outlined here. The details of ordination and other methods are readily traced from Gauch et al (1977). Computer programs for most methods discussed here plus others are available from H. G. Gauch, Jr., Division of Biological Sciences, Cornell University, Ithaca, NY, 14850. The only exception is Hill's (1973) true RA.

Principal Component Analysis. Conventional PCA's (centered, non-standardized or sample standardized) performed poorly for site ordination. PCA's involving species or double standardization performed well as long as monotonicity was preserved. In no case did PCA outperform other methods mentioned, but PCA appears useful for the purpose of grouping or classifying sites or species (Gauch et al, 1977).

Correspondence Analysis. Although the modified PCA solution to Correspondence Analysis, also called RA but actually PCA-RA, is superior to conventional PCA for ordinations, Hill (1974) indicates that it shares some of the latter's problems. The standardization of species scores is essential.

Then, the PCA-RA performs much like true RA. The only exceptions are when most species are very common or when samples vary greatly in species richness. In such cases, species standardization performs poorly—rare species and species-poor sites distort the results. Also, for reasons discussed previously, centering of data should be inappropriate.

The main strength of true RA is that it capitalizes upon a good species ordination to provide a good site ordination and vice versa. The method relates species of like site distribution and sites of like species composition independent of species abundance or site richness. There are also data strengths in reference to RA being a modified PCA. The data are non-centered and symmetrically standardized. Non-centering favors abundant species and rich sites but symmetric standardization favors unique sites, especially rare species confined to species-poor sites. Hill (1973, 1974) indicated superior theoretical properties and demonstrated excellent performance.

Turning to comparisons with other methods (Gauch et all, 1977; Fasham, 1977), even PCA-RA is an indirect method of ordination that does not depend upon any initial arrangement of the data. So is PCA. PO requires a choice of ordination endpoints, and MDS is somewhat dependent upon an initial arrangement of the sites and the dimensionality of the data.

In Monte Carlo studies PCA-RA provided the correct sequence of samples for all coenoclines examined. The coenoclines reflected low to high beta diversity, the high extending to a "worse possible" case. Although true RA was proposed for coenocline analysis, for coenoplanes of low beta diversity even PCA-RA outperformed both PO and MDS. For coenoplanes of moderate beta diversity, PCA-RA is somewhat better than PO.

RA is clearly the choice for species ordinations and for finding the first and major direction of sample variation. The main problem of RA appears to be the horseshoe effect. In some cases, RA will present a true 2D structure, but the 2nd RA axis can be polynomial and the true 2nd axis not appear until the 3rd or even a later RA axis.

Again: The so-called RA computer program used by Gauch and Fasham actually provides a modified PCA solution. Therefore, the poorer Correspondence Analysis algorithm was the only one tested.

Polar Ordination.
The major limitation of PO is the need for known endpoints. An incorrect choice of endpoints can be critical, even moderate misses of endpoints can destroy an ordination. When true endpoints are anchored, PO limits distortions from nonlinearity. It is not as subject as are PCA or RA to coenocline curvature because calculations of sample positions are independent of one another. Also, the calculations for any one axis tend to produce a linear ordination, so a true 2D community structure is presented by the first two PO axes.

The continued usefulness of this mathematically unsophisticated technique serves as a reminder to anyone prone to discredit quick-and-dirty methods. PO clearly is a useful technique that will serve well in the set of methods that should be applied to any ordination. In such an approach, an understanding of the sources of discrepancies among ordination techniques should serve well in understanding the actual community structure.

Nonmetric Multidimensional Scaling.
Usually, MDS performs as well or better than PCA-RA or PO and outperforms PCA. The major strength of MDS exists in the nonlinear and nonparametric features of the method. For coenoclines, MDS should place samples along a line. In actual coenocline ordination, MDS does produce a slight horseshoe effect, but smaller than that for other methods discussed. The horseshoe effect is the result of the similarity coefficients now in use. Perhaps better

coefficients soon will be available. The primary shortcomings of MDS are dependency upon knowing the true number of dimensions in the community structure and requiring some idea of the true ordination. Potential solutions to these problems were already mentioned.

In conclusion, active study of ordination methods implies a desire for better methods, perhaps a search for a single best method. In spite of this, many good methods exist and not all of the good methods were examined here. No single method should be the only device for attempting an analysis of community structure. Perhaps those presented are sufficient and perhaps current methods will provide satisfactory results for any ecological study. This can be answered only by future research. In any event, the state of the art is changing rapidly and the literature appears to be increasing exponentially.

II. Cluster Analysis

The purpose of applying cluster analysis is to segregate N individuals into g groups termed clusters. Naturally part of the intention is that individuals within any cluster are more closely related than are individuals in different clusters. In practice, it is recognized that a cluster can contain a single individual (a weak cluster) or all N individuals (a strong cluster). In the broad sense of evaluating relationships among N individuals, cluster analysis is similar to ordination. Most definitely cluster analysis shares the problems of ordination.

Cluster analysis is the generally accepted designation for the techniques to be reviewed; but, in reference to widespread applications in diverse disciplines, such descriptive terms as classification, clumping, grouping, numerical taxonomy, Q-analysis, typology, and unsupervised pattern recognition are often substituted. The applications are diverse. The aim might be simple data reduction, reducing N units to g groups to produce a more concise and hopefully more understandable portrayal of the individuals. The former purpose might be extended to generating hypotheses about the nature of the individuals and/or the clusters. These hypotheses should be tested by other approaches. In contrast to hypothesis generation, the objective might be to test hypotheses proposed from other methods.

A less errudite purpose is simple data exploration, essentially "fishing for results." Other goals include reducing a data set to types (clusters), model fitting, assignment of new individuals to previously established groups, and prediction based upon known clusters.

Most of the development of cluster analysis was linked with numerical taxonomy and can be attributed to the efforts of P. H. A. Sneath and R. R. Sokal (Sokal and Sneath, 1963; Sneath and Sokal, 1973). For biologists, the latter is the best reference. Other current texts are Everitt (1974) and Hartigan (1975), but many others exist. In fact, there are so many methods of cluster analysis that a text is required to do the subject justice. Therefore, further discussion is intended only to provide some familiarity with the overall approaches.

Methods of Cluster Analysis. Methods tend to follow a sequence of steps that are repeated until all individuals are placed within a cluster; but, procedurally, methods are of two types. Clusters can be formed by sequential addition of individuals or by sequential separation of subsets of individuals, i.e., by agglomerative or divisive techniques. For cluster analysis, agglomerative methods generally are preferred because most are more polythetic and less demanding of computer time. Divisive methods regularly utilize a single and different variable for each level of separation; agglomerative methods regularly apply all variables at each level. Although direct solutions exist for divisive methods, all algorithms are more time consuming than those for agglomerative methods. Even agglomerative methods tax computer time in the case of large N.

Agglomerative methods can be further characterized as hierarchical or nucleated. Hierarchical methods presume that $N - 1$ levels of groupings exist and all lower level groups are nested within higher levels, i.e., each group at level ℓ is part of a larger group at level $\ell + 1$, and all lower levels combine into a single group at level $N - 1$. Nucleated methods presume that each cluster is unique. Naturally a choice of one or the other, hierarchical or nucleated, should in part at least result from the structure of the data analyzed.

8.8 Hierarchical Clustering Methods

At the onset of examining the relationships among N individuals, a complex of decisions must be made. One must choose between ordination and cluster analysis. In either case a further choice is required, that of an association coefficient. Even if cluster analysis is preferred, clustering methods and their results can differ so greatly that a strategy is virtually mandatory. None of the association coefficients reviewed in section 8.1 might be most appropriate for a particular study, so other measures should be examined (Sokal and Sneath, 1973). Even if an agglomerative technique appears more appropriate, a nucleated method might be superior to a hierarchical one. In spite of all this, as a matter of treatment simplification, further discussion shall be limited to application of the Euclidean distance to hierarchical methods, a so-called SAHN approach. The acronym is descriptive of the approach as sequential, agglomerative, hierarchical, and non-overlapping. It will be seen that even in this simplified exposition, much room for diverse strategies exist.

Certain hierarchical methods are more appropriate since they are compatible and combinatorial. In addition, the manner in which clusters are distorted by each method is important. A compatible method provides likes measures of association at all levels of grouping. For example, both the original Euclidean distances between individuals and the distances calculated between groups at any higher level are statistics of the same kind. A combinatorial method is a compatible method that utilizes distances from level ℓ to compute distances for level $\ell + 1$. For example, consider groups i and j (with n_i and n_j members and d_{ij} distance between groups) becoming the larger group k ($n_k = n_i + n_j$) and the distance between the new group k and another point m. From the values d_{ij}, n_i, n_j, d_{im} and d_{jm}, a combinatorial method will provide d_{km} from

$$d_{km} = \alpha_i d_{im} + \alpha_j d_{jm} + \beta d_{jm} + \lambda |d_{im} - d_{jm}|.$$

A very important consequence of this combinatorial formula is that it will cause many distances between individuals to be altered from the original distances.

The different clustering methods to be discussed are compatible and combinatorial, but differ in values of the coefficients α_i, α_j, β, and γ. A primary concern is that when $\gamma = 0$ and $\alpha_i + \alpha_j + \beta \geq 1$, the successive minimum distances from the first through last level are monotonic.

The unavoidable consequence of forming groups at successive levels is space distortion. Even if original Euclidean space is defined accurately by the original d_{ij}, at higher levels the distances between groups do not pertain exactly to the same original space. This can be appreciated in reference to some original distances being altered by the combinatorial algorithm. Owing to differences in the coefficients of the combinatorial algorithm, some clustering algorithms tend to be space-contracting, as groups are formed some individuals are moved closer to adjacent groups. Other algorithms are space-dilating, causing newly formed groups to become more remote from one another. Space-contracting algorithms are prone to "chaining," linking successive levels in increments so similar that separate clusters are not distinguishable. Space-dilating algorithms have the opposite effect.

To demonstrate the potential of different possible clustering strategies, eight often seen combinatorial algorithms will be considered. Fortunately, the clustering computations are similar for the different algorithms.

Clustering Computations. Being compatible and combinatorial, any level $\ell + 1$, is obtained from the prior level, ℓ, by a sequence of four steps:

1. From the matrix of distances between groups, as they exist at level ℓ, find the smallest d_{ij}.
2. Combine groups i and j to form a new single group k.
3. Using the combinatorial algorithm, compute new distances, d_{km}, between the newly formed group, k, and the other m points (individuals at level 1, often groups at higher levels).
4. Redefine the d_{km} as d_{ij} and return to step 1 until $N - 1$ cycles are completed.

Clustering Algorithms. Recall that the primary differences between algorithms are in the values for α_i, α_j, β, and γ in the combinatorial formula. Changes in the coefficient values alter the distance as computed in the formula, hence the nature of the shortest intergroup distance used for joining groups into a growing cluster. The names of the eight methods to be discussed are descriptive of the

Table 8.5 Some characteristics of hierarchical clustering methods.

method	space	monotonic	a_i	a_j	β	γ	coef.
nearest neighbor	contracting	no	0.5	0.5	0	−0.5	many
fartherest neighbor	dilating	no	0.5	0.5	0	0.5	many
centroid	conserving	no	n_i/n_k	n_j/n_k	$-\alpha_i\alpha_j$	0	d
median	conserving	no	0.5	0.5	−.25	0	d
group average	conserving	yes	n_i/n_k	n_i/n_k	0	0	d?
simple average	conserving?	yes	0.5	0.5	0	0	d?
minimum variance	conserving?	yes	$\dfrac{n_m + n_i}{n_m + n_k}$	$\dfrac{n_m + n_j}{n_m + n_k}$	$\dfrac{-n_m}{n_m + n_k}$	0	d?

nature of the distances, but many synonyms and acronyms exist for the methods. The acronyms will be omitted.

The nearest neighbor, minimum, or single linkage method has coefficient values of $\alpha_i = \alpha_j = .5$, $\beta = 0$, and $\gamma = -.5$, causing distances between groups to be that between the two closest individuals. Perhaps obviously, the method causes space contraction, so chaining is a frequent result.

The furthest neighbor, maximum, or complete linkage method has equal values of .5 for all coefficients except $\beta = 0$. As a consequence, distances between two groups are between the two most distant individuals. For this reason, the method is space dilating.

The centroid, center of gravity, or unweighted pair group centroid method defines distances between centroids of the groups as they exist at any level. The combinatorial formula requires the use of d_{ij} and coefficient values of $\alpha_i = n_i/n_k$, $\alpha_j = n_j/n_k$, $\beta = -\alpha_i\alpha_j$, and $\gamma = 0$. Although the method is space-conserving, its and all previous methods lack of monotonicity can result in backward links in the formation of clusters. The consequences of backward links are to destroy the hierarchical nature of clusters and to make clusters difficult to interpret.

The median or weighted pair group method was proposed to remove a potential bias in the centroid method, but not the problem of backward links. In using the centroid method to join groups that differ appreciably in numbers of individuals, the larger group will have greater influence on the value of the new centroid. The median method assigns equal weight to each group, hence more appropriately pertains to the concept of medians rather than centroids. Like the centroid method, d_{ij} are required; but coefficients are $\alpha_i = \alpha_j = .5$, $\beta = -.25$, and $\gamma = 0$.

The group average method, or unweighted pair group method using arithmetic averages, joins groups i and j on the basis of mean similarity between pairs of individuals in the two groups $(1/n_in_j \sum d_{ij})$. In this most popular of the methods, the coefficients are $\alpha_i = n_i/n_k$, $\alpha_j = n_j/n_k$, and $\beta = \gamma = 0$, which satisfy the criteria for monotonicity. Although also space-conserving, the group average method is more space-distorting than the centroid method.

The simple average method, or weighted pair group method using arithmetic averages, also was devised to remove the potential bias of larger groups. To correct this, the simple average gives equal weight to groups by $\alpha_i = \alpha_j = .5$ and $\beta = \gamma = 0$, which maintain monotonicity. It requires emphasis that neither the simple average nor median methods are necessarily better than the so-called biased centroid or group average methods. Although technically space conserving, in taking account of small groups, median and simple average dilate distances between groups.

The minimum variance method joins the two groups that, when combined, provide the smallest increase in the within group sum of squares. It is combinatorial and the d_{ij} are monotonic when $\alpha_i = (n_m + n_i)(n_m + n_k)$, $\alpha_j = (n_m + n_j)/(n_m + n_k)$, $\beta = -n_m/(n_m + n_k)$, and $\gamma = 0$. The method has not been evaluated in depth.

The flexible method allows a user's strategy in selecting some coefficients within the limitations of $\alpha_i + \alpha_j + \beta + \gamma = 1$, where $\gamma = 0$, $\alpha_i = \alpha_j$, and $\beta \leq 1$. The strategy can emphasize extreme space-dilation $(\beta = -1)$ to extreme space-contraction $(\beta = +1)$. The value of β for space-conservation is dependent upon the values of d_{ij} but would tend to be a small negative value. Sneath and Sokal (1973) disapprove of this method since, by manipulation, results can be made to follow preconceived conclusions of the user.

Selection of Method.   No method consistently outperforms all others, but furtherest neighbor and group average are often accepted as better techniques. In spite of this, other methods cannot be ignored since the performance of any algorithm is data dependent. Consistent with the latter is the practice of applying a variety of methods and measuring their performance. The most used measure of performance is the cophenetic correlation, the correlation between the original d_{ij} (between pairs of individuals) and the d_{ij}^* after application of the combinatorial formula. Usually, cophenetic correlation coefficients above .75 are considered good fit and the method providing the largest coefficient is deemed best.

Numerical Example.   Consider a matrix of hypothetical distances between five individuals,

$$\mathbf{D} = \begin{bmatrix} 0 & 1 & 3 & 5 & 4 \\ 1 & 0 & 2 & 4 & 5 \\ 3 & 2 & 0 & 3 & 4 \\ 5 & 4 & 3 & 0 & 2 \\ 4 & 5 & 4 & 2 & 0 \end{bmatrix}$$

Using the above data, the nearest neighbor and furtherest neighbor methods will be demonstrated. However, "computations" will amount to actual selection of the nearest or most distant individuals rather than applying the combinatorial algorithm. The actual selections will demonstrate the nature of each of the methods. The reader can apply the particular combinatorial formula to verify procedures.

For nearest neighbor, the $N-1$ ($= 4$) levels are obtained sequentially by finding the shortest distance between groups as they exist, merging the closest groups, and calculating the new closest distances to other groups.

For level 1,

$\min(d_{ij}) = d_{12} = 1$, and

$d_{(12)3} = \min(d_{13}, d_{23}) = d_{23} = 2$

$d_{(12)4} = \min(d_{14}, d_{24}) = d_{24} = 4$

$d_{(12)5} = \min(d_{15}, d_{25}) = d_{15} = 4,$

which modifies $\mathbf{D}$ to $\mathbf{D}_2$,

$$\mathbf{D}_2 = \begin{array}{c} \\ 12 \\ 3 \\ 4 \\ 5 \end{array} \begin{array}{c} \begin{array}{cccc} 12 & 3 & 4 & 5 \end{array} \\ \begin{bmatrix} 0 & 2 & 4 & 4 \\ 2 & 0 & 3 & 4 \\ 4 & 3 & 0 & 2 \\ 4 & 4 & 2 & 0 \end{bmatrix} \end{array}$$

Note that the 3×3 submatrix at the lower right remains unaltered from the original. For level 2,

$$\min(d_{ij}) = d_{45} = 2.$$

Since d_{12} are already joined, $d_{(12)3}$ is arbitrarily delayed until level 3. Continuing,

$$d_{(12)(45)} = \min(d_{14}, d_{15}, d_{24}, d_{25}) = d_{15} \text{ or } d_{24} = 4$$

$$d_{3(45)} = \min(d_{34}, d_{35}) = d_{34} = 3,$$

which modifies $\mathbf{D}_2$ to $\mathbf{D}_3$,

$$\mathbf{D}_3 = \begin{array}{c} \\ 12 \\ 3 \\ 45 \end{array} \begin{array}{c} \begin{array}{ccc} 12 & 3 & 45 \end{array} \\ \begin{bmatrix} 0 & 2 & 4 \\ 2 & 0 & 3 \\ 4 & 3 & 0 \end{bmatrix} \end{array}$$

For level 3,

$$\min(d_{ij}) = d_{(12)3} = 2, \text{ and}$$

$$d_{(123)(45)} = \min(d_{14}, d_{15}, d_{24}, d_{25}, d_{34}, d_{35}) = d_{34} = 3,$$

which modifies $\mathbf{D}^3$ to $\mathbf{D}_4$,

$$\mathbf{D}_4 = \begin{array}{c} \\ 123 \\ 45 \end{array} \begin{array}{c} \begin{array}{cc} 123 & 45 \end{array} \\ \begin{bmatrix} 0 & 3 \\ 3 & 0 \end{bmatrix} \end{array}$$

For level 4, the $\min(d_{ij}) = 3$, the only remaining value. As a consequence of the above, distances d_{ij} are altered to d_{ij}^*

$$\mathbf{D}^* = \begin{bmatrix} 0 & 1 & 2 & 3 & 3 \\ 1 & 0 & 2 & 3 & 3 \\ 2 & 2 & 0 & 3 & 3 \\ 3 & 3 & 3 & 0 & 2 \\ 3 & 3 & 3 & 2 & 0 \end{bmatrix}$$

For furtherest neighbor, the only difference is in the use of maximum distance when a choice is available. The d_{ij} selected at any level still is the minimum.

For level 1,

$$\min(d_{ij}) = d_{12} = 1, \text{ and}$$

$$d_{(12)3} = \max(d_{13}, d_{23}) = d_{13} = 3$$

$$d_{(12)4} = \max(d_{14}, d_{24}) = d_{14} = 5$$

$$d_{(12)5} = \max(d_{15}, d_{25}) = d_{25} = 5, \text{ so}$$

$$
\mathbf{D}_2 =
\begin{array}{c c}
 & \begin{array}{cccc} 12 & 3 & 4 & 5 \end{array} \\
\begin{array}{c} 12 \\ 3 \\ 4 \\ 5 \end{array} &
\left[
\begin{array}{cccc}
0 & 3 & 5 & 5 \\
3 & 0 & 3 & 4 \\
5 & 3 & 0 & 2 \\
5 & 4 & 2 & 0
\end{array}
\right]
\end{array}
$$

For level 2,

$$\min(d_{ij}) = d_{45} = 2, \text{ and}$$

$$d_{(12)(45)} = \max(d_{14}, d_{15}, d_{24}, d_{25}) = d_{14} \text{ or } d_{25} = 5$$

$$d_{3(45)} = \max(d_{34}, d_{35}) = d_{35} = 4, \text{ so}$$

$$
\mathbf{D}_3 =
\begin{array}{c c}
 & \begin{array}{ccc} 12 & 3 & 45 \end{array} \\
\begin{array}{c} 12 \\ 3 \\ 45 \end{array} &
\left[
\begin{array}{ccc}
0 & 3 & 5 \\
3 & 0 & 4 \\
5 & 4 & 0
\end{array}
\right]
\end{array}
$$

For level 3,

$$\min(d_{ij}) = d_{(12)3} = 3, \text{ so}$$

$$d_{(123)(45)} = \max(d_{14}, d_{15}, d_{24}, d_{25}, d_{34}, d_{35}) = 5, \text{ so}$$

$$
\mathbf{D}_4 =
\begin{array}{c c}
 & \begin{array}{cc} 123 & 45 \end{array} \\
\begin{array}{c} 123 \\ 45 \end{array} &
\left[
\begin{array}{cc}
0 & 5 \\
5 & 0
\end{array}
\right]
\end{array}
$$

For level 4, $\min(d_{ij}) = 5$. Therefore,

$$
\mathbf{D^*} = \begin{bmatrix}
0 & 1 & 3 & 5 & 5 \\
1 & 0 & 3 & 5 & 5 \\
3 & 3 & 0 & 5 & 5 \\
5 & 5 & 5 & 0 & 2 \\
5 & 5 & 5 & 2 & 0
\end{bmatrix}
$$

As a next step, the elements of $\mathbf{D^*}$ are applied to the construction of treelike, hierarchical linkages from lower through higher levels of groupings. In this sense, $\mathbf{D^*}$ is called a tree matrix and the diagram, a dendrogram (Figure 8.5).

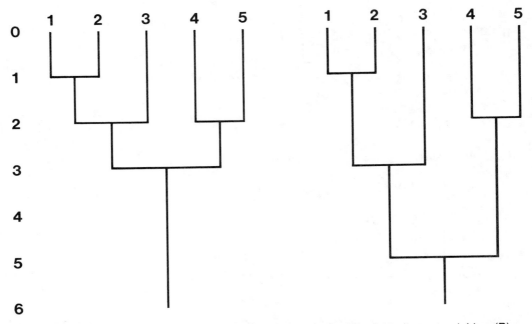

Figure 8.5 Dendrograms resulting from nearest neighbor (A) and furtherest neighbor (B) algorithms applied to hypothetical data.

Comment.   Dendrograms, often justified by a cophetic correlation coefficient, are the primary basis for evaluating results. The simplicity of the above example causes the nearest neighbor and furtherest neighbor dendrograms to be atypical. Although two divergent analyses are portrayed, the results are the same except for tree distances. For actual studies, nearest and furtherest neighbor methods could lead to different clusters and interpretation of relationships. On the other hand, the dendrograms do contain some indication of problems that can occur in defining different clusters. For example, either dendrogram might be considered to define three, two, or even a single cluster.

Although many studies disclose that clustering methods can provide similar diagrams and clusters can be readily definable, the above was stated to emphasize that a given set of cluster analyses can be subject to more than one interpretation. For this reason, cluster analysis can be highly dependent upon the expertise of the investigator.

Critique of Cluster Analysis.

The heirarchical, agglomerative clustering strategies outlined are the most widely applied in the literature. Recognized authorities on cluster analysis (Sneath and Sokal, 1973) consider the same methods to be better than those not discussed. Some of the reasons for this preference were indicated.

The potential problems of cluster analysis occur at every level of the strategy. The hierarchical feature limits the method to extracting a 1-D portrayal of relationships among OTU's. Farris (1977) has demonstrated that the single dimension extracted by cluster analysis need not be the most parsimonious representation of the OTU's. Whenever relationships among OTU's require more dimensions for clear portrayal, naturally cluster analysis will be imprecise. Next, recall that divisive methods theoretically are to be preferred over agglomerative methods. Also, the most widely applied resemblance measurements, the correlation coefficient and Euclidean distance, have both theoretical and practical problems. Perhaps most important is the practical problem of correlation between variables (Blackith, 1967). This is especially the case when two or more perfectly correlated variables ($r_{ij} = 1$) are present in the data matrix (Farris, 1977).

Since different strategies can result in drastically different dendrograms, various interpretations of a given study are possible. The practice of choosing the "best" strategy from several on the basis of the strategy producing the largest cophenetic correlation coefficient should be approached very carefully. Farris (1977) has clearly shown that the cophenetic correlation coefficient is not a reliable measure of the best strategy. Therefore, selecting a best strategy must be accomplished on a non-statistical basis. Most definitely this demands considerable expertise on the part of any user of cluster analysis. Unfortunately this does not mean that several experts in reference to a given study would select the same dendrogram as being most biologically meaningful.

In spite of the above difficulties, there seems to be little doubt that cluster. analysis can and has produced useful results. This would imply that in given circumstances the potential pitfalls either do not exist, or methodology is sufficiently robust to circumvent the pitfalls.

Chapter 9
Multivariate Analysis
of Variance and Covariance

Methodology now changes from examining independent patterns of variation and/or association of variables within a single population to examining contrasting variation between two or more populations. The first step in comparing samples from g populations is to test the hypothesis that the g population centroids are equal. This sounds like the univariate analysis of variance, anova. The methodology is that of the analysis of variance; however, since multiple measurements are made upon each individual, the method is the multivariate analysis of variance, manova.

The matter of testing the significance of population means, granted after their prior adjustment, is also the subject of the analysis of covariance. This analysis is also generalized to the multivariate case, the multivariate analysis of covariance or mancova. Owing to a role similar to manova, mancova also is a subject of this chapter. However, as was done in the chapter on ordination and cluster analysis, the manova and ancova parts are, for the most part, treated separately.

I: Multivariate Analysis of Variance

Since manova is only a first step in contrasting populations and other methods examine differences in some detail, only one model is needed, the fixed model one-way manova. In algebraic form, the model decomposes an individual data vector into grand centroid, treatment effect and error effect:

$$X_{ijk} = \mu_j + \alpha_{ij} + \epsilon_{ijk}$$

where $k = 1, 2, \ldots, N$ individuals, $j = 1, 2, \ldots, p$ observations on an individual, and $i = 1, 2, \ldots,$ g groups or populations. In matrix algebra form the model is

$$\mathbf{X}_{ik} = \mu + \alpha_i + \epsilon_{ik}.$$

It is assumed that the reader is familiar with one-way fixed model anova. Further discussion shall stress the assumptions of manova, the computations of manova, and shall return to the turtle example with a manova of males vs. females.

9.1 Assumptions and Their Implications

The assumptions of manova are those of anova extended to the multivariate case. Each sample is a random sample from a population following the multivariate normal distribution. The g populations have the same dispersion. The distribution of errors is random and independent. The main effects are additive, i.e., there is no interaction.

Some problems caused by invalid assumptions were considered with principal component analysis. They will be reviewed and their pertinence to manova will be mentioned.

Random Sampling.
Since temporal, ecological and/or behavioral responses may vary in different sexes, age classes, etc., it is not surprising that truly random samples of organisms probably cannot be obtained in nature. In essence, the above responses may differentially influence where organisms are at the time of collection, hence collections are not likely to be random samples.

Various steps can be taken to rectify the situation. The major effort should be placed in restricting the definition of the population sampled. For example, in taxonomic or ecological studies, populations designated by sex, age, time and habitats should provide more reliable interpretations. When these factors are ignored, unequal representation of the latter factors between samples might imply differences that do not exist between populations.

There is an additional advantage to explicit diagnosis of populations. Every factor of diagnosis becomes a major feature for recognizing affinities among populations. The conclusions of any study parallel the precise nature of the populations sampled.

In many investigations it is impractical and frequently impossible to define populations in much detail. Frequently, museum or herbarium specimens must be used, hence limit population definitions. Even when the investigator has control over collecting, specimens needed to define a population might not be available from all populations. In fact, in all studies one has to assume that individuals taken are representative of the populations sampled. Again, extreme measures for population definition might not be necessary. The abundance of results repeated within an ever-growing framework of detail supports conventional sampling and various degrees of limiting population definitions.

Normality and Homoscedasticity.
Multivariate normality and equality of population dispersions are tested simultaneously and so are inseparable as departures from assumptions. Either departure is naturally a function of the data. In chapter four, indices were said to be especially prone to both greater inaccuracy of measurement and departure from normality. Even single variables may not follow the normal distribution. Many texts on univariate analysis disclose the types of data that might not approximate normality. The most serious problems usually are skewed distributions or variances being functions of means. Tests of skewness and kurtosis, the Kolmogorov-Smirnov D_{max} test, and a test of equality of coefficients of variation focus on these problems in univariate analysis. They can be applied here and perhaps even lead to transformations that satisfy the assumptions.

When extreme caution is necessary, one can use the average of five, perhaps even three, single measurements as a single observation. This suggestion follows from the *central limit theorem:* if N independent variates have finite variances, their sum or mean tends to be normally distributed as N approaches infinity. However, most statistical textbooks present as accepted fact that normality is approximated by the average of a very few variates.

Recall that departure of single variables from normality probably is insufficient to invalidate statistical analysis. Both qualitative and quantitative biological data, where studied, typically do not depart too far from normal to invalidate results. As was stated and cited previously this also appears to be the case for multivariate data.

Multivariate Normality.
Only in the study of single populations is multivariate normality independent from equality of dispersions. The only multivariate methods discussed in this book that are applicable to single populations are principal component analysis, factor analysis and canonical corre-

lation analysis. Although each method can be applied without the requirement of multivariate normality, each becomes more limited. Any of the associated tests, e.g., an eigenvalue equals zero, requires normality. Also, the concepts of ellipsoids of variation and rotation of axes are linked to multivariate normality, so it probably should be approximated to assure valid interpretation of results.

Fortunately, the central limit theorem again is helpful. Because calculated component, factor or canonical coefficients are based upon many individuals, the coefficients are more prone to be normally distributed than are the original measurements.

Equality of Dispersions.

The assumption of homoscedasticity is unique to comparing populations. Assuming homogeneity of dispersions is dubious. In biology, as means increase often so do variances. In ecology, inequality of dispersions has lead to many prefixes, steno-, dys-, oligo-, eury-, etc. Perhaps we biologists are too impressed by comparing means. If populations have unequal dispersions, the populations are different even if their centroids are equal.

There is another reason for the assumption of homoscedasticity being dubious. The test is extremely powerful—even minor differences between group dispersions likely will be discovered. Large samples typically lead to rejecting the hypothesis of equality. Even the test of equality of male and female turtle dispersions with 24 individuals per group produced an F with 6 and 15,330 degrees of freedom, providing very little opportunity for a Type **II** error.

How can one compare centroids of populations having unequal dispersions? Perhaps the data can be transformed so it approximates normality and equality of dispersions. Familiar univariate tests, nature of data, and transformations could help here, but more likely they will not. However, transformations might not be required! Although the central limit theorem does not apply, investigations of unequal dispersions comparing two multivariate populations are encouraging (Ito and Schull, 1964).

Inequality of two dispersions is reflected in the true magnitude of Type **I** and Type **II** errors. If two population dispersions are equal,

$$\sum{}_1^2 \sum{}_2^{-2} = \mathbf{I}$$

and the Type **I** and Type **II** errors are correct. Parenthetically for later discussion, the eigenvalues of an identity matrix are all unity. Departures from the correct magnitude of the Type **I** error reflect sample sizes and which dispersion is greater. Consider a standard situation of samples N_1 and N_2. If $\sum{}_1^2 > \sum{}_2^2$ at least some of the eigenvalues of $\sum{}_1^2 \sum{}_2^{-2}$ will be greater than unity but the Type **I** error will tend to become smaller as N_1 increases over N_2. However, if $\sum{}_1^2 < \sum{}_2^2$, the eigenvalues are all less than unity and the Type **I** error increases as N_1/N_2 increases. On the other hand, if sample sizes are large and equal, inequality of dispersions has no real effect on the Type **I** and **II** errors. Also, other research implies that tests of equality of centroids are rather insensitive to moderate departures from normality and homoscedasticity. Therefore, when inequality of dispersions is likely and one wishes to test equality of centroids, large equal sample sizes should produce valid results.

Test of Homoscedasticity.

The test is Box's modification of Bartlett's test of homogeneity of dispersions. From the above, one can appreciate why many workers ignore the test. This should not be done since the test can aid interpretation. Acceptance of the hypothesis is important. Rejection justifies examining dispersions via multigroup component analysis which provides insight as to how populations differ.

In testing the equality of dispersions via an F test, p is the number of variables; g is the number of groups; $k = 1, 2, \ldots, g$ groups; and N_k is the sample size of the kth group, so sample sizes need not be equal. One first calculates

$$A_1 = \left(\sum_{k=1}^{g} \frac{1}{N_k-1} - \frac{1}{N-g} \right) \left(\frac{2p^2+3p-1}{6(g-1)(p+1)} \right)$$

$$A_2 = \left(\sum_{k=1}^{g} \frac{1}{(N_k-1)^2} - \frac{1}{(N-g)^2} \right) \left(\frac{(p-1)(p+2)}{6(g-1)} \right)$$

$$M = (N-g) \, log_e \, \mathbf{S}^2 - \sum_{k=1}^{g} (N_k-1) \, log_e \, \mathbf{S}_k^2$$

$$n_1 = \frac{(g-1)p(p+1)}{2}$$

If $A_2-A_1^2$ is positive, then

$$n_2 = \frac{n_1 + 2}{A_2 - A_1^2}$$

$$b = \frac{n_1}{1-A_1-(n_1/n_2)}$$

$$F = \frac{M}{b} \text{ with } n_1 \text{ and } n_2 \text{ degrees of freedom.}$$

If $A_2-A_1^2$ is negative, then

$$n_2 = \frac{n_1 + 2}{A_1^2 - A_2}$$

$$b = \frac{n_2}{1-A_1 + (2/n_2)}$$

$$F = \frac{n_2 M}{n_1(b-M)} \text{ with } n_1 \text{ and } n_2 \text{ degrees of freedom.}$$

Linear Models.   All the methods in this book involve linear models. However, no truly linear relationship is likely to exist in any set of variables over their full range of variation. The consequences of nonlinearity are to reduce the magnitude of the first eigenvalue and to change probabilities in tests of significance, e.g., one is more likely to accept null hypotheses of equality of group centroids, but to reject null hypotheses of equality of group dispersions. Under most circumstances, nonlinearity is unlikely to be serious. A possible exception occurs when organisms in a sample range from very young to fully mature. Then, the full range of variation of variables might be very nonlinear.

Various means of coping with this problem have been suggested:

1. Use a stepwise procedure to select the best set of variables, BMDO2R (a stepwise regression program) to obtain the "best" set of predictors of an assumed linear criterion or BMDO7M (stepwise discriminant analysis) to obtain the "best" set of discriminators (Dixon, 1973). The premise is that a set of linear variables will result. In practice the problems of stepwise procedures prevail, often good variables being deleted and poor variables retained.
2. In view of (1), ignore the condition especially since nonlinearity is unlikely to change results sufficiently to alter fundamental relationships.
3. Perform the analysis using original data. Then perform the analysis using logarithms of the original data. Compare the two results. Generally the results will be similar. If not, rely on the logarithms since they remove multiplicative effects.

Interaction.   Unmeasured interaction can be a serious source of error in any analysis of variance. However, in morphometrics manova is only one of many methods in discriminant analysis. Since discriminant analysis eventually unravels interactions, interactions can be ignored here.

9.2 Manova

For each data vector $\mathbf{X}$ of p measurements, the matrix algebra manova estimate model is

$$\mathbf{X}_{ki} = \overline{\overline{\mathbf{X}}} + (\overline{\mathbf{X}}_k - \overline{\overline{\mathbf{X}}}) + (\mathbf{X}_{ki} - \overline{\mathbf{X}}_k)$$

where $\mathbf{X}_{ki}$ is the ith individual of the kth group, $\overline{\overline{\mathbf{X}}}$ is the grand centroid, $\overline{\mathbf{X}}_k$ is the centroid of the kth group, $(\overline{\mathbf{X}}_k - \overline{\overline{\mathbf{X}}})$ is the between group effect of $\mathbf{X}_{ki}$, and $(\mathbf{X}_{ki} - \overline{\mathbf{X}}_k)$ is the within group or error term of $\mathbf{X}_{ki}$. If the grand centroid is subtracted from both sides of the model so $(\mathbf{X}_{ki} - \overline{\overline{\mathbf{X}}})$ becomes the total effect, then the model becomes

$$(\mathbf{X}_{ki} - \overline{\overline{\mathbf{X}}}) = (\overline{\mathbf{X}}_k - \overline{\overline{\mathbf{X}}}) + (\mathbf{X}_{ki} - \overline{\mathbf{X}}_k)$$

or total effect equals group effect plus residual effect for the ith individual. Summing over all individuals,

$$\sum(\mathbf{X}_{ki} - \overline{\overline{\mathbf{X}}}) = \sum(\overline{\mathbf{X}}_k - \overline{\overline{\mathbf{X}}}) + \sum(\mathbf{X}_{ki} - \overline{\mathbf{X}}_k)$$

results in total effect equaling group effect plus residual effect for all individuals of all groups. "Squaring" both sides provides the total crossproducts equal to the between group crossproducts plus the residual crossproducts, all $p \times p$ matrices

$$\sum_{k=1}^{g} \sum_{i=1}^{N_k} (\mathbf{X}_{ki} - \overline{\overline{\mathbf{X}}})^2 = \sum_{k=1}^{g} \sum_{i=1}^{N_k} (\overline{\mathbf{X}}_k - \overline{\overline{\mathbf{X}}})^2 + \sum_{k=1}^{g} \sum_{i=1}^{N_k} (\mathbf{X}_{ki} - \overline{\mathbf{X}})^2$$

or

$$\mathbf{T} \quad = \quad \mathbf{B} \quad + \quad \mathbf{W}$$

Then the between covariance matrix S_B^2, and the within covariance matrix, S_W^2, are obtained by dividing each element of each matrix by the appropriate degrees of freedom, a scalar (actually multiplication by reciprocals is involved),

$$S_B^2 = \frac{1}{g-1} \, \mathbf{B}$$

$$S_W^2 = \frac{1}{N-g} \, \mathbf{W} \text{ // where } N = \sum_{k=1}^{g} N_k.$$

Test of Equality of Population Centroids.   The test is a generalization of the anova F-test. The manova test uses Wilk's determinant ratio or Lambda test,

$$\Lambda = \left| \frac{\mathbf{W}}{\mathbf{T}} \right| \, .$$

Notice that as $\mathbf{T}$ increases relative to $\mathbf{W}$ (as centroids are more different), the ratio decreases in magnitude. This means that as Λ gets smaller there is greater chance of rejecting equality of centroids.

If the acceptance or rejection of manova on the basis of the nature of Λ is confusing, recall the nature of an univariate F-test in anova. The anova F is a ratio of between mean square divided by residual mean square. The manova considers total where total equals between (or treatment) plus residual. The only basic difference between anova and manova is that manova uses total rather than between. The fact that determinants are involved in manova should not prove perplexing.

Wilks (1932) originally described Lambda equal to eta-squared, the multivariate generalization of Fisher's correlation ratio

$$\eta^2 = 1 - \Lambda,$$

where the statistic eta-squared expresses the proportion of the treatment variance explained by the variables. For example, if $\mathbf{T} = \mathbf{W}$ there is no between group or treatment effect and no explanation of treatments by variables.

The test to be used is Rao's (1952) approximation of an F-statistic. Computations are as follows:

$$s = \sqrt{\frac{p^2(g-1)^2 - 4}{p^2 + (g-1)^2 - 5}}$$

$$n_1 = p(g-1)$$

$$n_2 = s\left[(N-1) - \frac{p+(g-1)+1}{2} \right] - \frac{p(g-1)-2}{2}$$

$$y = \Lambda^{1/s}$$

$$F = \left(\frac{1-y}{y} \right)\left(\frac{n_2}{n_1} \right) \text{ with } n_1 \text{ and } n_2 \text{ degrees of freedom.}$$

Tests After a Significant F. There is no manova test that determines how synergistic interplay of variables leads to a significant manova. Also, p univariate F-tests followed by multiple range tests for significant F-values need not disclose the nature of a significant manova. That is the function of discriminant analysis. However, p univariate anova can provide insight into a problem. For this reason, the p diagonal elements of **B** and of **W** can be changed to mean squares, p F-tests performed, and appropriate multiple range tests made.

9.3 Turtle Example

Table 9.1 summarizes basic statistics for the turtle data of Jolicoeur and Mosimann (1960).

The F-test for equality of dispersions provides an $F = 4.004$ with 6 and 15,330 degrees of freedom ($p = .0002$) so the hypothesis is rejected. However, the samples are sizable and equal for each group, so the test of equality of centroids is appropriate. Here, $F = 21.29$ with 3 and 44 degrees of freedom ($p < .000$). Table 9.2 summarizes the univariate F-tests

Since all female dimensions singly are significantly larger, there is no sense in transforming the data to logs to determine if calculated F-values would have even more remote probabilities of equality of dimensions. Again, substantially larger and equal sample sizes would not likely alter results appreciably.

The results at this point are trivial. Although manova discloses that females are longer, wider and higher than males, manova does nothing to contrast the body form of the sexes. Again, contrasting body form is the function of a complete discriminant analysis. This is the next step. However, the manova supports a previous morphometric analysis. Since manova confirmed inequality of male and female dispersions, the multigroup component analysis of males and females is justified. Review of that analysis will confirm that unequal dispersions might be involved with only minor differences in components.

Table 9.1 Basic statistics for 24 male and 24 female painted turtles

	males				females		
	length	width	height		length	width	height
means	113.75	88.29	40.71		136.00	102.58	51.96
standard deviations	11.78	7.07	3.36		21.25	13.11	8.16
T =	19716.38	11937.69	7794.00		13573.63	8057.38	4739.63
	11937.69	7552.00	4803.00	**W** =	8057.38	5100.88	2873.69
	7794.00	4803.00	3310.69		4739.63	2873.69	1791.92

Table 9.2 Univariate F tests for turtle dimensions

Variable	MS_B	MS_W	F	p(F)
length	6142.75	295.08	20.82	0.00004
width	2451.13	110.89	22.10	0.00002
height	1518.77	38.95	38.99	0.00000

II: Multivariate Analysis of Covariance

The aim of the multivariate analysis of covariance, *mancova,* is to use one or more variables to adjust a set of variables prior to a manova on the latter set. The adjusted set is further referred to as a dependent, predicted, concomitant, or criterion set. The adjustors are termed independent, predictor, covariate, control, experimental, or residual variables. Discussion returns not only to the terminology of regression but to its methodology.

The aim is to present the purposes, characteristics, limitations and computations of mancova without an example. Mancova is not emphasized. Its appropriateness in most discriminant analyses is questionable. Since discussion is brief, some of the mathematical considerations are above the general level assumed for the rest of the book. Rather than take the time to develop some of this mathematics, their implications are stated for those not following the level of discussion.

9.4 Purposes

Perhaps somewhat arbitrarily, a given mancova can be said to have one of five goals:

Increasing Precision.   Predictors might be selected for an otherwise manova of dependent variables so the adjusted dependent variables are more effective in differentiating among groups. Such things as sexual, temporal, behavioral, physiological, morphological and/or ecological differences might be of little experimental interest. In such cases, rather than defining treatment groups to account for these sources of variation, covariates that eliminate the unwanted variation are applied. Then, the variables of interest, after adjustment by covariates, become predicted variables devoid of the sources of experimental error.

Removing Experimental Bias.   During the course of an experiment, it is possible that a factor otherwise "designed out" inadvertently becomes important. For example, in a lab experiment equipment may fail, standards lost, or organisms may die. If for any reason, original controls cannot or might not be duplicated, it often is possible to quantify the original and new conditions. The one or more variables that do this then become covariates. Such correction of data is necessary so this would be a case where mancova would be used in discriminant analysis.

Modifying Treatments.   The biologist might wish to compare groups of organisms independent of some major feature. For example, the biologist might know that two or more groups differ in size. However, he may be interested in shape independent of size. In such cases, an indicator of size, perhaps length, can be used to rule out size as a differentiator among groups.

Since size differences can involve allometric growth, a nonlinear function, and mancova adjustments generally assume linear relationships, one should be sure that the regression function between the size variable(s) and predicted shape variables is linear. If not, a standard transformation, perhaps logarithmic, should satisfy the test of linearity.

Comparing Group Regressions.   Since mancova is based upon the regression of predicted variables on covariates, one can emphasize individual group regressions. Here, analysis would pertain to linearity or curvilinearity of regression for each group and equality of regressions.

Influence of Additional Variables.   It may be of interest to learn if one or more additional variables would add information about group differences. This sounds like stepwise regression but is not the same. First, a set of variables is used to contrast groups. Then, this same set is used as covariates in a second study. In the second study the covariates will remove their bases for separating groups from the additional variables. Therefore, any possible between group differences found in the second study can be attributed to unique aspects of the new variables.

9.5 Characteristics

The analysis of covariance is the most complicated of the basic parametric statistical methods. It can apply to a single adjusted variable, ancova, as well as to two or more, mancova. The methodology combines regression in the broad sense with anova or manova.

An ancova resembles simple linear regression when a single covariate is used to adjust a single criterion. All aspects of linear regression apply, especially the test of linearity of the regression function. However, the goal is to obtain a predicted y-value for each individual. Then, these adjusted y-values are analyzed by anova. Mancova resembles multiple regression when there is a set of predictors and a single criterion. Again, regression analysis applies prior to an anova using the adjusted y for each individual. When there is a set of covariates and a set of criteria, mancova resembles canonical correlation. However, in mancova sets are not interchangeable, one set consists of adjuster variables, the other of adjusted variables; and only adjusted variables are analyzed by manova. Also, mancova deals with many groups and canonical correlation treats individuals as a single group.

9.6 Limitations

First, mancova is limited by its assumptions. Aside from random sampling, for every value of covariates, the dependent variables are assumed to follow the multivariate normal distribution. All group multivariate normal distributions are assumed to have equal dispersions. Finally, covariates must be linearly related to dependent variables, a premise equivalent to assuming linearity of regression. The consequences of not satisfying assumptions shared with manova are the same as for mancova. Lack of linearity of regression can be very serious. Most definitely the additional assumptions can lead to problems in analysis.

Mancova is a very powerful but often a misleading technique. The researcher must know the properties of variables and precisely what is being done in a given mancova.

Essentially there are three basic situations in which mancova can be used:

Experimental Situation.   Covariates are measured. Only then are individuals assigned to groups at random. Therefore, covariates cannot be correlated with treatments, the critical feature to identify the experimental situation. In short, treatments do not influence the values of covariates.

The precision of the experiment is increased. In the sense of manova, adjustment of criteria reduces their within group variation and thereby increases the F-ratio. Naturally if there is no predictor-dependent variable correlation within groups, there is no adjustment and mancova is equivalent to manova without prior adjustment.

Observational Situation.   Here, individual assignment is determined by treatments or groups. This is the case when individuals are restricted to and treatments are localities. Then, covariates may be correlated with treatments. For example, if covariates are selected to "standardize" individuals for size and size differs among localities, size is correlated with localities. The correlation might be verified by a significant manova of covariates.

The purpose of mancova differs in this situation. Both dependent variables and treatments are adjusted by covariates. Therefore, mancova tests the relationship between adjusted variables and adjusted treatments. In general, interpretation of the analysis is difficult.

Biologists might consider this design an opportunity to introduce experimental controls on a nonexperimental study. However, statisticians might refuse to perform such a mancova because it amounts to partial correlation adjustment. I agree with the statisticians that consider partial correlation a poor substitute for proper design. The reasons follow from a reconsideration of partial correlation.

Partial correlation is applied to eliminate the effect of one or more variables upon other variables. For example, one might want to partial out the effect of variable one on the correlation of variables two and three. For readers wishing mathematical assurance, the procedure starts from a partitioned correlation matrix:

$$
\begin{bmatrix}
1 & \vdots & r_{12} & r_{13} \\
\cdots & \vdots & \cdots & \\
r_{21} & \vdots & 1 & r_{23} \\
& \vdots & & \\
r_{31} & \vdots & r_{32} & 1
\end{bmatrix}
=
\begin{bmatrix}
\mathbf{R}_{11} & \mathbf{R}_{12} \\
\mathbf{R}_{21} & \mathbf{R}_{22}
\end{bmatrix}
$$

Then, the matrix of residuals is

$$
\tilde{\mathbf{R}}_{22} = \mathbf{R}_{22} - \mathbf{R}_{21}\mathbf{R}_{11}^{-1}\mathbf{R}_{12} = \mathbf{R}_{22} - \hat{\mathbf{R}}_{22}
$$

which in the example equals

$$
\begin{bmatrix}
1 - r_{21}r_{12} & R_{23} - r_{21}r_{13} \\
r_{32} - r_{31}r_{12} & 1 - r_{31}r_{13}
\end{bmatrix}
$$

where $\hat{\mathbf{R}}_{22}$ is the predicted regression matrix of $\mathbf{R}_{22}$. Recall that the residual matrix $\tilde{\mathbf{R}}_{22}$ contains the parts of variables two and three unaccounted for (unrelated to) variable one.

Since

$$
r_{21}r_{12} = r_{12}^2 \text{ and } r_{31}r_{13} = r_{13}^2,
$$

the correlation between residual parts of variables two and three, the partial correlation coefficient, is

$$
r_{23.1} = \frac{r_{23} - r_{21}r_{13}}{\sqrt{(1-r_{12}^2)}\sqrt{(1-r_{13}^2)}}
$$

The problem occurs in an observational situation whenever the partialed variables are correlated with the adjustor. For example, if variable one is the covariate x, variable two the dependent variable

y, and variable three the treatment or group variable g; and correlation between covariate and treatments is zero, then the partial correlation reduces to

$$\frac{r_{23}}{\sqrt{(1-r_{12})^2}} = \frac{r_{yg}}{\sqrt{(1-r_{xy})^2}}.$$

This is relatively easy to interpret. It amounts to the cosine of the angle between y and g in their own space. It also is the cosine of the angle between y and the projection of y in a space at right angles to x. Since g is at right angles to x, this is the space of g. Therefore, y is directly projected to g without any modification of the latter. However, when x is correlated with g, it is both difficult to visualize and interpret the nature of simultaneous projection of y and g to a space at right angles to x. Therefore, when there is no correlation between x and g, partial correlation is very useful; but when there is, partial correlation is imperfect and dependent upon the magnitude of x and g correlation.

Mixed Situation.   It might be possible to assign individuals completely at random, experimental; however, treatments are correlated with covariates, observational. This arises whenever treatments influence covariates as well as dependent variables and can be expected in many laboratory studies. In such situations one might be able to unravel the dependent variables if comparisons are made of covariates before, covariates after, unadjusted dependent variables, and adjusted dependent variables. This might disclose the extent of treatment effect of dependent variables as adjusted by covariates correlated with treatments. However, the results might be so complex as to be virtually uninterpretable.

9.7 Mancova

Consider $\mathbf{X}$ a set of p-dimensional multinormal predictors and $\mathbf{Y}$ a set of q-dimensional multinormal criteria. Set pairs define individuals from g different populations. Assuming samples from all populations are of size N, the adjusted centroid for each group is

$$\hat{\mathbf{Y}}_j = \overline{\mathbf{Y}}_j - \mathbf{b}(\overline{\mathbf{X}}_j - \overline{\overline{\mathbf{X}}})$$

where $\overline{\mathbf{Y}}_j$ is the unadjusted criteria centroid for the jth group, $\mathbf{b}$ is a vector of regression coefficients, $\overline{\mathbf{X}}_j$ is the predictors centroid for the jth group, and $\overline{\overline{\mathbf{X}}}$ is the predictors grand centroid.

Equality of Adjusted Centroids.   The first step is to form the within $\mathbf{W}$ and total $\mathbf{T}$ crossproducts matrices:

$$\mathbf{W} = \begin{bmatrix} \mathbf{W}_{11} & \mathbf{W}_{12} \\ \mathbf{W}_{21} & \mathbf{W}_{22} \end{bmatrix} \text{ and } \mathbf{T} = \begin{bmatrix} \mathbf{T}_{11} & \mathbf{T}_{12} \\ \mathbf{T}_{21} & \mathbf{T}_{22} \end{bmatrix},$$

where $\mathbf{W}_{11}$ and $\mathbf{T}_{11}$ are sum of squares and crossproducts of independent variables; $\mathbf{W}_{22}$ and $\mathbf{T}_{22}$ are sum of squares and crossproducts of dependent variables; and $\mathbf{W}_{12}$, $\mathbf{W}_{21}$, $\mathbf{T}_{12}$ and $\mathbf{T}_{21}$ are crossproducts of dependent and independent variables of the two sets. Correcting the q-criteria in terms of the regression of the p-predictors, the within and total are adjusted to

$$\mathbf{W}_{2.1} = \mathbf{W}_{22} - \mathbf{W}_{21}\mathbf{W}_{11}^{-1}\mathbf{W}_{12}$$

$$\mathbf{T}_{2.1} = \mathbf{T}_{22} - \mathbf{T}_{21}\mathbf{T}_{11}^{-1}\mathbf{T}_{12}$$

Then, following the procedure of manova, Wilk's Lambda becomes

$$\Lambda = \frac{|\mathbf{W}_{2.1}|}{|\mathbf{T}_{2.1}|}$$

and if

$$z = g - 1$$

$$s = \sqrt{(q^2z^2 - 4)/(q^2 + z^2 - 5)}$$

$$n_1 = qz$$

$$n_2 = s\left[(N - p - 1) - \frac{q + g}{2}\right] - \frac{qz}{2}$$

$$y = \Lambda^{1/s}$$

then Rao's F-test is

$$F = \frac{(1-y)n_2}{(y)n_1} \text{ with } n_1 \text{ and } n_2 \text{ degrees of freedom.}$$

Tests After Mancova.   As is the case for manova, the effect of each adjusted variable separately can be determined by a multiple range test. The simultaneous effect of all adjusted variables can be determined by discriminant analysis.

Chapter 10
Discriminant Analysis

From the viewpoint of morphometrics, no area of multivariate analysis is so loosely designated as is discriminant analysis. In part, the terminology problems are the consequence of discriminant analysis generally being considered synonymous with classification (Cacoullos, 1973, in preface). The latter consideration is understandable, since from inception a major purpose of discriminant analysis was, still is, and should be classification. In morphometrics, the additional procedures for examining relationships between groups becomes remote from classification. Unfortunately, the refined procedures historically were proposed as different devices for classification purposes.

In developing terminology for morphometrics, terms will not be redefined but will be restricted or expanded, and mostly will follow Blackith and Reyment (1971) or Hope (1969). Some of the clarification of terms can start now. Although discriminant analysis often is synonymous with classification, in this text discriminant analysis shall be expanded to include the five major methods discussed in this chapter. Canonical analysis of discriminance shall be the only designation applied to one of the methods, the only one involving an eigenvalue-eigenvector solution. Although the model of the latter is called a discriminant function, the term discriminant function shall be restricted to Hope's (1969) model for "multigroup discriminant analysis." Even more confusing are erroneous applications of discriminant function analysis to a manova alone or even to a PCA of centroids representing several groups.

We shall treat manova and four new methods as comprising discriminant analysis. Only one of the latter four methods shall pertain to classification, but elsewhere the other three methods still are so applied. Partly for the latter reason, in developing the last three methods, their applications in classification will be shown.

Many of the sources of confusions connected with methodology, purposes, assumptions, and selection of variables are discussed in the first section. The next two sections examine a portion of the discriminant analysis methodology. Section two examines the classification of two populations with a single variable of measurement. Although a single variable study is inappropriate for a multivariate technique, the aim is to provide an introduction that might be familiar to the reader. Section three continues with the classification of two populations but extends analysis to many variables. Also, much of the other methodology of discriminant analysis is introduced in some detail and six models of canonical analysis of discriminance are outlined. This section contains many details, but hopefully clarifies many terms in reference to how they are currently applied in other works and here. The fourth section contains an example of a two group multivariate discriminant analysis—the turtles again.

Sections five through nine examine various facets of multigroup multivariate discriminant analysis. Section five introduces multigroup methodology. Sections six and seven examine the two main groups of morphometric analyses, one of which will be emphasized depending upon the nature of the data. Section eight adds certain interpretive aids to one of the former approaches. Section nine, reification, discloses how to use angular comparison of vectors from discriminant analysis and theoretical vectors to unravel the morphometric implications of a study. The final section is a critique of the overall methodology.

Before proceeding farther, let us consider another possible source of confusion, namely discriminant space. Appreciating its nature and the fact that it can be defined by fewer orthogonal dimensions than variable space must come later. Now, simply imagine discriminant space as one derived from a synergistic relationship among variables. Therefore, the space is one of total form that is greater than the sum of the variables defining the space.

A final point of introduction: Discriminant analysis resembles ordination by PCA in that a graph of individuals is involved in evaluating similarities and differences between individuals. The clear difference in applying the two methods pertains to the original association between individuals and groups, or populations. In ordination by PCA, in a sense, all individuals belong to a single group. One need not have any knowledge of how individuals might relate to the concept of samples from specific populations. For discriminant analysis, each individual must pertain to a sample from a recognizable population. Discriminant analysis must start with a set of populations and a sample from each population. Therefore, the goals of PCA ordination and discriminant analysis are similar, but the experimental designs differ markedly. Also, the design for a discriminant analysis leads to more comprehensive results.

10.1 Preliminary Considerations

To be considered here are the methods, purposes, assumptions, and selection of variables in discriminant analysis. Methods and purposes are obviously related in the sense of overall analysis, but their separation serves a purpose. The subsection on methods is used to name the procedures as they will be applied in morphometrics. The subsection on purposes considers methods in reference to morphometric goals and in reference to applications elsewhere. If these differences are appreciated at the end of this section, further discussions will be easier to follow.

Methods.   Of the five methods, the first in order of utilization is *manova*. The manova must establish a significant difference among groups for the analysis to proceed. Second is *classification* of each individual to the group it most closely resembles, a classification to an individual's own group being considered a hit and to any other group, a miss. Third is determination of the *generalized distances* or *Mahalanobis' distances* between group centroids and perhaps between individuals and group centroids. This amounts to examining distances, quantitative differences, as they exist in discriminant space. Fourth is *multigroup discriminant function analysis* which includes a vector for angular comparison of groups in discriminant space. The fifth, and last method, is *canonical analysis of discriminance*, used mostly for ordination including contrasting generalized distances in restricted orthogonal dimensions of discriminant space.

Purposes.   In morphometrics, a given discriminant analysis can have three purposes, or a single purpose. We will consider the three purposes situation and parenthetically add the single one. The purposes amount to areas of interpretation which really are not unique since they generally overlap. However, the five major methods accomplish three purposes in three steps.

The first step is *testing the equality of group centroids* by manova. Given a number of related measurements on each individual of samples from two or more populations, manova provides a single test of the null hypothesis that the group population centroids are equal. A valid discriminant analysis requires a significant difference among the population centroids.

The second step, *classification of individuals to groups,* is analagous to a multivariate multiple range test. A true multivariate multiple range test would determine which of the *g* population centroids are significantly different from one another and which are not; however, in discriminant analysis another approach is taken, namely assigning individuals into groups. Let us consider two of many possible statistical means of classification. First, one can classify individuals on the basis of their generalized distances from each group centroid. There are a variety of ways in which distances can be implemented. For example, they can be calculated directly as a generalized distance statistic, indirectly as a related statistic from canonical analysis of discriminance, or indirectly as a related statistic from multigroup discriminant functions. Although the actual classification procedure to be used here, the second method, is somewhat more complicated than classifying by generalized distances, these distances are involved. For the present, imagine that an individual will be classified to the population having the closest centroid. Generally, one hopes for perfect discrimination, all hits.

Classification is extremely helpful in systematic studies of variation among populations within a species and among species within a genus. Hits versus misses provide some idea of relationships and how different is different.

Parenthetically, the single purpose study, *diagnosis,* can be inserted here since statistically it uses the same methods as classification. After a study involving a complete discriminant analysis, individuals not used in the original study can be classified on the basis of statistics derived from the original analysis. Therefore, diagnosis is nothing more than classification of new individuals by applying statistics derived from a previous study. Again, this can be extremely useful in taxonomy or in any classification problem. For example, diagnosis by discriminant analysis is commonly a basis for medical diagnosis at large medical centers. A patient's physical records, laboratory measurements and symptoms are applied to diagnose his illness.

The third step following true classification is studying the *relationships among populations.* Manova and even classification rarely provide a clear picture of how different one population is from another or how populations are interrelated. Now applied independent of classification, generalized distance analysis, multigroup discriminant function analysis, and canonical analysis of discriminance determine how physiology, anatomy, behavior and/or ecology of one population of organisms contrasts with other populations. In other words, the third step stresses morphometric analysis.

The above three steps or purposes of morphometrics are another possible source of confusion when the morphometric approach is compared with other uses of discriminant analysis. Step one is the least perplexing since manova either is the only method applied for testing the null hypothesis of equality of population centroids, or is not included in a discriminant analysis.

Step two, classification, is a much greater source of confusion. The procedure preferred here is Geisser's classification probabilities. However, since generalized distance, multigroup discrimination scores and canonical variates (scores for individuals from canonical analysis of discriminance) historically and still are applied to classification, their different roles in morphometrics must be appreciated.

In step three, studying relationships among populations, there are two alternatives, either studying relationships in total discriminant space, or in canonical space. Canonical space, defined by canonical analysis of discriminance, amounts to a precise, controlled examination of a few (generally two or three) dimensions of discriminant space at any one time, a type of ordination.

If discriminant space contains few enough dimensions to be examined in total, the generalized distance is applied as a quantitative measurement of differences between groups. The generalized distance amounts to discriminant space distances in standard deviation units. Also, when discriminant space is sufficiently simple, group discrimination vectors (obtained as part of each group's multigroup discriminant function) are subjected to angular comparisons. Groups having small angular departures between their discrimination vectors display minor qualitative differences and those having large departures display large qualitative differences. Angular comparisons can be possible even when discriminant space is too complex for generalized distance analysis.

When discriminant space is too complex, as often is the case, generalized distance analysis is restricted but still possible within the framework of canonical analysis of discriminance. An integral part of canonical analysis is a canonical graph which portrays the aforementioned precise, restricted views of discriminant space. Naturally, ordinations becomes involved with the graphs.

The study of relationships among groups will be extended even farther. In the sense of unravelling the variation of each group, a principal component analysis of each group, i.e., a multigroup principal component analysis, also is appropriate for studying group relationships.

Assumptions. The assumptions for discriminant analysis are those of manova. Only a few additional points need be emphasized.

The sampling procedure defines the populations studied and the biologist must know the populations he is studying. (In cases where populations are not known, some ordination must suffice.) If the purpose of the study is to segregate taxa, one need not know the taxonomic status of any sample. Rather, population definitions might be based upon space, time, sex, age, habitat, and perhaps behavior and ecology.

Recall that departures from a linear model, multivariate normality and/or homoscedasticity probably will not be too serious to invalidate a manova. This will be discussed further later. Other aspects of data vectors are of interest. It is assumed that each individual is represented by a random vector of p elements. However, it is further assumed that elements of random vectors are somewhat but not perfectly interdependent, i.e., measurements are intercorrelated! The methods of discriminant analysis use rather than remove intercorrelations between variables.

Assumption of Known Parameters. If multigroup discriminant functions or distances are applied to the classification of individuals to groups, the parameters of group centroids and dispersions must be known and all groups must have the same dispersion. In most studies these criteria would make classification questionable if other methods were not available.

Selection of Variables. The biologist has the right to select variables. However, the addition or deletion of variables might change results. This is the reason that no two groups can be proven identical. If one or more variables were added, they might segregate the groups. Therefore, the biologist's goal must be to "cover" the domain of his study with variables. The way to accomplish this is to summarize the shape of individuals.

For painted turtles, length, width, and height provide a crude outline of shape. However, the analysis of these dimensions leads to insight into the nature of painted turtle form. The other extreme would be more measurements than needed to describe the essence of painted turtles. There really is no answer to the number of variables necessary. About ten measurements, well selected to outline form, have produced excellent results in studies of various groups. So have fewer variables. I believe that about ten measurements are minimal to outline form and to be sure the domain is covered. Perhaps twenty measurements are not too many.

With the above in mind, there should be about eight criteria for selecting variables:

1. *Theoretical:* Knowledge of the literature should suggest measurements that might be correlated with group differences. Most definitely measurements of previous studies should be considered.
2. *Accuracy, precision* and *cost:* One should obtain measurements that are close to true values and are repeatable. One must judge this in terms of money, equipment, facilities, time, and effort.
3. *Nature of variation:* A good variable will vary more between groups than within any group.
4. *Direct:* Avoid taking a single measurement, e.g., oxygen consumption, for an entire sample and applying it to all individuals of that sample. Parenthetically, this criterion was ignored in the example of canonical correlation. However, there and in most like situations separate weather data were not available for each plant. There was no choice. If weather data are used in one set, they should be from a weather station both close to the individuals and "typical" of their site. Such data problems should not occur in a discriminant analysis.
5. *Inconsistent variables:* No variable can change in meaning or specification from individual to individual. This refers back to the definition of each population. Each population must be so defined and sampled so all sets of corresponding observations have the same basis in all individuals. For example, measurements of male and female gonads are not of the same variable.
6. *Complete:* Each individual must consist of a complete set of observations. Missing observations virtually destroy morphometrics.
7. *Variables:* No measurement can be a constant for all groups, even if the value (measurement or state) or the constant differs from group to group. The methodology applies only to variables.
8. *Artificial linear constraints:* As was the case in canonical correlation, the inverse of matrices is required. Therefore, ipsative and related measures must be avoided so matrices are of full rank.

10.2 Two Populations, One Variable

This and the following two sections comprise a unit on the special case of two populations. The first two sections, 10.2 and 10.3, introduce some basic concepts of discriminant analysis in reference to classification. The classical role of the generalized distance and canonical analysis of discriminance in classification, rather than in studying relationships, will be shown. The last section, 10.4, presents a two group example that does include the morphometric approach. Along with the two group example some useful angular comparisons are developed; but group discrimination vectors will not be introduced until later.

The present section pertains to certain properties of z-scores and the chi-squared statistic. Since the section considers only a single variable, multivariate analysis is not involved. On the other hand, the discussion will place the generalized distance into a familiar framework, e.g., the generalized distance between two means is the difference between their z-scores. Later it will be learned that in the multivariate case, generalized distances are not z-score differences.

Consider two populations with means, μ_1 and μ_2, and a common standard deviation, σ for this single variable. If X is a measurement for a new specimen, to which population does X belong? If $\mu_1 < \mu_2$, a reasonable diagnosis (a posteriori classification) rule is to assign X to population **I** if

$$X < (\mu_1 + \mu_2)/2,$$

or to population **II** if

$$X > (\mu_1 + \mu_2) \, 2$$

Therefore, the average of the two population means is the boundary, or cutoff point, for assignment.

An important consideration is how often will misses occur? If a specimen belongs to **I**, a miss exists whenever

$$X > (\mu_1 + \mu_2) \, 2.$$

However, this form of indicating a miss is inconvenient—the units are those of the variable. Statistically, it is much more convenient to deal with standard deviation units. In standard deviation units, a miss exists whenever

$$\frac{X - \mu_1}{\sigma} > \frac{(\mu_1 + \mu_2) \, 2 - \mu_1}{\sigma} = \frac{\mu_2 - \mu_1}{2\sigma} = \frac{\delta}{2\sigma}$$

where δ, the absolute difference between the two sample means, is

$$\delta = \mu_1 - \mu_2 \, .$$

But note that

$$\frac{X - \mu_1}{\sigma} = z,$$

where z is a standardized normal score, i.e., follows the normal distribution with mean equal to zero and standard deviation equal to unity.

The generalized distance, Δ, between two means is defined as their difference in standard deviation units,

$$\Delta = (\mu_1 - \mu_2) \, \sigma$$

so $\delta/2\sigma$, the cutoff point, is half the generalized distance between the two means. If all X_1 and X_2 are transformed to z-scores based upon deviations from the general mean, then

$$\Delta = \mu_1 - \mu_2 = \delta$$

where μ_1 and μ_2 now are z-score means that deviate from the general mean of zero, the cutoff point.

The square of the generalized distance can be used as a classification chi-squared. Eliminating the z-scores and returning to raw data,

$$\Delta^2 = \delta^2/\sigma^2 = (\mu_1 - \mu_2)^2/\sigma^2 = \delta\sigma^{-2}\delta$$

but

$$\chi^2 = (\mu_1 - \mu_2)^2/\sigma^2.$$

This can be extended so the two Δ^2 for an individual, one for μ_1 and the other for μ^2, are two chi-squared values whose probabilities are those of group **I** and group **II** membership.

Let us reconsider the problem in reference to two samples from two populations of unknown means and equal but unknown variances. The difference between the two sample means, d, is

$$d = X_1 - X_2.$$

From the pooled variance, s^2, from the samples of size n_1 and n_2, the standard deviation, s, is obtained. Then, the generalized distance, D, between the two sample means is

$$D = \frac{d}{s} = \frac{\overline{X}_1 - \overline{X}_2}{s}$$

and $1/2 \, D$ is the cutoff point between the two samples. The square of the generalized distance between the sample means is

$$D^2 = \frac{d^2}{s^2} = \frac{(\overline{X}_1 - \overline{X}_2)^2}{s^2},$$

which only approximates a classification chi-squared to test the hypothesis that $\mu_1 = \mu_2$, or to obtain the probability of a given X belonging to **I** or **II**. Since s^2 rather than σ^2 is used, D^2 only approximates the chi-squared distribution. This is a reason that D per se is not preferred for classification purposes.

10.3 Two Populations, Many Variables

Consider samples from two populations and p measurements on each individual. To introduce methodology, canonical analysis of discriminance is developed as a classification device. In canonical analysis, a model of the type

$$\mathbf{y} = \mathbf{v}'\mathbf{x}$$

is introduced. In the model, $\mathbf{x}$ is an individual's vector of raw score deviations from the grand centroid, $\mathbf{v}$ is a canonical vector of weights providing optimum assignment of individuals to their own group, and $\mathbf{y}$ is a canonical variate (a score of an individual on a canonical axis). The vector $\mathbf{v}$ will be replaced by other vectors ($\mathbf{w}$, $\mathbf{a}$, $\mathbf{b}$, $\mathbf{c}$ and $\mathbf{u}$) in different models but all are canonical vectors pertaining to transformation to the same discriminant space (actually canonical space).

Two group classification by canonical variates, $\mathbf{y}$, is quite simple. Much the same as before, the cutoff point between the two groups is the average of all canonical variate scores, $\overline{y} = 0$, which is the value of the canonical variate for the grand centroid. Also, the canonical variate scores for both group centroids have the same absolute values, one group will have a positive sign and the other a negative sign. Since the cutoff point is zero, all individuals having canonical variate scores less than zero will be classified into the group whose transformed centroid provides the negative canonical variate score and all others into the other group.

When more than two groups are studied, canonical analysis of discriminance utilizes the characteristic equation. This will be demonstrated in section 10.6. In the two group situation, a direct solution of canonical vectors is possible.

In this section, further discussion of two group canonical analysis of discriminance pertains to the six models. The models are examined in some detail. Their mathematical relationships plus their unique properties and interpretations are stressed. Within this framework, generalized distances and classifications are included. The treatment of generalized distances introduces their later role in morphometrics, but that of classification does not introduce the final method to be used. Rather, the topic of classification serves both to expand upon properties of canonical analysis and to preview certain concepts of classification.

Six Models.   In addition to the confusion of discriminant analysis including five major methods and six purposes, six models for canonical analysis of discriminance are in general use. More models are possible. Fortunately the models are closely related in the sense that one can be derived from the other. Although this allows simple algebraic manipulations that disclose the interrelations of models, the fact that there are six models can be perplexing. However, it should help if one realizes that all models pertain to and define the same discriminant space. The primary differences among them pertain to whether the original data vectors are first transformed to z-scores and to the nature of the different variances of the canonical variates. For this reason, each model is interpreted somewhat differently. Eventually, concern will be limited to only three of the models. Still later, it will be seen that much that is stated here applies directly to morphometrics.

Canonical Analysis of Discriminance.   In view of prior mathematical discussion, for two groups, the canonical function probably is best understood in the form

$$\mathbf{y}_w = \mathbf{w}'\mathbf{x}.$$

In this model the canonical vector $\mathbf{w}$ is the product of the inverse of the pooled dispersion matrix, $\mathbf{S}^2$, and the difference between centroids, $\mathbf{d}$,

$$\mathbf{w} = \mathbf{S}^{-2}\mathbf{d} = \mathbf{d}'\mathbf{S}^{-2}.$$

Note that the vector $\mathbf{w}$ corresponds to the vector of multiple regression weights in

$$\mathbf{b} = \mathbf{R}_{11}^{-1}\mathbf{R}_{12}.$$

Therefore, the inverse of the dispersion matrix is similar to the inverse of predictor intercorrelations and the difference between centroids to predictor-criterion intercorrelations. In essence, $\mathbf{w}$ is the best predictor vector of differences between groups in terms of the covariances of within group variables. More important is the effect of multiplying $\mathbf{d}$ by $\mathbf{S}^{-2}$. This transforms Euclidean space to discriminant space, a space based upon intercorrelation between variables, a synergistic space.

The above canonical function is linked to the square of the generalized distance. It is of interest to develop the square of the generalized distance as if it were to be the basis for assignment of individuals to groups. The difference, $\mathbf{d}$, now between an individual and a group centroid, can be calculated for each individual, first for the first and then for the second centroid,

$$D^2 = \mathbf{d}'\mathbf{S}^{-2}\mathbf{d} = \mathbf{d}'\mathbf{w};$$

and the smaller D^2 is the basis for assigning an individual to a particular group. Since D^2 approximates the chi-squared distribution, assignment can involve probabilities derive from that distribution.

The square of the distance is an interesting statistic. It is analogous to the multiple correlation coefficient,

$$R^2 = \mathbf{R}_{21}\mathbf{R}_1^{-1}\mathbf{R}_{12} = \mathbf{R}_{21}\mathbf{b}.$$

In this sense, D^2 is a measure of the relationship between the groups. Also, the square of the generalized distance is the variance of the canonical function,

$$\mathbf{y}_w = \mathbf{w}'\mathbf{x},$$

since

$$\mathbf{s}^2_{y_w} = \mathbf{w}'\mathbf{S}^2\mathbf{w}$$

$$= (\mathbf{d}'\mathbf{S}^{-2})\mathbf{S}^2(\mathbf{S}^{-2}\mathbf{d})$$

$$= \mathbf{d}'\mathbf{S}^{-2}\mathbf{I}\mathbf{d}$$

$$= \mathbf{d}'\mathbf{S}^{-2}\mathbf{d}$$

$$= D^2.$$

In other models, the variance of canonical variates is not equal to D^2.

Normalized Canonical Analysis of Discriminance. The above canonical vectors can be normalized, i.e., changed so that the sum of squares of the coefficients of the canonical vector equal unity, without losing the power of discrimination or changing the fundamental nature of the discriminant space. The way to do this is to divide each element of $\mathbf{w}$ by the square root of its sum of squares,

$$\mathbf{v} = \mathbf{w}/\sqrt{\sum \mathbf{w}^2} \text{ so}$$

$$\mathbf{v}'\mathbf{v} = \mathbf{1}.$$

The reason for this is to create $\mathbf{v}$ as a fundamental vector from which other models can be derived.

Then the normalized model becomes

$$\mathbf{y}_v = \mathbf{v}'\mathbf{x} \text{ where}$$

$$\mathbf{v}'\mathbf{v} = 1,$$

$$s_{y_v}^2 = \mathbf{v}'\mathbf{S}^2\mathbf{v}, \text{ and}$$

$$\mathbf{y}_v = s_{y_w}^{-1}\mathbf{w}'\mathbf{x}$$

For later use note that the grand or total covariance of $\mathbf{y}_v$ is

$$\mathbf{S}_{y_T}^2 = \mathbf{v}'\left[\frac{1}{N-1}\right]\mathbf{T}\mathbf{v}' = \mathbf{v}'\mathbf{S}_T^2\mathbf{V},$$

where N is the total sample size, $\mathbf{T}$ is the total *SSCP* matrix and $\mathbf{S}_T^2$ is the total dispersion matrix of $\mathbf{y}_x$.

The canonical vectors $\mathbf{v}$ are useful for examining relationships in a manner analogous to that for PCA eigenvectors, a "unit stimulus of discrimination" providing a certain synergistic relationship among variables. Therefore, the elements of $\mathbf{v}$ must and do provide the cosine of the angle of rotation of individual variable axes to the canonical axes.

Standard Canonical Analysis of Discriminance for Raw Variables.
The foregoing models have a major disadvantage in ordination. Units on a graph are not related directly to a convenient statistic. Greater convenience would exist if a unit on a graph were one standard deviation of the canonical variate.

Standardization, causing the variance of $\mathbf{y}$ to be unity, can be accomplished by within or total variation and by appropriate scaling of $\mathbf{v}$. Since $\mathbf{v}$ is an unscaled vector, it is readily manipulated to derive any other vector. For standardization by within

$$\mathbf{y}_a = \mathbf{a}'\mathbf{x} \text{ where}$$

$$\mathbf{a}_j = s_{y_{v_j}}^{-1}\mathbf{v}_j$$

and $s_{y_{v_j}}^{-1}$ is a vector of reciprocals of standard deviations obtained from the normalized model, so

$$\mathbf{y}_a = \mathbf{y}_v s_{y_v}^{-1} \text{ and}$$

$$\mathbf{a}'\mathbf{S}^2\mathbf{a} = 1.$$

Therefore, **a** is derived simply by dividing each coefficient of **v** by the scalar standard deviation of $_{ia}$. which causes the variance of $\mathbf{y}_a$ to be unity.

For the standardized by total model,

$$\mathbf{y}_b = \mathbf{b}'\mathbf{x},$$

as one might expect, the scaling factor for **v** is the reciprocal of the standard deviation for total of $\mathbf{y}_v$, a scalar. Therefore,

$$\mathbf{b} = \mathbf{s}_{y_T}^{-1}\mathbf{v}, \text{ so}$$

$$\mathbf{y}_b = \mathbf{y}_v \mathbf{s}_{y_T}^{-1} \text{ and}$$

$$\mathbf{b}'\mathbf{S}_T^2\mathbf{b} = 1.$$

Neither of the above models allow consideration of a canonical vector derived from a unit stimulus. Interpretation is relative to the disperison of raw variables, absolute variation. The advantages of the vectors **a** and **b** are best appreciated when individuals are plotted in canonical space. Then, units of canonical discriminant space are in standard deviations of dispersion or total variation. Standardization by within is especially useful since the units of each canonical axis can be assumed to be standard deviations of any group along that axis. The effect of standardization by total is more data dependent so the model is not as useful in morphometrics.

Standardized Canonical Analysis of Discriminance of Normal Scores.   These standardized variation or z-score models are often called canonical factor models. Again standardization can be by within or by total dispersion, but since z-scores are involved, the within correlation matrix, **R**, or total correlation matrix, $\mathbf{R}_T$, is involved. Ths standardization by within, the canonical factor for within model, can be derived from

$$\mathbf{y}_a = \mathbf{a}'\mathbf{x}.$$

Since

$$\mathbf{x} = \mathbf{S}_{diag}\mathbf{z},$$

where $\mathbf{S}_{diag}$ is the vector of standard deviations for within for original variables in **x**,

$$\mathbf{y}_a = \mathbf{a}'\mathbf{S}_{diag}\mathbf{z};$$

and, defining

$$\mathbf{c}' = \mathbf{a}'\mathbf{S}_{diag},$$

$$\mathbf{y}_a = \mathbf{y}_c = \mathbf{c}'\mathbf{z}, \text{ and}$$

$$s_{y_a}^2 = s_{y_c}^2 = \mathbf{c}'\mathbf{R}\mathbf{c} = 1.$$

Therefore we see that either model for within produces equal canonical variates, $\mathbf{y}_a = \mathbf{y}_c$, hence identical canonical graphs. Later it will be seen that $\mathbf{a}$ and $\mathbf{c}$ can be compared in reference to a single canonical graph. The comparison is one of absolute vs. standardized variation.

The canonical factor for total model is derived from

$$\mathbf{y}_b = \mathbf{b}'\mathbf{x}$$

where

$$\mathbf{x} = \mathbf{S}_{Tdiag}\mathbf{z}.$$

$\mathbf{S}_{Tdiag}$ is the vector of standard deviations of original variables for total, so

$$\mathbf{y}_b = \mathbf{b}'\mathbf{S}_{Tdiag}\mathbf{z};$$

and, defining

$$\mathbf{u}' = \mathbf{b}'\mathbf{S}_{Tdiag},$$

$$\mathbf{y}_b = \mathbf{y}_u = \mathbf{u}'\mathbf{z}, \text{ and}$$

$$s_{y_b}^2 = s_{y_u}^2 = \mathbf{u}'\mathbf{R}_T\mathbf{u} = 1.$$

These z-score models pertain to standardized variation just as the corresponding raw variable models pertain to absolute variation. The models for total are applied mostly to studies of psychological and/or educational data where canonical graphs are less likely and the morphometric approach is not applied. Again, the within models are more appropriate for morphometrics since canonical graph units, within group standard deviations, are directly translatable. Another major benefit occurs from the fact that within canonical graphs for both models are identical for a given set of data. This also is the case for both total models.

Recapitulation of Models.   Each of the above is a model for canonical analysis of discriminance. In all, $\mathbf{y}$ is a *canonical variate,* a value projected to a $\mathbf{y}$-*canonical axis*. Parenthetically, the $\mathbf{y}$-values for the canonical factor models often are called canonical factor scores, and the $\mathbf{c}$ or $\mathbf{u}$ coefficients, canonical factor coefficients. All models provide the same set of $\mathbf{y}$ axes, the only difference being the "units" of some of the different $\mathbf{y}$'s (variances of $\mathbf{y}$.) Projection to a $\mathbf{y}$-axis can lead to both classification and study of relationships of individuals. Each model contains a *canonical vector* ($\mathbf{w, v, a, b, c,}$ or $\mathbf{u}$) whose coefficients provide the synergistic relationship among variables that maximizes discrimination by transforming data vectors to the same discriminant space. The vector $\mathbf{w}$ is interpreted on a $\mathbf{y}_w$-axis where the square of the distance between groups is equal to the variance of $\mathbf{y}_w$. The $\mathbf{w}$-vector is convenient for classification or visualization of methodology but is not very useful for ordination or heuristic analysis. The vector $\mathbf{v}$ is unique in that it allows interpretation of the maximum synergistic discrimination of raw variables as a response to a stimulus of discrimination. Since it is the only vector whose coefficients represent cosines of angles of rotation of original variable axes to axes in discriminant space it might be considered the fundamental vector of any canonical analysis. Although $\mathbf{y}_v$ in this model is also subject to graphing, the $\mathbf{y}_v$ variance, hence units of canonical space, is data dependent. In this two group case the single vector $\mathbf{v}$ contrasts the patterns of variation in the two groups. The coefficients can be "read" in reference to contrasting patterns of variation, form, between groups. Patterns of variation most definitely are not independent as they are in principal component analysis.

Of the vectors $\mathbf{a, b, c}$ and $\mathbf{u}$, additional preference goes to $\mathbf{a}$ or $\mathbf{c}$ for graphing, since on a graph a circle of unit radius about each centroid can be used to indicate one within group standard deviation. Also, when the within standard deviation is assumed to be unity in all directions about a centroid, each parametric or population centroid can be estimated by a confidence interval about its sample centroid.

Since tests of homogeneity of dispersions often are not accepted, the within models use of a pooled dispersion matrix, $\mathbf{S}^2$ or $\mathbf{R}$, to estimate a common population dispersion might seem dubious. However, again evidence indicates that methodology is sufficiently robust to take advantage of the concepts of unit standard deviations and confidence intervals.

In later discussion only three models are emphasized, the normalized model $\mathbf{y}_v = \mathbf{v}'\mathbf{x}$ for heuristics, plus the standardized for within model $\mathbf{y}_a = \mathbf{a}'\mathbf{x}$ and the canonical factor for within model $\mathbf{y}_c = \mathbf{y}'\mathbf{z}$ for graphic comparisons. Again, since $\mathbf{y}_a = \mathbf{y}_c$, both of the later models lead to exactly the same canonical graph. Therefore, a comparison of $\mathbf{a}$ and $\mathbf{c}$ in terms of the graph permits a simultaneous evaluation of absolute and standardized variation of the original variables.

Two Populations, Classification.   Two models are frequently used elsewhere to examine assignment of individuals to groups. These are the basic and total factor models. In the basic model where

$$\mathbf{y}_w = \mathbf{w}'\mathbf{x},$$

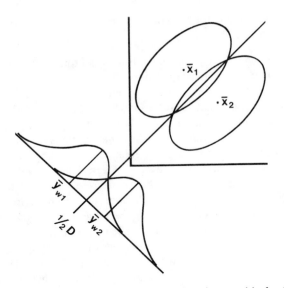

Figure 10.1 Projection of bivariate raw data vectors and centroids for two samples from raw data axes X_1 and X_2 to a *single canonical axis* $y_w = w'x$. D is distance between $\bar{y}_{w1}$ and $\bar{y}_{w2}$ ($\bar{X}_1$ and $\bar{X}_2$ in canonical space).

if $\bar{y}_{w_1} < \bar{y}_{w_2}$ and the cutting point is to be $1/2\, D$, half the distance between the two group centroids, the cutting point is

$$\bar{y}_{w_1} + 1/2\, D = \bar{y}_{w_2} - 1/2\, D.$$

This can be visualized as follows: Once the canonical vector is known, y_w can be calculated for all individuals of both samples. Assuming large samples, the canonical scores y_w will cluster into two groups approximating normal curves (Figure 10.1). For each group, the mean of y_w will be $y_w = \mathbf{w}'\bar{\mathbf{x}}$. The cutoff point is $1/2\, D$, where

$$\bar{y}_{w_1} - \bar{y}_{w_2} = D = (\mathbf{d}'\mathbf{S}^{-2}\mathbf{d})^{1/2} = ((\bar{\mathbf{X}}_1 - \bar{\mathbf{X}}_2)'\mathbf{S}^{-2}(\bar{\mathbf{X}}_1 - \bar{\mathbf{X}}_2))^{1/2}.$$

The implications of Figure 10.1 will be clarified later. The figure portrays the transformation of individuals from a 2-dimensional Euclidian space to a single canonical axis in discriminant space. In the Euclidian space the frequency of individuals is not shown as a third dimension. However, on the canonical axis the frequency of individuals is represented by two normal curves, one for each group. Actually both spaces correspond and have the same number of dimensions—discriminant space may or may not have fewer canonical axes than dimensions.

Other Classification Considerations. The cutoff point between two groups can be adjusted from halfway by a quantity r, the ratio of the larger over the smaller probability of membership,

$r = p_i/p_j$, where $p_i > p_j$ and $p_i + p_j = 1$.

Then,

$log_e r/D$

is subtracted from the group less likely to be represented. If group one has the lower probability, then

$$y_{w_1} + 1/2\ D - log_e\ r/D = y_{w_2} - 1/2\ D + log_e\ r/D$$

become the cutoff point.

In addition, one can calculate the probability of an individual's membership to each group via approximate chi-squared values, but these do not directly allow for unequal probabilities of group membership. The probabilities are obtained from

$$\chi_i^2 = \mathbf{d}_i{'}\mathbf{S}^{-2}\mathbf{d}_i$$

with p degrees of freedom for the ith group. Again, the chi-squared value is equal to Mahalanobis' square of the distance.

10.4 Two Groups, Example

To illustrate the techniques of two group morphometric analysis, the data for male and female painted turtles shall be contrasted. In the previous chapter, the first step in discriminant analysis, manova of the turtle data, was shown. Since the manova was significant, we can proceed.

Further discussion will be in reference to three topics: classification, generalized distance and canonical analysis.

Classification. First, consider generalized distances for classification purposes. Table 10.1 contains the generalized distance of each individual from each group centroid. The square of the distances could be used as classification chi-squareds to provide the probabilities of group membership—consistently the closer an individual is to one group centroid than to the other, the higher the probability of it belonging to the closer group.

Classification of individuals to groups (Table 10.1 and 10.2) is on the basis of the group having the higher probability. However, the actual probabilities shown in Table 10.1 are Geisser classification probabilities rather than those from classification chi-squareds. Note that the generalized distances would lead to the same classification. All individuals are classified to the group having the closer centroid. The only misses are of five females Three of these individuals, 1, 2 and 4, are relatively small. It is not surprising that they might display insufficient sexual dimorphism to segregate them from males.

Table 10.1 Generalized distances of individuals from group centroids and Geisser classification probabilities of group membership (misclassifications are indicated by an asterisk)

Case		Distance from Group		Probability of Membership	
		Females	**Males**	**Females**	**Males**
Female	1*	2.30	1.80	0.28	0.72
	2*	2.52	0.99	0.08	0.92
	3	2.32	2.99	0.82	0.18
	4*	2.13	1.53	0.27	0.73
	5	1.82	2.81	0.88	0.12
	6	2.31	3.94	0.98	0.02
	7	1.16	1.25	0.53	0.47
	8	0.75	2.50	0.93	0.07
	9	0.49	2.31	0.91	0.09
	10	0.49	2.31	0.91	0.09
	11*	1.89	1.22	0.27	0.73
	12*	1.74	1.34	0.36	0.64
	13	1.98	2.75	0.83	0.17
	14	1.87	2.54	0.79	0.21
	15	0.51	2.25	0.90	0.10
	16	1.15	3.40	0.99	0.01
	17	1.56	2.67	0.89	0.11
	18	2.43	3.27	0.88	0.12
	19	2.80	5.14	1.00	0.00
	20	1.70	3.75	0.99	0.01
	21	1.99	4.30	1.00	0.00
	22	2.21	4.54	1.00	0.00
	23	2.82	4.03	0.96	0.04
	24	3.07	4.69	0.99	0.01
Male	1	2.79	2.73	0.47	0.53
	2	2.83	1.32	0.06	0.94
	3	3.07	1.27	0.03	0.97
	4	2.31	1.84	0.29	0.71
	5	2.71	1.52	0.10	0.90
	6	2.77	0.75	0.04	0.96

7	2.13	1.07	0.17	0.83
8	2.28	0.84	0.11	0.89
9	2.87	1.02	0.04	0.96
10	2.65	0.66	0.05	0.95
11	2.66	0.34	0.04	0.96
12	2.87	1.05	0.04	0.96
13	1.63	0.83	0.28	0.72
14	2.75	0.60	0.04	0.96
15	2.81	0.72	0.03	0.97
16	3.26	1.32	0.02	0.98
17	3.48	1.94	0.24	0.76
18	1.70	0.70	0.24	0.76
19	3.17	1.43	0.03	0.97
20	2.03	1.14	0.21	0.79
21	2.28	0.84	0.11	0.89
22	2.49	1.17	0.10	0.90
23	2.69	1.76	0.14	0.86
24	3.59	2.60	0.08	0.92

Table 10.2 Classification table, rows are actual groups and columns are predicted groups.

Groups	Females	Males	% hits
Females	19	5	79.1
Males	0	24	100.0
Total	19	29	89.6

The 5 misses can be appreciated by referring back to the multigroup component analysis. First, in Table 10.4 note that coefficients of the canonical axis are very much the same as were those for the third component axes of females, of males, and of dispersion. This will soon be verified by angular comparison. In examining these components (Tables 5.2 and 5.3) or the coefficients of the canonical axis (Table 10.4), note that all deal with an increase in height at the expense of relationships with length and width, mostly length. In the three small females, there are small positive or negative component scores for the third component. In other words, there are three small females that hardly can be said to display female characteristics. The two larger females, 11 and 12, are the two outlier females that showed very high variation, 27.5 and 77.5 percent of their totals, in the third component. Since these variations are tied in with very high negative scores for the third component, it is not surprising that they should misclassify; in fact, they clearly are outliers.

If one wished to pursue the subject of classification by examining probabilities in detail, a comparison of male, female and dispersion component analyses with Tables 10.1–10.6 would allow this.

Generalized Distance. Table 10.3 presents the generalized distance between groups based upon all variables and each variable alone. In the sense of individual variables, distances disclose the importance of height as a discriminator. However, both length and width are also important discriminators. The total distance, 2.358, is in standard deviation units. Therefore, one can expect, from tables of the normal curve, approximately 88% correct assignment of males and females to their correct group. For this reason, 89.6% hits might appear remarkable in view of two outliers and inequality of group dispersions.

Canonical Vectors. Table 10.4 provides three forms of the canonical vector plus the structure, or correlation between original variables and canonical variates. The latter "shows" that all variables are "important" in the canonical analysis.

All canonical vectors disclose that discrimination involves contrasting height with length and width. Since the single normalized canonical vector departs less than 8° (see Table 10.6) from the third dispersion component, it can be assumed that discrimination contrasts an approximation of a fundamental pattern of variation that discloses sexual dimorphism. The means of calculating these angles are provided later.

It should not be surprising that discrimination relates to the minor component axis of turtles. In most cases, the line connecting group centroids (which approximates the canonical axis) is unlikely to relate to general growth, i.e., the first component. Primary differences between groups tend to be shape differences. Also, the basis for discrimination might approximate a component pertaining to small within-group variation, perhaps even the last component. In other words, discrimination would pertain to features displaying greater between than within group differences. Naturally, all this must be considered in reference to components being in Euclidean space and canonical vectors in discriminant space.

Table 10.5 provides canonical scores, $y_a = a'x$, for group centroids and individuals, all as deviations from the cutoff point, the grand centroid. Again, $y_a = y_c = c'z$. Figure 10.2 presents the same data in graphic form, a canonical graph. The table and graph disclose the distances from the cutoff point of all individuals, including the misclassified females.

In reference to later use of canonical analysis of discriminance, note that here, where only one canonical axis can exist in discriminant space, the difference between canonical variate scores for centroids, $1.179 - (-1.179) = 2.358$, is equal to the total distance between groups in discriminant space.

Table 10.3 Comparison of distances between groups based upon all and each variable

	Variables			
	All	Length	Width	Height
Distance	2.358	1.317	1.357	1.802
%	100.0	55.4	57.5	76.4

Table 10.4 Coefficients of normalized, standardized for within, and factor for within canonical vectors, plus structure (vector-variable correlations)

Variable	normalized	within	within factor	structure
length	−0.227	−0.116	−1.534	0.725
width	−0.080	−0.041	−0.336	0.740
height	0.970	0.495	2.683	0.880

Table 10.5 Distance of centroids and individuals, expressed as deviations from the grand centroid (the cutoff point) in one dimensional canonical space

Females (centroid = 1.179)

Individuals							
1	−0.437	7	0.049	13	0.777	19	3.932
2	−0.140	8	1.200	14	0.620	20	2.365
3	0.758	9	1.077	15	1.016	21	3.089
4	−0.464	10	1.077	16	2.177	22	3.354
5	0.970	11	−0.443	17	0.996	23	1.761
6	2.152	12	−0.262	18	1.027	24	2.663

Males (centroid = −1.790)

Individuals							
1	−0.065	7	−0.720	13	−0.419	19	−1.700
2	−1.335	8	−0.952	14	−1.526	20	−0.596
3	−1.649	9	−1.522	15	−1.567	21	−0.952
4	−0.413	10	−1.400	16	−1.881	22	−1.027
5	−1.065	11	−1.475	17	−2.328	23	−0.880
6	−1.512	12	−1.509	18	−0.511	24	−1.300

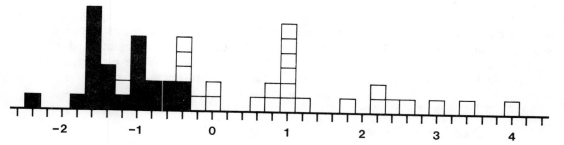
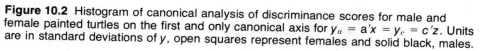

Figure 10.2 Histogram of canonical analysis of discriminance scores for male and female painted turtles on the first and only canonical axis for $y_a = a'x = y_c = c'z$. Units are in standard deviations of y, open squares represent females and solid black, males.

Table 10.6 Angles between canonical axes and variables, female components, male components, and dispersion components

variables		female		male		dispersion
length	−76.844	1	85.812	1	−89.493	1	86.856
width	−85.407	2	−81.079	2	88.089	2	−82.951
height	13.962	3	9.870	3	1.976	3	7.725

Since angles between canonical and component vectors were important, they were discussed. For this reason, two formulae for angular comparison will be presented. These formulae are generalized to the case where more than one canonical axis exists, but they apply equally to the present single canonical axis situation.

Table 10.6 provides useful angular comparisons between variables and the canonical vector. These angles are derived from

$$\text{cosine } \theta_{ij} = v_{ij}$$

which is the cosine of the angle of the jth variable axis from the ith canonical vector—in this case i equals one. Since in the turtles we are dealing with a single canonical axis, there is a single set of angles. These angles disclose the not surprising fact that length and width are farther removed from the canonical axis than is height. In other words, in the transformation from original variable **x**-axes to the **y**-axis, height generally contributes more to the value of **y**. However, length and width cannot be declared less important than height, since discrimination is in terms of a synergistic relationship among the three variables.

The aforementioned angular comparisons between components and canonical axes must be accomplished by vectors composed of direction cosines as elements. Since **v** is the only canonical vector having direction cosines as elements, $\mathbf{v}'\mathbf{v} = 1$, and all component eigenvectors are direction cosines,

$$\text{cosine } \theta_{ij} = \mathbf{v}_i'\mathbf{a}_j$$

provides the cosine of the angle between the canonical axis $\mathbf{v}_i$ and component axis $\mathbf{a}_j$, both of which involve p elements. (Note that here $\mathbf{a}_j$ is a component and not a canonical axis of the standardized for within model.) Therefore, $\mathbf{v}_i'\mathbf{a}_j$ is the sum of the products of p elements. We have already referred to the fact that all third component axes are close to the canonical axis. For this reason, it is not surprising that all first and second components are nearly orthogonal to the canonical axis.

10.5 Multigroup Discriminant Function Analysis

When more than two groups are involved, discriminant analysis becomes more complex. After a significant manova, how are we to classify? The boundaries between groups can no longer be determined with sufficient accuracy by a single canonical axis. Only in the two group case is a single axis, parallel to the line between centroids, capable of maximizing the differences between groups. In studies of more than two groups, for maximum discrimination, one discrimination vector is calculated for each group. However, each vector is part of a unique function derived for each group.

Historically, the primary purposes of multigroup functions were classification and diagnosis of many groups. Such functions became useful since canonical functions have different properties when more than two groups are involved. In fact, canonical functions become so different that their application to classification is unsatisfactory. In this section we shall finalize the procedure for classification as recommended here. After first showing how multigroup functions and distance can be applied to classification, Geisser classification probabilities are presented. Then, multigroup discriminant functions and generalized distances will be considered under their new roles as agents for studying relationships. Finally, the turtle example will be joined to methodology.

Classification. In terms of the square of the generalized distance being a classification chi-squared, each group's region of discriminant space can be recognized by its probability of occurrence. The probabilities theoretically correspond to the density function of each population. Since all populations are assumed to follow the multivariate normal distribution and have the same dispersion, the density function is assumed to be that for the *mnd*.

In the following discussion, details will be presented about the multigroup discriminant function model and an l-vector within that model in reference to their historical roles in classification. The entire model is not used in morphometrics. Although the l-vector is important in morphometrics, l has no role in classification. Rather, l will be involved in examining relationships between groups. In spite of this, the historical roles of both the model and l-vector provide important insight into methodology and the final role of the l-vectors.

Nature of Regions. The features of hyperspace regions can be appreciated in a 3-group, 3-region example. When three regions occur, individuals can be projected to a plane, a discriminant plane. Then each group can be imagined to exist in a circle, and the radius of each circle is the boundary for that region of discriminant space. However, the circles can overlap creating difficulty for discrimination (Figure 10.3).

The discriminant space and its regions are not automatically defined by canonical axes—discriminant space and a discriminant plane can be defined without orthogonal reference axes. Later we shall see how canonical axes are added. Even without canonical axes discriminant space is created by the inverse of the dispersion matrix, and each group's region can be defined by multigroup discriminant function.

Model. In matrix form, the model for multigroup discriminant analysis is

$$y_{ij} = \mathbf{X}_{jk}{}'l_i\overline{\mathbf{X}}_i - 1/2\overline{\mathbf{X}}_i{}'l_i + P_i$$

where $i = 1, 2, \ldots, g$ groups of classification; $j = 1, 2, \ldots, g$ groups of actual membership; $k = 1, 2, \ldots, N_g$ individuals; $\mathbf{X}$ is an individual's data vector; l is a group discrimination vector; $\overline{\mathbf{X}}$ is a group centroid; P is the probability of group membership; and y is a discrimination score for an individual.

Terminology again requires consideration. Each y_{ij} has been named a *classification score, discrimination score, discriminant score* or *discriminant function score*. The term $\mathbf{X}_{jk}{}'l_i\overline{\mathbf{X}}_i$ is equivalent to the *discriminant function* as originally described by Fisher (1936). However, Fisher treated the two

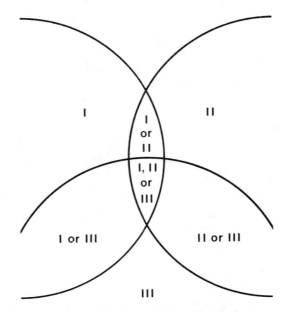

Figure 10.3 Three groups projected onto a 2-dimensional plane to emphasize areas of clear classification to the correct group, areas that could be classified into two groups, and an area that could be classified into three groups.

group case and used only a single function which provided a single canonical axis. Perhaps for this reason, when extended to the multigroup case, Fisher's function is termed a *classification function* in both a BIOMED (Dixon, 1973) and a SPSS (Klecka, 1975) computer program. Mathematical and statistical dictionaries are of little help in clarifying the situation (Kendall and Buckland, 1957; James and Beckenback, 1968). In essence, it seems that "discriminant function" could apply to any method for classifying individuals. It has substituted for manova; for the generalized distance; and for the vectors, functions, and scores of canonical analysis of discriminance. It might seem best either to abandon, or to restrict discriminant function to include both multigroup discriminant function analysis and canonical analysis of discriminance; but I believe that both possibilities avoid the issue. Since canonical analysis, except for the two group case, involves the characteristic equation and eigenvector function to develop canonical form, it appears that canonical analysis of discriminance is most appropriate for descriptive purposes, but canonical discriminant function also is acceptable.

In agreement with some usage, here discriminant function is restricted to $\mathbf{X}_{jk}'\,{}_{\ell}\,{}_{i}\overline{\mathbf{X}}_{i}$ but *classification function* is an appropriate synonym. In addition, consistent synonyms for multigroup discriminant analysis would be *classification function analysis* or *discriminant function analysis*; and *discriminant analysis* will be applied to include all methods presented in this chapter.

In the literature, the vector ℓ is called a (*group*) *classification, discrimination,* or *discriminant (function) vector*. The terms in parentheses may or may not be included. The individual elements of ℓ are ℓ-*weights* or ℓ-*coefficients*. Although canonical vectors have been called discrimination or discriminant vectors, this shall not be followed here. Also, ℓ-vectors shall not be called classification vectors since they shall provide a major source of heuristic analysis and shall never be used for classification. The ℓ-vectors shall be called *(group) discrimination vectors*.

The term $1/2\,\mathbf{X}_i{}'\ell_i$ is regularly called the *discriminant function constant, classification constant*, or simply *constant* and is not confused with other procedures. The term P_i is the *a priori probability of group membership*, i.e., to the ith group, that is selected by the investigator on the basis of some preconceived judgment. The elements of P_i frequently are referred to as *priors* and $\mathbf{P}$ as a *vector of priors*.

In the sense of the above, the model for multigroup discriminant analysis will be called the *(multiple) discriminant (function) score model*; the first term, the (multiple) discriminant function; the second term, the (multiple) (discriminant function) constant; and the last, the (multiple) (discriminant function) priors.

I must admit that the above represents personal preference in an attempt to clarify terminology. There is no uniformity of application. In fact, I have ignored some previous attempts at uniformity that I thought still allowed confusion.

Returning to the model, discriminant scores for an individual can be used to assign the individual to a region (group) in discriminant space. For each individual, a score is calculated for each group and the individual is assigned to the group producing the largest score. To appreciate this, more details about the formula for multigroup discriminant function analysis are necessary.

The vector of ℓ-weights can be defined further,

$$\ell_i = \mathbf{S}^{-2}\overline{\mathbf{X}}_i.$$

Note the similarity to the two group canonical vector $\mathbf{w} = \mathbf{S}^{-2}\mathbf{d}$. However, $\mathbf{w}$ is obtained by multiplying the inverse of the dispersion matrix by the difference in centroids, not the particular centroid as in ℓ. On the other hand, a closer relationship between $\mathbf{w}$ and ℓ can be imagined if $\overline{\mathbf{X}}$ is considered to be the difference between the ith centroid and the origin. In these terms, $\mathbf{w}$ can be visualized as a vector between group centroids, and each ℓ as a vector passing through its group centroid. Furthermore, both $\mathbf{w}$ and ℓ are analogous to regression slope coefficients; but $\mathbf{w}$ and ℓ serve to maximize scores of data vectors for their own groups rather than to estimate a point on the line of regression. Also, ignoring the relationship between centroid and origin, ℓ amounts to the rotation of the centroid from Euclidean into discriminant space.

The dispersion of discriminant scores $\mathbf{y}$ for any group is equal to

$$\ell_i{}'\overline{\mathbf{X}}_i = \overline{\mathbf{X}}_i{}'\ell_i,$$

which is the square of the generalized distance of that group from the origin. This follows from

$$\mathbf{S}^2{}_{y_i} = \ell_i{}'\mathbf{S}^2\,\ell_i$$

$$= (\overline{\mathbf{X}}_i{}'\mathbf{S}^{-2})(\mathbf{S}^2)(\mathbf{S}^{-2}\overline{\mathbf{X}}_i)$$

$$= \overline{\mathbf{X}}_i{}'\mathbf{S}^{-2}\overline{\mathbf{X}}_i = D^2$$

$$= \overline{\mathbf{X}}_i{}'\,\ell_i.$$

The effect of

$$\mathbf{X}_{jk} \, \ell_i \overline{\mathbf{X}}_i$$

is to cause the value to be a maximum when $\mathbf{X}_{jk} = \overline{\mathbf{X}}_i$. The greater the departure of an individual data vector, $\mathbf{X}_{jk}$, from a centroid, the smaller the discrimination score for that centroid.

The constant

$$1/2 \, \overline{\mathbf{X}}_i{}' \ell_i$$

is analogous to $\sum xy$ of regression. Together with the discriminant function, the constant is designed to give a maximum discriminant function score for correct assignment to a group. It amounts to subtracting half the square of the distance of a centroid to the origin and is a constant for each group.

The probability of group membership, if priors are equal is

$$P_i = 1/g,$$

but if probabilities are unequal

$$P_i = log_e P_i^*$$

where P_i^* is the prior probability of membership to the ith group. Unequal priors change locations of classification cutoff points between centroids, but naturally not the position of centroids.

The net effect of the three multigroup discriminant analysis terms is to provide a score that is a maximum for a particular centroid, other scores decreasing for individuals progressively farther away from the centroid.

Assignment to Groups.   Although each individual can be classified into the group having the largest discriminant score, this procedure suffers from poor statistical estimates, unequal group dispersions, overlapping regions, and doubtful regions of classification. For this reason, classification probabilities for membership to each group are useful. One source of such probabilities is from classification chi-squareds, the square of the generalized distance of each individual from each group centroid. However, recall that these probabilities are prone to mathematical error and are based upon equal priors of membership.

Geisser Classification Probabilities.   Geisser (1964) developed classification procedures superior to discrimination scores even when the latter are accompanied by probabilities based upon distances.

The shortcomings of discrimination scores were mentioned and various problems exist in generalized distances. If two groups have unequal dispersions, the group with the greater dispersion will be underassigned because it occupies more discriminant space. Also, distance alone allows no provision for a priori probabilities of group membership. Therefore, distance per se is reliable only when all groups have the same dispersion and all priors are equal.

A more stable classification rule is to accept the hypothesis (H_j) of individual r belonging to group j (Pr_{jr}) if the probability of belonging to group j is greater than that for belonging to any of the $i \neq j$ groups (Pr_{ir})

H_j if $Pr_{jr} > Pr_{ir}$, where

$$Pr_{jr} = \frac{P_{jr}}{\sum P_{ir}} \text{ and}$$

$$P_{jr} = q_j \left[\frac{N_j}{(N_j+1)} \right]^{1/2P} \frac{\Gamma 1/2(N-g+1)}{\Gamma 1/2(N-p-g+1)} \left[1 + \frac{N_j{}^2 D^2}{(N_j+1)(N-g)} \right]^{-1/2(N-g+1)}$$

In the above formula q_j is an element of a vector of prior probabilities, N_j is the sample size for the jth group, p is the number of variables, D^2 is the square of the generalized distance and still a classification chi-squared (but not to denote classification probabilities), and P_{jr} is the multivariate normal density function of individual r in group j. Fortunately, the center ratio of gamma functions is a constant for all P_{jr} and need not be calculated.

A further innovation is possible. If one is concerned about group dispersions being unequal, the vector of priors can take this into account. The element q_j can be based upon the determinant of the jth group,

$$q_j = \mathbf{S}_j^{-2} / \sum \mathbf{S}_j^{-2}.$$

Since groups with larger dispersions tend to be underassigned, the priors based upon the proportion of the sum of all generalized variances might be helpful.

Interpretation of Group Relationships. Although dismissed as methods of classification, generalized distance, group discriminant vectors, and canonical analysis of discriminance are critical devices for morphometric interpretation of relationships. How they are used is determined by the complexity of discriminant space. In the simplest case, all aspects of discriminant space can be evaluated by generalized distance and group discriminant vectors. In the most complex cases, only canonical analysis of discriminance might be applicable. The complex case is considered in the next section. The methodology here is primarily that of Blackith (see cited works).

Generalized Distance Analysis. The generalized distance among groups is a quantitative measure of the degree of resemblance among groups; a scalar contrast of form. Distance is easiest to analyze when a constructed 3-dimensional model approximates discriminant space distances between groups. The model can be constructed of any appropriate material, rods cut to lengths indicating all possible distances between groups (any arbitrary scale can be used for a unit of distance) and clay or styrofoam balls to indicate centroids. Using the completed distance model, every effort should be made to locate arbitrary axes that define biological relationships, e.g., axes of sexual dimorphism, taxonomic differences, geographic trends, ecological relationships, altitudinal gradients, etc. might be discovered. For example, a distance model of members of a genus might disclose axes of species, of sexes, and of ecological differences. The latter axis might disclose phenotypic plasticity or genetic differences. In either event, the axis would relate morphometric trends to ecological differences. When models cannot be constructed in 3-space, the value of the distances is related to their approximation of a 3-space fit. The approximation is a function of the complexity of the discriminant space. However, when distance analysis cannot summarize relationships, canonical analysis of discriminance serves this purpose.

Group Discrimination Vector Analysis. The discriminant function for each group provides a synergistic relationship of form that maximizes the distance, in standard deviation units, of a group from all other groups. Maximizing distances might appear to be another example of data grinding; but it seems logical to use a synergistic relationship, assuming that the whole is greater than the sum of its parts. Otherwise, one must assume that organisms are no more than an amalgamation of independent parts. However, the entire discriminant function is not proposed for studying relationships; the group discrimination vectors are more useful.

Each discrimination vector, since it expresses a centroid conditioned by variable covariances, provides a synergistic measure of a group's form. Actually, even ℓ rotates the corresponding centroid from Euclidean space to discriminant space. In more general terms, ℓ rotates raw variable axes such that the cosine of the angle between any pair of variable axes is equal to their correlation coefficient. The angle between the kth raw variable axis and the ith group discrimination vector is derived from

$$\text{cosine } \theta_{ik} = \ell_{ik}/\sqrt{\sum \ell_{ik}^2}$$

The angles allow evaluation of the participation of each variable in the synergistic definition of each group; however, the synergism again prevents any precise interpretation of importance of the contribution of each variable.

For morphometric comparison between groups, two features of ℓ are helpful. First, the magnitude of a particular measurement is implied by the magnitude of the corresponding coefficient of ℓ for each group. Unfortunately, such direct quantitative appraisal of size differences among groups is possible only if group dispersions approximate equality. In any event, this feature cannot be extended to quantitative comparison of total shapes—the synergistic interplay of variables prevents this. Second, since group dispersions usually are unequal, ℓ can be the basis for contrasting the relative response of each group to shape when dispersions are unequal,

$$E_i = \sum \ell_i^2$$

indicates how each group's dispersion corresponds to the pooled dispersion.

Qualitative comparison between the shape of groups often is more rewarding. This is done by evaluating the cosine of the angle between the ith and j^{th} groups' discrimination vectors,

$$\text{cosine } \theta_{ij} = \sum \ell_i \ell_j / \sqrt{E_i E_j}.$$

The greater the angular departure the greater the difference in shape changes in the two groups. This is the case because θ evaluates the equality of space orientation of the group dispersions, when dispersions are equal, $\theta = 0°$. (An important side issue enters here. When angles between vectors are similar, even if dispersions are unequal, canonical vectors are unlikely to be warped and canonical analysis of discriminance will clearly summarize most of intergroup differences in a few vectors.) Application of angular comparisons is facilitated by a diagram of angular differences between group vectors. When a 2-dimensional figure can be constructed with little error, (see Figure 10.4) the diagram is termed coplanar and is most useful. Such minor error often is the consequence of unmeasured differences between groups. On the other hand, in many cases groups conflict too much to allow a simple summary. Then, clusters of coplanar groups might be defined. When angular discrepancies are so great that clustering is arbitrary, the angular orientation of centroids in discriminant space is too complex to portray by this method. Any such discriminant space is termed hypermultivariate. A truly hypermultivariate discriminant space cannot be characterized in detail.

The distinction between quantitative and qualitative differences should now be clear. Quantitative differences are distances between centroids. Qualitative differences are angular departures in the orientation of group centroids in discriminant space and reflect differences in group dispersions.

The group discrimination vectors applied here are those of Hope (1969) and not those of Blackith (1960). Blackith's work pertained to a hierarchy involving natural pairs within the total number of groups examined. For each of the natural pairs, a single two-group canonical vector was calculated. For example, the sexes of two species of grasshoppers constituted four of the groups studied. For the sexes of each species, a single canonical vector was calculated. Finally, the angle between the two canonical vectors was computed and evaluated as discussed above. Therefore, usage here represents a generalization of Blackith's method to the case where no special relationship between pairs of groups is assumed. Naturally, when such dichotomies do occur within the groups, the original method of Balckith is preferred.

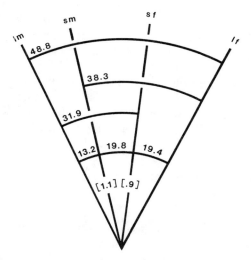

Figure 10.4 Angles between discrimination vectors for four groups of painted turtles (**lm**, large males; **sm**, small males; **sf**, small females; and **lf**, large females). Note the small angular departures (in brackets) across single vectors and that the angular discrepancy between large males and large females is 3.6°, an additional 1.6° greater than across a single vector.

Turtle Example. Since there are only two groups, a single dimension in discriminant 3-space will allow consideration of all relationships. Table 10.7 discloses that size and shape differences between turtle sexes lead to group centroids being a distance of 2.358 standard deviations apart in discriminant space. Therefore, one might suppose that about 12% of each group would overlap the other group, in classification about six misses. Although there were only five misses, we can review and expand previous comments on how few there were. First, we can eliminate two, now familiar, females (#11 and #12) that were classified as males as being outliers. The three remaining females which classified as males, since they were small, are likely misses. Therefore, classification actually is better than might be expected. In reference to theoretical restrictions, this is surprising because the two group dispersions are unequal.

Turning to discrimination vectors (Table 10.7) the coefficients would imply that males are wider and longer than females because both male coefficients are large in a positive sense. Although this is true if males and females of like sizes are compared, it is not true in an absolute sense because females get larger. This latter situation also is indicated by the analysis. The sum of the squares of female coefficients is greater than that of male coefficients, so females have a greater response to shape than do males. The presence of such a difference in response to shape is consistent with an appreciable angular departure, 19.5°, between the discrimination vectors. In essence this is a qualitative sexual differentiation which is further emphasized by comparing corresponding coefficients of the two discrimination vectors. The vectors clearly imply that males of a given length and width are distinctly lower in height than females of comparable length and width.

In spite of the 19.5° between the sexes both of their vectors associate well with the second, or "turtleness," components of dispersion and both groups (Table 10.8). This might seem to conflict with the single canonical axis of canonical analysis of discriminance being related to the third components. Rather than being confusing, hopefully this again emphasizes the differences between canonical axes and discrimination vectors. Canonical axes provide quantitative, not qualitative, differences between groups. Therefore, it should not be surprising that the canonical axis is 86.7° from the female and −73.7° from the male discrimination vectors.

Once more we can reevaluate the turtle components. The third component can now be called a component of quantitative sexual dimorphism. It is quantitative in that it creates a dichotomy corresponding to distance between the sexes. On the other hand, the second component can be termed a component of qualitative sexual dimorphism. It presents a unique pattern of variation that can be said to differentiate between the two sexes when individuals are of the same size, but it creates no true dichotomy for all individuals.

Parenthetically, do not expect discriminant analysis of your own data to lead to clear cut heuristics—it generally does not happen.

A Four Group Example. For simplicity, the male and female painted turtle groups will each be subdivided (arbitrarily and granted invalidly) into two groups, the smaller 12 and larger 12 individuals. This is done because it allows continuation from most of the previous discussions on turtles. For example, the four group multigroup principal component analysis differs little from before. The three components naturally are much the same, but those for both groups of males and for small females form a slight dichotomy with large females. Another result is that the small female second and third components are inverted in reference to the other groups. For small females, the second component is the third of other groups and the third is the second.

Table 10.7 Generalized distance, discrimination vectors, normalized discrimination vectors, and angle for male and female painted turtles

	females	males
distance	2.358	
vector		
length	−1.445	−1.171
width	3.139	3.236
height	.122	−1.045
normalized		
length	−.418	−.326
width	.908	.900
height	.035	−.291
angle	19.5°	

Table 10.8 Angles between discrimination vectors and components of male and female painted turtles

components		discrimination vectors	
		females	males
dispersion	1	83.3	84.6
	2	−7.9	−24.3
	3	−85.9	−66.4
females	1	83.0	84.7
	2	−9.3	−26.2
	3	−83.9	−64.4
males	1	84.1	84.1
	2	−7.9	−15.5
	3	84.8	−75.7

In the manova for the four groups, the tests of equality of dispersions and centroids are both significant. Classification of individuals (Table 10.9) discloses 40 hits (83%) and 8 misses (17%) but note that small females miss only to males, large males miss only to small females, small males miss to large males, and large females have no misses. This is consistent with the premise that the extremes of sexual dimorphism are more the consequence of female development rather than male.

Table 10.10 contains the group centroids and Table 10.11, the generalized distances between groups. Note the similarity in size and proportions of small females and large males. From a quantitative point of view, note that the distance between male groups is hardly closer than the distance between large males and small females. In a distance model (not shown, but the reader is encouraged to construct one), the male groups and small females approximate an equilateral triangle in 2-space with the large females being in a third dimension by themselves. Naturally one can orient the model to recognize male vs. female and small vs. large individuals axes, but any orientation of the model consistently emphasizes the uniqueness of the large females.

The group discrimination vectors are presented in Table 10.12 and the angles between the vectors in Figure 10.4. Again, they are closely allied to the second dispersion component. The angles disclose that, qualitatively, individuals of the same sex are more alike than like members of the other sex. Also, the greatest qualitative difference exists between large males and large females. Of course this is not surprising since in nature larger individuals are easier to diagnose as to sex.

Table 10.9 Classification of 12 individuals from each of four groups of painted turtles

actual group	group classified into			
	small females	large females	small males	large males
small females	7	0	3	2
large females	0	12	0	0
small males	0	0	10	2
large males	1	0	0	11

Table 10.10 Centroids for four groups of painted turtles

	small females	large females	small males	large males
length	119.417	152.583	103.750	123.000
width	93.083	112.083	82.750	93.833
height	45.667	58.250	38.083	43.333

Table 10.11 Generalized distances between four groups of painted turtles

	small females	large females	small males	large males
small females	0.0	3.433	2.045	2.210
large females	3.433	0.0	5.308	4.242
small males	2.045	5.308	0.0	2.033
large males	2.210	4.242	2.033	0.0

Table 10.12 Discrimination vectors for four groups of painted turtles

	small females	large females	small males	large males
length	-0.595	-0.259	-0.509	-0.133
width	2.444	2.034	2.560	2.505
height	0.465	1.134	-0.421	-0.898

10.6 Multigroup Canonical Analysis of Discriminance

We return to familiar methods. The multigroup models are the same as the two group models, but now the models deal with matrices of two or more canonical vectors rather than with single vectors **v, a, b, c,** or **u.** For the normalized model

$$y_V = \mathbf{V}'\mathbf{x} \text{ where } \mathbf{V}_i'\mathbf{V}_i = 1 \text{ and } s_{y_{v_i}}^2 = \mathbf{V}_i'\mathbf{S}^2\mathbf{V}_i.$$

For the standardized for within model

$$\mathbf{y}_A = \mathbf{A}'\mathbf{x} \text{ where } s_{y_{a_i}}^2 = \mathbf{a}_i'\mathbf{S}^2\mathbf{a}_i = 1 \text{ and } \mathbf{a}_i = s_{y_{v_i}}^{-1} \mathbf{V}_i$$

so the ith eigenvector of **A** is obtained by multiplying each element by the reciprocal of the standard deviation for $\mathbf{y}_v$. For the standardized for total model,

$$\mathbf{y}_B = \mathbf{B}'\mathbf{x}$$

where $s_{y_{b_i}}^2 = \mathbf{b}_i'\mathbf{S}_T^2\mathbf{b}_i = 1$ and $\mathbf{b}_i = s_{y_{T_i}}\mathbf{V}_i$.

For the canonical factor for within model,

$$\mathbf{y}_C = \mathbf{C}'\mathbf{z}$$

where $s_{y_{c_i}}^2 = \mathbf{C}_i'\mathbf{R}\mathbf{C}_i = 1$ and $\mathbf{C}_i = \mathbf{S}_{diag}\mathbf{a}_i$,

and for the canonical factor for total model,

$$\mathbf{y}_U = \mathbf{U}'\mathbf{z}$$

where $s_{y_{u_i}}^2 = \mathbf{u}_i'\mathbf{R}_T\mathbf{u}_i = 1$ and $\mathbf{u}_i = \mathbf{S}_{Tdiag}\mathbf{b}_i$

The basic model

$$\mathbf{y}_W = \mathbf{W}'\mathbf{d}$$

is not too useful here. In fact since both standardized for total models are of little use in morphomet-rics, they also shall be ignored in further discussion.

The canonical vectors do more than maximize distance and portray the differences among groups in as few canonical axes as possible. The axes accomplish this by maximizing relationships among variables. For example, one direction of an axis involves increases in certain variables and decreases in others.

In this section three topics are explored. First is the transformation from Euclidean to discrimi-nant space and definition of the latter by canonical axes. Although transformation from Euclidean to discriminant space cannot be diagrammed simply, some appreciation of what happens comes from considering, somewhat simplistically, the standardized for within models. The second topic, results of canonical analysis, augments previous discussions of the nature of the analysis. Finally, computa-tions briefly summarize the eigenvalue-eigenvector solution.

Euclidean to Discriminant Space.

In the case of two variables and many groups, the transfor-mation is from a 2-dimensional Euclidean space to a 2-dimensional discriminant space. Also, within the discriminant space two canonical axes can be placed to summarize the complete relationships between groups.

We can visualize the transformation as follows:

Imagine a rubber sheet stretched in a very special way. On the stretched sheet are plotted all individuals on the 2-variable axes. Since the samples are from multivariate normal distributions hav-ing equal dispersions, each sample can be summarized by an ellipse indicating the group dispersion. Since group dispersions are equal, each ellipse is of the same size, shape, and orientation as all others. This is Euclidean space.

What happens in the transformation to discriminant space? For a somewhat simplistic interpreta-tion, imagine that the rubber sheet is allowed to contract and that three things happen simultaneously. First, the 2-variable axes rotate until the cosine of the angle between them is equal to the within group correlation between the two variables and the ellipses become circles of the same size and reflect the like orientation of the original ellipses. Second, the rotation of original axes changes the positions of groups and individuals but groups and individuals maintain their relative positions to one another. Finally, there is an overall reduction in dimensions, a contraction, from Euclidean to discriminant space. All distances in the new space are generalized distances.

This transformation creates discriminant space, a necessary and important step prior to develop-ing the canonical axes that best summarize discrimination between groups. Then, the first canonical axis can be defined as the single axis that emphasizes the maximum possible differences between groups. Since there can be only one more canonical axis because the axes must be orthogonal, the first axis defines the second. Once the canonical axes are defined, the original variable axes can be located in canonical space, the discriminant space now augmented by canonical axes. Location of original variable axes allows applying variable relationships in the canonical space to examine group relation-ships.

Since the transformation to discriminant space alters group ellipses to circles, in canonical space defined by **a** vectors, a circle of unit radius about each group centroid approximates each group's disper-sion because $S_{y_a}^2 = 1$. However, even if population dispersions are equal, the "circles" will neither have exactly unit radii nor be exactly equal because the ellipses, since they were from samples, were neither perfect ellipses nor equal. In fact, one could calculate and graph the nature of the "circles"—they would tend toward ellipses. In spite of this, the average of the squared radii of the circles would be unity.

The distance between the centers of any two circles, the centroids $\mathbf{y}_{a_i}$ and $\mathbf{y}_{a_j}$, is the generalized distance between the ith and j^{th} group,

$$D_{ij} = |\bar{\mathbf{y}}_{a_i} - \bar{\mathbf{y}}_{a_j}|.$$

A final note on the transformation. If contraction of the rubber sheet had not rotated the original variable axes, the transformation would have been to z-scores. This could occur only if all the correlation coefficients between the original variables were zero.

Standardized for Within Model.
Since graphic representation for discrimination of absolute differences among variables uses the standardized for within model, consider the model and the simple case where only two canonical axes exist,

$$\mathbf{y}_{a_1} = \mathbf{a}_1'\mathbf{x} \text{ and } \mathbf{y}_{a_2} = \mathbf{a}_2'\mathbf{x}$$

for three groups and three variables. Figure 10.5 portrays each individual as a point in 3-space, Euclidean space, defined by 3-axes, X_1, X_2 and X_3. Each group centroid also is a single point in the 3-space, and is the center of gravity of the sample of points (individuals) in its group. Assuming the multivariate normal distribution, each group could be represented by an ellipsoid. Assuming equality of dispersions, all ellipsoids are of the same size, shape and orientation.

After a significant manova, it becomes convenient to summarize differences on two canonical axes, the maximum that exist in the discriminant 3-space. Plots of individuals on the canonical axes and the vectors per se allow a deeper appreciation of the relationships of the individuals and the groups in the study.

Results of Canonical Analysis.
Canonical analysis provides the best reduced rank, or dimension, model to simply but effectively indicate measured differences among groups. Generally each successive canonical axis portrays less between group differences than the former and the first two or three axes summarize enough among group differences so the reamining axes can be ignored. The number of "important" canonical axes is clear from the amount of between group variation summarized and the structural correlations between variables and axes.

The groups and individuals can be mapped in canonical space. Mapping and ordination may be in reference to two axes at a time, usually 1st vs. 2nd, 1st vs. 3rd, and 2nd vs. 3rd; but one can map according to three axes. This also allows heuristic evaluation of the results.

The number of dimensions in discriminant space is equal to the number of variables. However, the number of canonical axes, r, depends upon the number of groups, g, and variables, p

$$r = min(g-1,p).$$

Consider $g-1$, the centroids of two groups define a line, of three a plane, and of four a 3-space.

Since successive canonical axes are based upon maximum between group differences, one can appreciate why the first two or three axes often summarize the major among group differences. Often meaningfulness of axes is lost long before their lack of significance. However, when discriminant

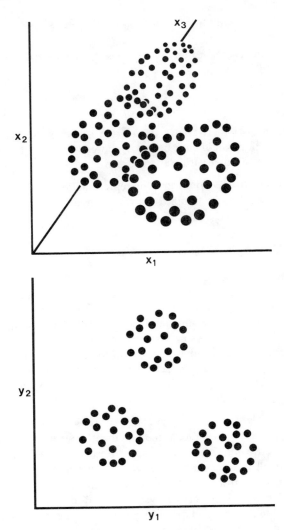

Figure 10.5 Transformation of three groups having three variables from Euclidean to canonical space. Above, three groups in 3-dimensional Euclidean space; below, three groups in 2-dimensional canonical space.

space is hypermultivariate, three axes provide an inadequate proportion of the total discrimination among groups.

Computations.   The method of multidimensional canonical analysis of discriminance, being canonical, involves eigenvalues and eigenvectors. Also as one might expect the end result is the normalized model, the normalized eigenvectors providing v_i rather than any of the scaled vectors. Also as one might expect the characteristic equation and eigenvector function select successive canonical axes that portray maximum possible differences among the groups as limited by orthogonal axes. However, the scores for individuals on different axes, canonical variates, are correlated.

For the computations, differences between groups are expressed in the form of the inverse of the within *SSCP* matrix, $\mathbf{W}$, times the between, or among, *SSCP* matrix, $\mathbf{B}$,

$$\mathbf{W}^{-1}\mathbf{B}.$$

Again, the fundamental or normalized model for the linear canonical function is

$$\mathbf{y}_V = \mathbf{V}'\mathbf{x},$$

which has within group *SSCP* of canonical variates $\mathbf{y}_V$ of

$$\mathbf{W}_y = \mathbf{V}'\mathbf{W}\mathbf{V}$$

and among group *SSCP* of $\mathbf{y}_V$ of

$$\mathbf{B}_y = \mathbf{V}'\mathbf{B}\mathbf{V}.$$

The aim of canonical analysis is reached by solving

$$\lambda_i = \left.\frac{\mathbf{v}_i'\mathbf{B}\mathbf{v}_i}{\mathbf{v}_i'\mathbf{W}\mathbf{v}_i}\right| \text{maximum}$$

by the characteristic equation,

$$\mathbf{W}^{-1}\mathbf{B} - \lambda_i\mathbf{I} = \mathbf{0}.$$

The canonical vectors are solved by the eigenvector function,

$$(\mathbf{W}^{-1}\mathbf{B} - \lambda_i\mathbf{I})\mathbf{v}_i = \mathbf{0},$$

subject to $\mathbf{v}_i'\mathbf{v}_i = \mathbf{1}$. The method leaves shapes to be contrasted and often the contrasted shapes can be associated with independent patterns of variation.

The other models can be derived in the manner indicated previously.

10.7 Interpretation of Canonical Analysis of Discriminance

The interpretation of canonical vectors is much different than that of component vectors. This is emphasized by the fact that, although any $\mathbf{v}_i'\mathbf{v}_i = \mathbf{1}$, any $\mathbf{v}_i'\mathbf{v}_j \neq \mathbf{0}$. Note also that no $\mathbf{v}_i'\mathbf{v}_j$ formed from Table 10.16 would be equal to zero. This does not mean that canonical axes are oblique axes. Rather, it is the consequence of the rotation of the original variable axes. Actually, the canonical axes are orthogonal; and like the original variables, canonical variates are correlated. Any confusion on this point might be alleviated by realizing that the characteristic equation operates on $\mathbf{W}^{-1}\mathbf{B}$, which can be considered to define between group differences in discriminant space. Also, recall the nature of the transformation of Euclidean space original variable axes to oblique axes (the cosine of whose angles are correlation coefficients) in discriminant space. Therefore, the nature of original variable axes in discriminant space precludes canonical variates being uncorrelated. Naturally, the purpose of discriminant analysis is to maximize variable relationships to maximize differences between different areas of discriminant space. If groups are separated, differences between groups are maximized.

In the sense that component scores are uncorrelated and said to represent independent patterns of variation, canonical variates are correlated and are said to represent contrasting patterns of variation. For example, plots of two canonical variates pertain to two canonical axes whose units of distance are in terms of contrasting form. This concept of form changing in all directions of discriminant space is the basis for calling discriminant space a synergistic space.

Coefficients of the canonical vectors disclose the nature of the synergistic relationship among the original variables that discriminates between the groups. These relationships are analogous to those of principal component coefficients. If all canonical coefficients are of the same sign, an unlikely situation of a general canonical vector, they indicate a "size" relationship. If of unlike sign, a bipolar canonical vector, the coefficients pertain more to shape. Also, the magnitude and sign of coefficients of the latter cannot be assumed to imply the independent importance of variables in discrimination. This naturally follows from coefficients representing a synergistic relationship. However, a variety of statistics help to judge the importance of vectors and give some idea of the contribution of variables to group differences. Again, nothing is absolute. As was the case with component analysis, the best criterion for any decision is that it makes good biological sense.

Certain of these interpretive devices already were discussed for the two group case. The models and their resultant canonical vectors pertain here. Also useful are canonical variates, angles between variables and canonical axes, and angles between canonical and component axes.

In interpretation the standardized for within model is emphasized; however, each statistic is easily modified to apply to other models.

Contribution to Distance Between Groups.   The distance between groups k and ℓ based upon the jth canonical variate alone is

$$D^*_{jk\ell}{}^2 = d_{k\ell}{}' a_j,$$

the percentage contribution of the jth canonical variate to the total distance between groups k and ℓ is

$$100(\mathbf{d}_{k\ell}{}' \mathbf{a}_j / D_{k\ell}{}^2),$$

and the cumulative distance between groups as r successive canonical variates are added is the sum of the r distances.

Canonical Structure.   The structure or correlation between original variables and the jth canonical variate is

$$\mathbf{s}^*_j = \mathbf{R}_T \mathbf{u}_j$$

where $\mathbf{R}_T$ is the total correlation matrix and $\mathbf{U}$ is a $p \times r$ matrix of total factor canonical vectors $\mathbf{u}_j$. Therefore, $\mathbf{S}^*$ is a $p \times r$ matrix of correlations between p variables and r canonical variates.

Structure can indicate highly correlated, relatively invariant, and irrelevant original variables. Highly correlated original variables will act as a single variable—they will act in like manner from axis to axis. In fact, each truly redundant variable (recall linear dependence) will reduce discriminant space by one dimension. Invariant variables will act the same way from axis to axis. Irrelevant variables will have near zero canonical vector coefficients and structure.

Variable Percentage Contribution.   Although the synergistic actions of variables must always be recognized, the percentage contribution of each original variable to each canonical variate does imply the relative role of each original variable to each canonical variate. Again this is measured by the percentage of each original variable's variance in each canonical variate,

$$100(\mathbf{v}_{ij}{}^2\lambda_i)/\sum \mathbf{v}_{ij}{}^2\lambda_i,$$

for the jth coefficient of the ith canonical vector.

Canonical Correlation.   The canonical correlation between the jth canonical function and the groups is

$$R_{c_j} = \sqrt{\frac{\lambda_j}{1 + \lambda_j}}\,.$$

To obtain this correlation, groups are coded as a set of binary dummy variables, unity for the particular group and zeros for all other groups. The correlation follows from the fact that canonical analysis of discriminance can be a special case of canonical correlation between binary dummy group and measurement variables.

As for canonical correlation analysis the square of the canonical correlation coefficient measure overlap, but here between canonical variates and hypothetical dummy variables pertaining to groups. For each group the dummy variables define a unit vector, the value of unity being for the group of the vector and the zeros for other groups. Therefore, a canonical correlation coefficient pertains to the overall goodness of fit of canonical variates to their group, but again is subject to the problems of any correlation coefficient.

Communalities for Each Variable.   When the number of canonical axes equals the number of variables p, the total variance of each variable is accounted for; however, whenever the axes equal one less than the number of groups, the total variances for variables need not be extracted. A low communality implies that the variable contributes little to discrimination. The communality of the ith variable for all j axes is

$$K_i = \sum \mathbf{s}^*{}_{ij}{}^2.$$

Trace of R.   The percentage of the trace of the total correlation matrix sums to 100 percent only when the axes equal p. When axes equal $g - 1$ the percentage of the trace of $\mathbf{R}_T(=p)$ for the jth axis is

$$T_j = 100\ \mathbf{s}_j^{*\prime}\mathbf{s}_j^{*}/\mathbf{tr}(\mathbf{R}_T).$$

When the 100 is removed, the above formula corresponds exactly to a portion of the formula for redundancy in canonical correlation analysis, the proportion of the variance of one set extracted by the other set. In the present case, the percentage of the trace of $\mathbf{R}_T$ amounts to 100 times the proportion of the variance between groups. Naturally, either the percentage or proportion provides an evaluation of total discrimination by all axes.

Redundancy.   The canonical correlation approach leads to redundancy in reference to each and all canonical axes. In contrast to canonical correlation, however, only one redundancy is pertinent, the measure of overlap between data vectors and the vectors of dummy variables for groups,

$$R_d = \mathbf{s}_j^{*\prime}\mathbf{s}_j^{*}R_j/p.$$

Therefore, it provides an overall evaluation of the uniqueness of groups.

Canonical Graphs.   Plots for individuals and groups are most useful. However, rough appraisal of differences and similarities among groups might be better appreciated if confidence circles are applied to each sample centroid to estimate each population centroid (Seal, 1969). The radius of the confidence circle of the ith group is

$$t_{(1-\alpha, N_i-1)}/\sqrt{N_i},$$

where t is the tabulated value of Student's t, α is the magnitude of the Type **I** error so $1-\alpha$ is the confidence coefficient, N_i is the sample size of the ith group, and N_i-1 is the degrees of freedom for entering a table of Student's t.

Confidence circles are crude approximations of differences that might exist between groups. Confidence ellipses are based upon individual group dispersions, are preferred statistically, and announce fewer significant results than do confidence circles. Since confidence ellipses become simple to use only when their computation is interfaced with graphics and/or plotters, and would alter conclusions infrequently, they are not presented here.

Also useful on canonical graphs are the vectors of variables in canonical space. The origin for all vectors is at the grand centroid, the zero coordinates of the graph. For example, consider a two-dimensional canonical graph of the ith and j^{th} axes. Using an arbitrary scale, corresponding i^{th} and j^{th} axis coefficients of either **a** or **c** for each variable can be plotted as points about the origin. Then, vectors can be drawn from the origin to each point and the vectors labeled as to variables. Usually an arbitrary scale for plotting can be found so the vectors can be placed at the origin, the grand centroid. Of course the original arbitrary coordinate axes for plotting vectors must be placed parallel to the corresponding canonical axes.

Again the vectors recommended for plotting pertain to standardized for within models. Since

$$\mathbf{y}_A = \mathbf{A}'\mathbf{x} = \mathbf{y}_C = \mathbf{C}'\mathbf{z},$$

a plot of $\mathbf{y}_A$ on the ith and j^{th} axes corresponds exactly to a plot of $\mathbf{y}_C$ for the same axes. Therefore, either one copy of a plot can contain **A** vectors and another copy **C** vectors, or both sets of vectors can be shown on a single graph. Then, the manner which absolute variation $(\mathbf{a}_i,\mathbf{a}_j)$ and standardized variation $(\mathbf{c}_i,\mathbf{c}_j)$ differ or agree in discrimination can be appraised.

Vectors of variables may or may not imply the direction in the magnitude of a particular variable's increase. For example, a group centroid might be at a point that is a direct extension of a single variable's vector because the centroid has a large value for that variable. However, the centroid can be at the same point because it has an extremely low value for another variable whose vector is extending in the opposite direction. More specifically, the placement of any group centroid is the result of the complex synergistic relationship among variables. Often the vectors of variables barely indicate the synergism that places a centroid where it is.

The relationships between the two standardized for within models is of special significance in reference to their canonical graphs. Since both models produce identical canonical variates for each individual, both canonical graphs, except for vectors of variables, are identical. In both models, the ith canonical axis passes exactly through the same path in the same discriminant space, the units of canonical axes are identical, the dispersion of individuals and centroids is the same, and confidence circles for centroids correspond exactly. Also, as one might expect, the angles between variables in both sets of vectors of variables are identical and there is no angular difference in a single variable between sets, i.e., the two sets of vectors of variables superimpose precisely. However, there is a difference between the two sets of vectors of variables, the only difference between the two canonical graphs. For any single variable, the lengths of the two vectors, one on each graph, are not likely to be the same. This stems from vectors of variables being plots of canonical vectors, a_i or c_i. Recall that the jth coefficient of c_i, c_{ij}, is the product of a_{ij} and the jth within group standard deviation. Since no within group standard deviation is likely to be equal to unity, no a_{ij} and c_{ij} are likely to be equal. Also, considering possible differences in magnitudes of standard deviations, proportional lengths of vectors for individual variables might change, i.e., a variable's length might appear quite long within one set and short in the other. All this means that a single canonical graph can be used to summarize both absolute and relative variation. The only difference between the two models could be shown on a single set of vectors of variables containing one vector for each variable—on each vector of a variable both original and z-score variation lengths could be indicated.

Four Group Example, Continued. Table 10.13 discloses that only the first two of the three canonical axes provide significant discrimination and summarize 99.8% of the differences between groups. The canonical structure (Table 10.14) plus redundancy and related statistics (Table 10.15) stress the major importance of the first canonical axis and its 86.43% of the total discrimination possible. Since the second axis also provides significant discrimination, it also will be considered. The other interpretive aids are not presented since their use has been shown elsewhere and they are not required here.

Note the normalized (Table 10.16), standardized for within for raw variables (Table 10.17), and standardized for within for z-scores (Table 10.18) canonical vectors. The first two canonical vectors both approximate the 3rd dispersion components, sexual dimorphism; but again the canonical vector refers to a contrasting pattern of variation. The canonical graph of group centroids on the first two canonical axes (Figure 10.6), the only significant ones, summarizes 99.8% of the differences between groups so not surprisingly portrays what was said about the generalized distances between groups. Also note that no confidence circles overlap so it can be assumed that all groups are significantly different from one another. In addition, the vector of variables disclose the difference between raw and z-score variables influence in canonical space. The z-score vectors were obtained by multiplying the raw variable coefficients by within standard deviations, length by 10.390, width by 7.067 and height by 3.920. Therefore, the marked increase in the z-score vector for length, especially in contrast to height, is not surprising. Note that the height vector is the longest for raw variables, but length is longest for z-scores, and that the vectors of variables support what was said about form differences in

Table 10.13 Chi-squared tests of significance of eigenvalues (canonical axes) and related statistics for four groups of painted turtles

λ removed	R_c	R_c^2	λ	χ^2	d.f.	$P(X^2)$	trace	% trace
0	0.899	0.807	4.194	93.825	9	0.0000	86.43	86.43
1	0.627	0.394	0.649	22.161	4	0.0002	13.37	99.80
2	0.097	0.009	0.009	0.409	1	0.5234	0.19	100.00

Table 10.14 Canonical structure, correlations between canonical variates and variables, for four groups of painted turtles

	1	2	3
length	0.946	−0.306	0.109
width	0.921	−0.241	0.305
height	0.992	−0.045	0.117

Table 10.15 Percent of trace of R, proportion of variance extracted and redundancy for canonical variates for four groups of painted turtles

variate	trace	variance extracted	redundancy
1	90.906	.909	.817
2	5.126	.051	.032
3	3.968	.040	.004
Total	100.000	1.000	.853

Table 10.16 Normalized canonical vectors for four groups of painted turtles

	1	2	3
length	0.084	−0.391	−0.407
width	−0.291	−0.008	0.871
height	0.953	0.920	−0.274

Table 10.17 Standardized canonical vectors for within for raw variables for four groups of painted turtles

	1	2	3
length	0.030	−0.228	−0.181
width	−0.104	−0.005	0.386
height	0.340	0.537	−0.122

Table 10.18 Standardized canonical vectors for within for z-scores for four groups of painted turtles

	1	2	3
length	.312	−2.369	−1.881
width	−.735	−.035	2.728
height	1.333	2.106	.478

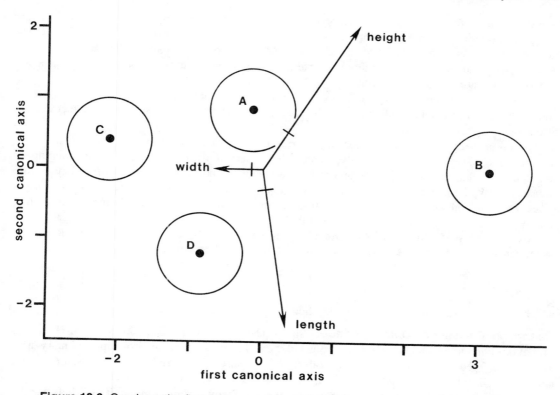

Figure 10.6 Graph on the first and second canonical axes of the centroids for four groups of painted turtles (**A**, small females; **B**, large females; **C**, small males; and **D**, large males). Large dots indicate group centroids, circles around each dot are confidence circles for population centroids, and the three vectors are vectors of variables as labeled. The length of vectors in terms of raw variables are indicated as cross bars and in terms of z-scores as the points of the arrows terminating each vector.

females and males. A distinct male-female axis is apparent. Note that even small females are distinct from large males. Also, an axis of age, young-old, is recognizable. Finally, the uniqueness of older females is quite apparent. As should be the case, canonical analysis of discriminance provides an excellent summary of all said previously.

10.8 Reification

This section reflects closely the cited works of Blackith. For heuristics, angular comparisons among vectors of all types might unravel relationships. All possible contrasts of group components, dispersion components, group discriminant vectors and canonical vectors can unite patterns of variation and bases for discrimination. Again, relationships may be so complex as to prevent this methodology from being very useful. A further possibility is the use of theoretical vectors. These can be derived by the biologist proposing a vector to test some preconceived concept. Also, the theoretical vectors can come from canonical correlation analysis or predicted components. For example, analysis might disclose a meaningful high correlation between precipitation in set one and morphology in set two. Then, the morphological or predicted component vectors of set two can be reified with the various group component and discriminant analysis vectors. Angular comparisons would associate patterns of variation and groups with precipitation.

This procedure could be especially useful to study geographic variation. For example, a morphological vector associated with latitude, when reified with groups, could lead to morphometric rates of change over geographic distance.

When properly applied, reification might bring together all facets of a morphometric analysis. Therefore, it is a separate means of analysis and included here only for convenience. Angular comparison should emphasize discrimination and canonical vectors as points of reference since they are the statistics of discriminant space. Components should be contrasted with the reference, but components should be examined with one another by multigroup component analysis as well. Some further indication of possible reifications might prove helpful. First, it must be pointed out that for reification, all vectors must be normalized. All but group discrimination vectors already are normalized. The j coefficients of the ith group's discrimination vector are normalized by

$$\ell_{ij}^* = \ell_{ij}/\sqrt{\sum \ell_{ij}^2}$$

which causes the sum of squares of the coefficients of ℓ_i^* to be unity, $\ell_i^{*\prime}\ell_i^* = 1$.

Canonical Analysis of Discriminance.
In the usual case where generalized distance models cannot be constructed, canonical graphs are mandatory. Then, canonical axes or supplementary arbitrary axes defined on graphs might reflect collections of taxa; the sexes and/or age groups; altitudinal, latitudinal and/or longitudinal sampling; temporal and/or ecological units; etc. In some cases the group differences will correspond directly to one or more of these features. For example, if one finds that the two sexes of each species maintain the same general relationship on a canonical graph, sexual dimorphism could be assumed identical in the various species. On the other hand, the basis for separating groups might be more complex and appear as a disruptive pattern of like and unlike units. However, even the latter patterns can be unravelled if they are expected. In essence, one must seek and evaluate possible bases for group differences.

Although canonical vectors and vectors of variables provide some idea of the bases for discrimination, the synergistic relationships require other procedures for their identification. Since canonical vectors represent contrasts, the origin of these contrasts might be found in components. Upon reification, a canonical axis might represent a single component as was the case in the turtle example. However, a canonical axis also can be derived from two or more components, a condition implied by a fairly uniform canonical axis departure of about 45° from two or more components.

Reification of canonical and component axes is much less direct and often less detailed than comparing group discriminant vectors and components. This is true since discriminant vectors pertain to the entire discriminant space. In spite of this, even when discriminant space is extremely complex, reification of canonical and component axes can be helpful. Also, an attempt might be made to reify canonical and group discrimination vectors; however, the results generally are meaningless since the two vector types appraise qualitative versus quantitative differences.

Multigroup Component Analysis.   Interpretation of group components should stress degree of relevance rather than presence alone of patterns of variation. It is entirely possible that a particular shape contributes major variation to one group and minor variation to another. The percentage of variance in components determines this.

Since first components for all groups likely represent growth, they might be quite similar. Therefore, other group components are more likely to diverge and thus indicate morphometric differences between groups. If any group has a component approximating 90° from all components of all other groups, this implies a unique pattern of variation. However, like vectors need not result from identical patterns of variation per se. Rather they can come from identical physiological stimuli, perhaps hormonal, controlling the phenotype. In practice, this is unlikely to be far removed from what the biologist is seeking.

In comparing components among groups, one can judge angular equality on the basis of the test of equality of a theoretical and an observed vector. However, recall that the magnitude of elements of a given vector influences the test more than the angle between it and the reference vector. For this reason, one might pay more attention to angles between vectors alone. As a quick-and-dirty rule, one could consider angles less than 15° rather small and not be too impressed by angles as much as 20°.

The comparison of a group's discrimination vector with the same group's components might reference the former to a most closely approximated pattern of variation. This would show which component is emphasized and which others are subdued in discrimination. However, no such component-discrimination vector relationship might occur. Again, discrimination provides a contrast, one that might apply to more than one component.

Dispersion Component Analysis.   Dispersion components give absolute discriminating capability, but perhaps not actual discrimination of individual variables. The components pertain mostly to shared variation and so might not be involved in discrimination. In essence, the components are more likely to emphasize major variation shared by groups than one group's unique limited variation.

As in the case of individual group components, large samples are most meaningful. With large samples the test of sphericity is more likely to be rejected. Then, one can place reliance on many patterns of variation, both for group, and for dispersion components.

Group Discrimination Vectors.   When these vectors are not coplanar, clustering group vectors into coplanar subgroups can be very difficult. Reification with components might be helpful. If it is possible to associate each discrimination vector with a dispersion component, group vectors can be clustered on the basis of their dispersion component alliance. Naturally, the degree of coplanarity discovered in any study deserves comment. Rarely should this method provide no association whatsoever. Even when difficult, reification tends to be productive. I have seen this even in hypermultivariate distributions.

10.9 Critique of Discriminant Analysis

Like other statistical analyses mentioned so far the methods of discriminant analysis are very powerful statistically, are far from being objective, require assumptions about the nature of the data, and are based upon decisions on the part of the user.

Methodology.   When using the methodology, one presumes that the best way to compare groups is by overall form of individuals. On the other hand, form might be assumed to be a function of each variable independent of all others or of equal weighting of variables acting together. However, as was the case with components analysis, the approach to discriminant analysis appears consistent with biological thought. Biologists generally accept that the whole of a phenomenon is greater than the sum of its parts. For those not accepting this, when used in conjunction with component analysis, discriminant analysis should not be objectionable at all.

Variables.   All said previously about variables pertains here. Of primary importance in discriminant analysis, variables must cover the domain of form to be contrasted but must be linearly independent. Extraneous variables cannot be included and important ones must be included. For this reason the worker must have some idea of the phenomenon under study.

Assumptions and Transformations.   Conventional transformations of univariate data often do not satisfy most multivariate assumptions so it might be best to use no transformations. How then can we use discriminant analysis? There are three bases for justification. First is the apparent robustness of manova. Since a critical assumption is equality of group dispersions, recall that for two groups, if sample sizes are large and equal, inequality of dispersions has no real effect on manova Type **I** and **II** errors. Therefore, there is justification for performing a discriminant analysis. However, the influence of heteroscedasticity upon discrimination vectors, canonical axes and generalized distance is poorly known. In spite of this, studies indicate that reliance can be placed upon these statistics when used for morphometric interpretations. Their shortcomings appear to exist mainly in the precise classification of individuals to groups. This leads to the second justification. Classification is by Geisser classification probabilities which Monte Carlo studies indicate perform well. The third area of justification can come from reification of vectors, especially discrimination vectors and canonical axes, with other vectors not so subject to assumptions. For example, turtle discrimination vectors were reified with component axes and a correspondence was found. In essence, reification related the two kinds of variation, contrasting and independent.

Evidence consistently supports the robustness of the methodology. However, there is a device for checking on the influence of heteroscedasticity. In addition to a raw variable analysis, another analysis is made after transforming all original variables to common logarithms. Recall that this also acts to remove nonlinearity of variables. If the log analysis does not deny the raw data analysis, the raw data analysis generally is considered acceptable.

Still another check on methodology is possible. Especially in studies involving a few groups, successively the inverse of each group's dispersion matrix can be substituted for the inverse of the pooled within group dispersion matrix. The resulting g discriminant analyses can be compared for consistency of performance. If the g analyses disagree, the results should be fairly easy to interpret. Each analysis refers to generalized distances from the group of reference in standard deviation units of that group.

Canonical Vector Choice.   Aside from which model of canonical analysis of discriminance and which vectors are most useful for specific items of interpretation, one must determine how many vectors of a given model one should interpret. As was stated previously, this usually is much simpler than the analogous problem in components analysis.

Objectivity of Methodology.   Discriminant analysis like all multivariate analyses was shown to be based upon rigid assumptions about the nature of variation. Also, an attempt was made to verify the appropriateness of such approaches to examining variation. However, if one accepts methodology as being realistic, individual studies in a sense become objective. The investigator cannot influence the results once data are collected. The data determine the outcome of the analysis. It is on this basis that I consider morphometrics to be an objective procedure for studying biological phenomena.

Computer Programs for Morphometrics

Two statistical programs were developed for morphometrics. Both programs run on an IBM 360/40, G-level compiler, 200K available. There is a complex overlay structure in each program which normally would require about 550K. DISANAL performs all the operations discussed in reference to multigroup discriminant analysis, multigroup component analysis, and component analysis. CORANAL provides all computations indicated in the chapter on canonical correlation.

The documentation of the programs was written with two assumptions in mind: First, the reader is familiar with canned computer programs and all the pitfalls in making a data deck. Those unfamiliar with the real problems that can exist should first consult the introductory material in either the BIOMED (Dixon, 1973) or SPSS (Nie et al, 1975) computer programs. Further pertinent commentary is in Harris (1975). Second, the reader is familiar with the text. Commentary here is brief and restricted to the most critical items that validate procedures. Also, a few univariate statistics not really discussed in the text are included with only a brief indication of their application and a reference. Finally, an item not mentioned in the documentation is that the probability of obtaining the calculated value of a statistic in any test of a null hypothesis is provided.

DISANAL

Output (maximum possible):

Header including various items to verify correctness of data deck:
> *PROBLM card*—reproduction of input card #1.
> *Problem description*—reproduction of title cards.
> *Parameters*—number of groups, number of variables, etc.
> *Group labels*—six character alphanumeric codes or abbreviation for groups.
> *Variable labels*—six character alphanumeric codes or abbreviation for variables.
> *Sample sizes*—for each group, even if equal.
> *Variable format*—for data input, especially important since an incorrect format can negate the analysis.

Data including raw, transformed, z-score and/or rearranged sequence of variables—a useful check for correct input. An option calls for data previously stored on file 1.

Tests of normality for each variable—the grand centroid is added to the deviation of each data vector from its group centroid prior to output for each variable of:

Grand mean and 95% confidence interval

Total variance

Total standard deviation

> g_1 *statistic* (a measure of skewness) *and 95% confidence interval* (which should bracket zero in a normal distribution)—especially useful for ascertaining the need for transformation of a variable, a confidence interval greater than zero indicating need for a log transformation (see Sokal and Rohlf, 1969).

g_2 *statistic* (a measure of kurtosis) *and 95% confidence interval* (which should bracket zero in a normal distribution)—see Sokal and Rohlf, 1969.

Multivariate analysis of variance
  Basic statistics for each group
    Centroid
    Standard deviations
    SSCP matrix
    Group dispersion matrix
    Group dispersion determinant (computed from matrix inversion)—value of zero not critical here.
    Group correlation matrix
  Basic statistics for total
    Grand centroid
    Pooled standard deviations
    Total SSCP matrix
    Between groups SSCP matrix
    Within groups SSCP matrix
    Dispersion matrix
    Dispersion determinant—should be greater than zero; if ≤ 0, check principal component analysis of dispersion (below) for a more reliable value.
    Pooled correlation matrix
    Tests of equality of group dispersions (includes both an *F*-test and a chi-squared test, results can differ slightly)—since group dispersion determinants are used, check for their accuracy in the principal component analysis of each group (if called for).
    Univariate F-tests of equality of population means for each variable
    Multivariate F-test of equality of population centroids
    Check on inverse of dispersion—the dispersion times the inverse of dispersion must closely approximate an identity matrix to assure sufficient accuracy for discriminant analysis calculations. If this cannot be satisfied, the program may or may not stop here.
    Univariate Student-Newman-Keuls multiple Range tests—each variable is tested with a Type I error of 5%—see Sokal and Rohlf, 1969.

Multivariate analysis of covariance
  Output parallels that for the multivariate analysis of variance.
  Adjusted variables result—an option provides a data deck of adjusted variables.

Multigroup principal component analysis
  Principal component analysis of dispersion matrix
    Centroid and dispersion matrix
    Dispersion determinant (product of eigenvalues)—check for critical value greater than zero
    Eigenvectors
    Percentage of the variance of each variable's contribution to each component
    Angle of deviation and tests of equality of the first four components to an hypothetical isometric vector
    Component scores

Distance of individuals from origin in space of all components and percentage of each individual's variance accounted for by up to 14 components

Eigenvalue, percent trace, cumulative percent trace, degrees of freedom, chi-squared, and probabilities for Bartlett's and Anderson's sphericity tests (Anderson's is based upon large sample theory) for each component.

Principal component analysis for each group

Same output as for principal component analysis of dispersion

Graphs of dispersion and group component scores—1st vs. 2nd, 1st vs. 3rd, and 2nd vs. 3rd.

Geisser Classification

Header including prior probabilities

Classification of individuals

Number and percent hits and misses

Classification tables by numbers and percents

Generalized distance analysis

Generalized distances of individuals from group centroids

Group centroids

Euclidean distances between groups

Generalized distances between groups

Generalized distances between groups based upon each variable

Discrimination vector analysis

Discrimination vectors

Direction cosines of discrimination vectors

Percentage contribution of each direction cosine to each group

Percentage contribution of each direction cosine among groups

Angles between group discrimination vectors

Canonical analysis of discriminance

Canonical R, R^2, eigenvalue, chi-squared, degrees of freedom, probabilities of accepting equality of centroids, lambda, percent trace, and cumulative percent trace for each canonical variate

Normalized canonical vectors for raw data—V

Angles between canonical vectors and original variables

Percentage of the variance of each variable in each canonical variate

Standardized canonical vectors for within for raw data—A

Canonical factors for within—C (standardized canonical vectors for within for z-scores)

Canonical structure

Communalities for canonical vectors

Percent of trace of R, proportion of variance and redundancy for each canonical axis

Plots of canonical variates for all and each group—1st vs. 2nd, 1st vs. 3rd, and 2nd vs. 3rd.

Addenda to multigroup component analysis

Test of goodness of fit of group to dispersion components

Angles between group and dispersion components

Reification

Angles between discrimination, component and canonical vectors

Also includes theoretical vectors of input

Input (numbers refer to computer card columns)

1) *A Control Card*

1-6	PROBLM (Mandatory)
7-8	p, total number of variables input per case (= $m + n$ in ancova); maximum = 50
9-11	N_i, sample size if the same for all groups; otherwise 0 which calls for SAMSIZ card(s), card type 7.
12-15	N, total sample size (sum of all group sample sizes); maximum = 3000.
16-17	g, number of groups; maximum = 75; note: $g = 1$ calls for principal component analysis of a single group—in such cases columns 7-8, 9-11, 16-17, 18, 19, 20, 21-22, 28, 29, 31, 36, and 37 must be defined or are obvious options; also, TITLE, LABELS, GROUPS (for one group), and FINISH cards must be used.
18	*1* calls for transformation card type 5 which contains numeric codes for possible standard transformation of individual observations; *8* calls for data from tape file 1 rather than a data deck; *9* also calls for data from file 1 but adds transformation as per card type 5. Note: for values of 8 or 9 a blank variable format card (type 6) must be included.
19	*> 0* calls for z-score transformation of data after any other transformation; 2 calls for punched output of z-score data deck.
20	*1* calls for rearranging the order of the p variables according to the variable order card, type 9—can be used to segregate m and n ancova variables into successive subsets.
21-22	m, number of independent variables for ancova only; must be variables 1 through m.
23-24	n, number of dependent variables for ancova only; must be variables m + 1 through p.
25-26	number of title cards (type 2) to describe the problem; at least one must be used.
27	number of variable format cards (type 6) used to format input of each data vector; at least one must be used.
28	*1* calls for printed output of raw data *2* calls for printed output of raw data following an identification field of six characters (the first field). Variable format card must start with "A6" to represent the field, i.e., (A6, . . .)
29	*1* calls for printed output of transformed data
30	*1* calls for printed output of z-scores
31	*1* calls for a test of normality for each variable
32	*2* calls for tests of homogeneity of group dispersions and all group output; *1* calls for group dispersion and correlation output; *0* suppresses all the above.
33	*2* calls for punched output of the group centroids and inverse of dispersion for future diagnosis studies; *1* calls for punched output of group centroids only.
34	*1* causes punched output of covariance adjusted data.
35	*1* calls for Student-Newman-Keuls multiple range test for each variable.
36	*1* calls for principal component analysis of dispersion using centered data; *2* same as *1* but for uncentered data; *3* adds PCA of each group's dispersion using uncentered data—should not be done unless group sample sizes exceed 24; *4* same as *3* but for uncentered data;

           5 adds tests of significance of all group vs. dispersion components using centered data;

           6 same as *5* but for uncentered data;

           0 eliminates all principal component analyses.

37   *1* calls for component scores, distances of individuals from the origin, and percentage of each individual's variance accounted for by up to 14 components for dispersion if col. 36 > 0;

           2 adds the above to each group's components if col. 26 > 1.

38   *1* calls for angular comparison of group vs. dispersion if col. 36 > 1.

39   sets a priori classification probabilities:

           0 equalizes priors;

           1 sets priors according to sample sizes;

           2 calls for prior probability card(s), type 10, for input of "arbitrary" priors based upon previous knowledge;

           3 sets priors in terms of group dispersion determinants.

40   *1* calls for norming group sample sizes, card type 11 (sample sizes from a previous identical experiment which also had the priors)—generally not used; *5* calls for a diagnosis only study which needs:

          1) control card defining g, p, N_i, and N plus appropriate values for cols. 18, 20, 25-26, 39, 40 and 41

          2) title card(s)

          3) label card(s)

          4) transformation card(s) if required

          5) variable format card(s) for data vectors

          6) sample size card(s) if $N_i = 0$

          7) data vectors

          8) variable order card if col. 20 = 1—and from the previous study:

          9) group centroids (10X, 5E14.7) obtained previously by col. 33 = 2

        10) inverse of dispersion (10X, 5E14.7) obtained previously by col. 33 = 2

        11) priors card(s) (containing priors from previous study)

        12) norming group sample sizes (=group sample sizes from previous study)

        13) finish or control card.

41   *1* calls for distances of individuals from group centroids

42   *1* calls for comparisons of distances between groups based upon each variable singly

43   > *0* calls for from 1 to 9 canonical variate scores for each individual

44   > *0* calls for cumulative distances among groups based upon successive canonical variates; maximum = 9—do you need more than 3?

45   *1* calls for canonical graphs of centroids and all individuals.

46   *1* calls for canonical graphs for each group singly

47   > *0* calls for input of that many sets of theoretical vectors, each of p elements for reification; input of vectors as per card(s) type 12

48   > *0* calls for graph data to be output on magnetic tape for 3-D graphics on a PDP 11/45 via the program DISGRAPH.

49   > *0* calls for output of the group label and that number of eigenvectors for each PCA called by column 36.

2) Title Card(s)—cols. 25-26 (card 1) of them
   Note: even if cols (25-26) = 0, at least one title card must be used—card may be blank
   1-80   any alphanumeric description of the problem
3) Labels Card(s)—to contain p labels
   1-6   LABELS (mandatory)
   10-15   label for measurement 1
   20-25   label for measurement 1
   . . .   . . .
   70-75   label for measurement 7
              as many cards exactly as above as needed for p measurements
4) Group Identification Card(s)—to contain g groups
   1-6   GROUPS (mandatory)
   10-15   abbreviation for group 1
   20-25   abbreviation for group 2
   . . .   . . .
   70-75   abbreviation for group 7
              as many cards exactly as above as needed for g groups
5) Transformation Card—p values called by col. 18 (card 1) = 1
   Note: data are transformed in reference to the original order of data input and not the
              sequence after rearranging the variables
   Transformation codes:
   0   no transformation
   1   absolute (X)
   2   square root (X)
   3   square root $(X + 1/2)$
   4   square (X)
   5          $\log_{10} (X)$
   6          $\log_{10} (X + 1/2)$
   7   arcsine (square root (X)) where X is proportion 0 to 1
   8   arcsine (square root (X)) where X is 0-100%
   9   1/X
   Input
   1-6   TRANSF (Mandatory)
   11   transformation code for variable 1
   12   transformation code for variable 2
   . . .   . . .
   60   transformation code for variable 50
6) Variable Format Card(s)—col. 27 (card 1) of them for p variables
   1-80   as appropriate for each case of p variables
   Note: start card (A6,. . .if col. 28 (card 1) equals 2
7) Sample Sizes—g of them if $g \neq 1$ and $N_i = 0$
   1-6   SAMSIZ (mandatory)
   11-15   size of sample for group 1
   16-20   size of sample for group 2
   . . .   . . .

76-80   size of sample for group 14
          as many cards exactly as above as needed for g samples

8) Data Deck
    1-80   punched as per above variable format card(s)

9) Variable Order Card(s)—for p variables

Generally this option is for convenience only; one might wish to rearrange variables into an order
    more convenient for interpretation

For the analysis of covariance, variable order must place all m variables before all n variables
    1-6   XORDER (mandatory)
    11-12   new number for input variable 1
    13-14   new number for input variable 2
    . . .   . . .
    79-80   new number for input variable 35
          as many cards exactly as above as needed for p variables
    Note: no variable can be allowed to have a number of zero

10) Prior Probabilities Card(s)—required if col. 39 (card 1) = 2
    1-6   PRIORS (mandatory)
    11-15   for group 1
    16-20   for group 2
    . . .   . . .
    76-80   for group 14
          as many cards exactly as above as needed for g groups

11) Norming Group Sample sizes—if g >1, g of them if col. 40 (card 1) = 1
    Note: when not provided are set equal to present sample sizes
    1-6   NORSAM (mandatory)
    11-15   size of sample 1
    16-20   size of sample 2
    . . .   . . .
    76-80   size of sample 14
          as many cards exactly as above as needed for g groups

12) Reification Vectors—col. 47 (card 1) sets of the following three card types:
  A. Parameter Card
      1-6 REIFIC (mandatory)
      9-10 NV, number of vectors to be input in this set (maximum = 75)—each vector must
           contain p (if no covariance adjustment) or n (if covariance adjustment elements)
      11-12 NT, number of title cards to identify this set of vectors.
  B. Title Card(s)—NT of them
      1-80 any alphanumeric description of this set of NV vectors.
  C. Vector Cards—for NV vectors, each as follows:
      1-6 VECTOR (mandatory)
      11-20 value of coefficient 1
      21-30 value of coefficient 2
      . . .   . . .
      71-80 value of coefficient 8
          as many cards exactly as above as needed for p elements.

13) Finish or Control Card
      1-6   PROBLM or FINISH
      Note: a new control card starts a new problem which requires all the above input. FINISH
              in cols. 1-6 stops the program

CORANAL

Output (maximum possible):
Header including reproduction of PROBLM card, problem description, variable labels, sample sizes
    and variable format
Data including raw or transformed
Canonical correlation
    Basic statistics
    Computation checks including inverse of R_{11} check and determinant of R_{11}
    R_{11} *canonical vectors then* R_{22} *canonical vectors,* each followed by:
        Normalized canonical vectors
        Percentage of the variance of each variable's contribution to each canonical vector
        Canonical structure
        Variance extracted and redundancy
    Canonical R, R^2, chi-squared, degrees of freedom, probability and lambda prime for each
        canonical R
    Graph of canonical variates —1st vs. 2nd, 1st vs. 3rd, and 2nd vs. 3rd
Reification
    Angular comparison between first set of canonical vectors and total R_{11}, predicted R_{11} and
        residual R_{11} components
    Angular comparison between second set of canonical vectors and total R_{22}, predicted R_{22} and
        residual R_{22} components
    Angular comparisons between all possible pairs of R_{11} total, predicted and residual components
    Angular comparisons between all possible pairs of R_{22} total, predicted and residual components
Principal component analysis of total R
    Output as per DISANAL
Principal component analysis of R_{11}
    Output as per DISANAL
Principal component analysis of R_{22}
    Output as per DISANAL
Principal component analysis of R_{11} predicted by R_{22}
    Output as per DISANAL
Principal component analysis of residual R_{11}
    Output as per DISANAL
Principal component analysis of R_{22} predicted by R_{11}
    Output as per DISANAL
Principal component analysis of residual R_{22}
    Output as per DISANAL

Canonical correlation of R_{11} vs. R_{22} component scores

    Output as per above Canonical Correlation

Principal component analysis of $R_{11}-R_{22}$ component scores

Output as per DISANAL

    Input:

1) A Control Card

1-6	PROBLM (mandatory)
7-8	p, number of variables in set one; must be the first p variables (xorder card(s), type 7, can accomplish this); maximum = 50, but maximum $p + q = 75$
9-10	q, number of variables in set two; must be the last q variables; maximum = 50, but maximum $p + q = 75$
Note:	in addition to all p and q variables having to be together, it is required that p $\geqslant$ q — the program will automatically adjust the latter.
11-15	N, sample size; total individuals of $p + q = t$ measurements per individual; maximum = 3000
16-20	g, number of groups; to each observation is added the deviation of its group mean from the grand mean—use with care as a crude device to remove group differences
21-25	number of title cards (type 2) used to identify the problem
26-28	number of variable format cards (type 6)
29	*1* calls for variable order cards (type 7) to resequence the variables
30	*1* calls for transformation card (type 5) which allows standard transformations of individual observations
35	*1* causes punched output of predicted R_{11} components
36	*1* causes punched output of residual R_{11} components
37	*1* causes punched output of predicted R_{22} components
38	*1* causes punched output of residual R_{22} components
39	*1* causes punched output of set one normalized canonical vectors of raw data
40	*1* causes punched output of set two normalized canonical vectors of raw data
41	*1* causes punched output of principal components of $R_{11}-R_{22}$ component scores
42	*1* causes punched output of set one normalized canonical vectors of R_{11} component scores
43	*1* causes punched output of set two normalized canonical vectors of R_{22} component scores
44	*0* limits the analysis to only basic canonical correlation
	1 calls for all other steps
	2 adds component graphs if N is not greater than 3000
45	*1* calls for data already stored on file 1; neither data deck nor variable format card(s) are then used; all other options are available

2) Title Cards—cols. 21-25 (card 1) of them

1-80	any alphanumeric description of the problem

3) Labels Cards—to contain t labels in the original data sequence

1-6	LABELS (mandatory)
10-15	label for measurement 1
20-25	label for measurement 2
. . .	. . .

| 75-80 | label for measurement 7 |

as many cards as needed for t measurements

4) Sample Size Cards—required even for 1 group

1-6	SAMSIZ (mandatory)
11-15	sample size for group 1
16-20	sample size for group 2
. . .	. . .
76-80	sample size for group 14

as many additional cards as needed for g groups; the sum of the g samples sizes = N

5) Transformation Card—t values called by col. 30 (card 1) = 1; in original data sequence

1-6	TRANSF (mandatory)
11	transformation code for variable 1
12	transformation code for variable 2
. . .	. . .
80	transformation code for variable 70 if needed, cols. 11-15 of a second card is used to include the maximum number of variables, 75
Note:	codes as per DISANAL

6) Variable Format Card(s)—cols. 26-28 (card 1) of them

| 1-80 | variable format for t variables of a case |

7) Variable Order Card(s)—for all t original variables

Note:	to run the program, the p variables must all precede the q variables
1-6	XORDER (mandatory)
11-12	new number for original variable 1
13-14	new number for original variable 2
. . .	. . .
79-80	new number for original variable 35

as many cards as needed for t variables

8) Data Deck—data deck of N individuals, each formated as per card 6

9) Problm or Finish Card

| 1-6 | PROBLM or FINISH |
| Note: | a new control card starts a new problem which requires all the above input. Blanks or FINISH in cols. 1-6 stops the program. |

Glossary

The glossary contains informal, essentially concept-oriented explanations of terms rather than precise, formal mathematical definitions. Terms pertaining to extensive topics of the text are defined very briefly, perhaps too briefly. About one-fourth of the terms are not in the text. The purpose of the glossary, in part, is to summarize the concepts of morphometrics and to review the basic language of multivariance analysis. Note that one term frequently leads to another.

abscissa—the horizontal coordinates in a 2-dimensional coordinate system, usually rectangular, but sometimes oblique; horizontal coordinates of a 2-dimensional graph.

accuracy—the closeness of a measure to its true value.

achieved communality—the communality of any variable resulting from analysis of a set of data; loosely the same as communality.

additive process—a random process or function in which all increments, e.g., the differences between adjacent ranked values of a variable, are independent.

absolute variation—pertaining to the dispersion of variables in terms of their original scales of measurement, in multivariate analysis the variation measured by a covariance matrix.

adjusted variable—a dependent variable generally as predicted by regression methods as in the analysis of covariance.

agglomerative methods—cluster analysis or other methods that start by creating pairs of ''most similar'' individuals and proceeds by pairing pairs, etc. on the basis of similarity; roughly, the opposites of divisive methods.

algorithm—any procedure for solving a mathematical formula.

allometry—allometric growth or differential growth rate among structures; differential rates of size increase thereby providing a change in shape; also see isometry.

alternate hypothesis—the hypothesis accepted after a significant statistical test (rejection of a null hypothesis); generally the hypothesis that the investigator wishes to establish by statistical proof; also see statistical inference.

among (groups) (effect)—pertaining to the variation among means or centroids of different populations or to the different groups as a whole in contrast with that within each group; syn.: between, between groups, classes, treatment effect, treatments.

analysis of covariance—the analysis of variance of dependent variate(s) after their adjustment by regression and the analysis of the equality of the regressions among the groups.

analysis of variance—a procedure for contrasting among group variation with within group variation to ascertain if population means are significantly different from one another; syn.: variance analysis.

ancova—abbreviation for analysis of covariance.

anova—abbreviation for analysis of variance.

a posteriori probability—here a probability applied after establishment of a basis for its use, e.g., applying a multiple range test after a significant anova.

a priori probability—here a probability determined, often somewhat arbitrarily, by the investigator before a trial (analysis) is performed; e.g., a Type I error selected prior to a statistical test of hypotheses; also see priors.

Aristotelean logic—inference based upon the refinement of the logical syllogisms of Aristotle and criticized primarily when used for nonstatistical judgment of the nature or implication of data per se; also see hypothesis and syllogism.

array—an orderly listing of numbers, e.g., any arrangement of a series of items from smallest to largest, or of elements of a vector or matrix; in a matrix, any row (a horizontal array) or any column (a vertical array).

association—any accessment of the degree of dependence existing between two or more variables or between any two entities.

association matrix—a symmetric similarity matrix containing measures of association between pairs of individuals as off diagonal elements and measures of perfect association of individuals with themselves along the principal diagonal; values generally range from zero (no association) to unity (perfect association; example, a Q-type correlation matrix).

assumption (statistical)—an axiom or statement of basic mathematical properties from which all other properties can be derived; generally a statement presented without proof, but many statistical assumptions are subject to tests of null hypothesis (acceptance does not establish statistical proof of the assumption); the consequences of invalid assumptions need not negate a particular statistical analysis; see robust method.

axis (coordinate)—a line along which a coordinate is measured; pairs of intersecting axes at right angles to one another are rectangular axes and at any other angle are oblique axes, the point of intersection being the origin and the coordinates being rectangular or oblique coordinates respectively; the rectangular or oblique lines used to locate Cartesian coordinates; also see vector.

b-weights—partial regression coefficients, often restricted to a raw variable or absolute variation model.

beta-weights—partial regression coefficients, often restricted to a z-score or standardized variation model, then called standard partial regression coefficients.

between or between groups—same as among groups.

bias (statistical)—generally a systematic distortion, increase or decrease, of a statistical result in contrast to random error, e.g., a systematic increase or decrease involved in measurements taken through time.

biometry—the principles and practices of statistics as applied to biological research; loosely, statistical methods applied to the study of biological phenomena.

bipolar (eigenvector)—one containing both positive and negative coefficients which imply a change in shape; especially such an eigenvector from PCA.

CA or **canona**—abbreviation of canonical analysis.

canonical—used only in the sense of reduction of some quantity, especially a matrix, to a simple form; see canonical form.

canonical analysis—loosely, the reduction of square matrices to canonical form; usually restricted to such reduction in canonical correlation analysis or canonical analysis of discriminance.

canonical analysis of discriminance—a procedure for defining in discriminant space a set of orthogonal axes such that the first and each successive axis indicates the maximum possible generalized distance between groups.

canonical axis—a reference axis in discriminant or canonical correlation space.

canonical coefficient—an element of a canonical vector.

canonical correlation analysis—given a sample of N data vectors, each vector composed of two sets of variables, a procedure for obtaining successive maximum possible but independent correlations between the two sets.

canonical correlation coefficient—in canonical correlation analysis or canonical analysis of discriminance, the correlation coefficient between the two canonical variates (each variate resulting from transformation of a set of original variables) in reference to a specific pair of canonical axes; in canonical analysis of discriminance, one of the axes of each pair is implied.

canonical form (of a matrix)—the reduction of a square matrix to a diagonal matrix; also see Jacobi canonical form; syn.: normal form.

canonical graph—a portrayal of individuals in 2-or 3-dimensions of canonical space, usually in reference to discriminant space defined by canonical axes but also to canonical correlation space; also see canonical space.

canonical space—in canonical analysis of discriminance, discriminant space as defined by canonical axes; in canonical correlation, the space of each data vector to produce maximum but independent correlations between subvectors.

canonical structure—the correlation between canonical variates and the original variables from which the former were derived.

canonical variates—the scores of individual raw data vectors after transformation from original reference axes to canonical axes; the vectors of scores of individuals on a set of canonical axes.

canonical vector—an eigenvector that defines a transformed (canonical) axis in canonical correlation analysis or canonical analysis of discriminance.

Cartesian coordinates—the set of numbers that locate a point by its distances from two intersecting straight lines (axes), the distance from each axis being measured along a line parallel to the other axis; generally such numbers pertaining to a maximum of 3-dimensional space; also see axis.

Cartesian space—the space defined by Cartesian coordinates, rectangular Cartesian space (that defined by orthogonal axes) being the same as Euclidean space.

case—same as individual.

CCA—abbreviation of canonical correlation analysis.

CDA—abbreviation of canonical analysis of discriminance.

central limit theorem—simplistically, the sum or mean of all possible samples of size N of a variate will approach the normal distribution as N increases without bound; also extends to transformed variates such as component scores, canonical variates, etc., approaching normality when original data vectors are not normal.

centroid—a vector with each element being the mean value for an observation of a sample or population; a vector containing the means for each variable for a sample or population.

characteristic determinental equation (of a matrix)—same as characteristic equation.

characteristic equation (of a matrix)—a mathematical statement that transforms a square matrix $\mathbf{A}$ of order p to a diagonal matrix $\mathbf{L} = \lambda\mathbf{I}$ containing diagonal elements of λ, by $\mathbf{A} - \lambda\mathbf{I} = 0$; important properties are $\sum a_{ij} = \sum \lambda_i$ and $\Pi\lambda_i = \mathbf{A}$; equation of the type providing canonical form in multivariate analysis; syn.: characteristic determinental equation.

characteristic root or value—same as eigenvalue.

characteristic vector—same as eigenvector.

classes—same as groups.

classification—in discriminant analysis the assignment of each individual of each group's sample to the group it most closely resembles, an assignment of an individual to its own group being called a *hit* and to any other group, a *miss;* also see cluster analysis and ordination.

classification chi-squared—the square of the generalized distance (which approximates the chi-squared distribution) as often designated when used for classification.

classification constant—the second term in the multigroup discriminant function.

classification function—same as discriminant function.

classification function analysis—same as multigroup discriminant (function) analysis.

classification score—same as (multigroup) discriminant (function) score.

classification vector—same as (group) discrimination vector.

cluster analysis—computational methods, usually on a similarity matrix, used for classifying OTU's into groups; any method for grouping contiguous units of a sample of OTU's.

coefficient—in a mathematical function, a constant in contrast to a variable.

coefficient of determination—the same as the square of the multiple correlation coefficient.

coefficient of variation—the standard deviation divided by the mean, often multiplied by 100; used to standardize and then compare the variation between organisms of different sizes since variation tends to be a function of size; sensitivity to statistical error, especially in the mean, limits its use.

coenocline—a 1-dimensional community gradient generally reflecting an environmental gradient.

coenoplane—a 2-dimensional community pattern reflecting environmental conditions.

common (factor) variance—in factor analysis, that part of the variance of a theory matrix which is shared with other variables; see theory variance.

communality—the proportion of the total variance of a variable extracted by an analysis, usually pertaining to standardized variation hence variances and covariances that actually are correlation coefficients; in a PCA of a correlation matrix, communalities are unity for all variables since the total variance of each variable is extracted; in most factor analyses (involving a common factor matrix, a specific factor matrix and a residual matrix that collectively sum to a correlation matrix) the diagonal elements of the common factor variance are the achieved communalities of the variable.

component—generally pertaining to an eigenvector of PCA.

component axis—an axis defined by an eigenvector in PCA.

component correlations—the structure or correlations between component scores and original variables; syn.: component structure.

component distribution—the distribution of individuals in reference to any arbitrary linear function of the vector variables; in a principal component distribution, orthogonal axes including the major and minor axes are defined.

component pattern—numerically identical to component correlations; the regression coefficients providing predicted z-score data vectors from factor scores; also called factor loadings.

component scores—the scores of original data vectors after transformation to component space.

component space—a Euclidean space defined by rotating original axes to a set of orthogonal axes defining new variables (component scores) that are uncorrelated with one another; also the axes define successive maximum possible variances or distances between individuals.

component structure—same as component correlation.

component vector—an eigenvector in PCA.

concomitant variable—a dependent variable.

conditional distribution—the predicted distribution of a variable from the known distribution of one or more other variables, e.g., a line of regression.

confidence circle—on a canonical analysis of discriminance graph an estimate of a group's population centroid in terms of the number of axes (dimensions) shown on a canonical graph; in a strict sense a 2-dimensional estimator.

confidence coefficient—the probability associated with the estimation of a parameter.

confidence interval—a range within which a particular univariate parameter can be expected in the proportion of cases defined by a confidence coefficient.

confidence limits—the values which form the upper and lower limits of a confidence interval.

confidence sphere—the 3-dimensional estimator equivalent to a confidence circle.

conform—pertaining to matrices whose orders permit addition, subtraction or multiplication.

constant—in contrast with a variable, a single number or a symbol that does not change in value throughout an analysis.

continuous variable—a variable that can assume any decimal fraction, theoretically an infinite number of values, between consecutive whole numbers.

coordinate—any number of a set used to locate a point in space, essentially a variable used for this purpose.

coordinate axis—see axis (coordinate).

coplanar—in the same plane; referring to vectors that can be diagrammed in terms of their angular deviations from one another in a single plane.

correlation—in the general sense, any method to examine association, interdependence, or relationship between variables; also pertaining to methods using correlation coefficients, or to an implied or statistically proven meaningful correlation; loosely, also a correlation coefficient.

correlation coefficient—a statistical measure of the interdependence between two variates; usually varies from -1 (perfect negative) through zero (no correlation) to $+1$ (perfect positive); as used here, the Pearson product moment correlation coefficient (see text).

correlation matrix—a symmetric matrix containing diagonal elements of perfect correlation of entities with themselves and off-diagonal elements of correlation coefficients between pairs of entities (intercorrelations), corresponding columns and rows pertaining to the same entity; R-type: a $p \times p$ correlation matrix pertaining to p-variables, a dispersion matrix measuring standardized variation between variables; Q-type: an $N \times N$ correlation matrix pertaining to N individuals or centroids, generally called OTU's.

correlation structure—the correlation between original variables and eigenvectors; also see component structure and canonical structure.

correspondence analysis—an ordination method for arranging character states of each of two variables into a sequence such that the two sequences are maximally correlated.

cosine (of an angle)—in a right angle triangle, the ratio of the side adjacent to the angle and the hypotenuse.

covariance—a measure of simultaneous variation between two variates; generally applied to usual scales of measurement of variables but can apply to z-scores of variables then becoming a correlation coefficient; see cross products.

covariance analysis—same as analysis of covariance.

covariance matrix—same as variance-covariance matrix.

covariate—same as independent variable.

CP—abbreviation of cross products.

criterion—same as dependent variable.

critical region—the area of a statistical probability distribution, e.g., normal distribution, chi-squared distribution, Student's t distribution, F distribution, etc., that represents the magnitude of the Type I error.

cross products—the sum of the products of the deviations from their means of the first through last of a pair of variables, each pair representing an individual of a sample or population; when divided by the degrees of freedom, a covariance.

cross products matrix—same as sum of squares and cross products matrix.

curvilinear regression—a regression function producing a curve rather than a line; generally a polynomial regression.

data analysis (multivariate)—the principles and applications of multivariate analysis to stress the heuristic evaluation of phenomena rather than statistical inference.

"data grinding"—colloquialism used in the broad sense to denote the collection and/or analysis of data; in another sense implying the improper analysis of data.

data matrix—a matrix whose columns each consist of p observations (a data vector) for an individual of a sample and whose rows each consist of a sample of N measurements of one of the p variables (a vector of a variable); a variables $\times$ individuals matrix.

data vector—a vector whose p elements are variables, each an observation on the same individual; syn.: vector of variables.

degree of an equation—the degree of its highest order term, which is equal to the sum of the exponents of its variables, e.g., $2 + x^4y^2$ has a degree of six from the sum of the x and y exponents, and $3 + xy^2 + x^4y$ has a degree of five from the sum of the exponents of x^4y; one also can refer to degree with respect to a variable per se, both expressions having a degree of four in x and a degree of two in y.

degree of a variable—see degree of an equation.

degrees of freedom—in a sample of N individuals the number of free observations of a variable entering into calculating a statistic; e.g., $\sum_{i=1}^{N} (X_i - \overline{X})^2$ has $N-1$ degrees of freedom because $\overline{X}$ is fixed and $\sum_{i=1}^{N} (X_i - \overline{X}) = 0$, i.e., when all but the last value of X is known the last value is determined by the sum being equal to zero.

deme—see population.

density dependent ecological factor—any biological and/or physical feature whose effect becomes increasingly severe as the density of individuals increases.

density function—an expression providing the frequency of a value or x or range of x as a function of x, a function of a variable that generates the frequency distribution of that variable; a statistical distribution function that provides the frequency of a value of a variable, the frequency between two values of the variable and/or the cumulative frequency from the lowest value up to a given value of the variable; syn.: distribution function.

dependent variable—a variable that although observed is also calculated or predicted from a function of one or more other so-called independent variables; syn.: concomitant variable, covariate, criterion, predictand, regressand.

derived variable—a variable formed by addition, subtraction, multiplication and/or division of two or more original variables; therefore, any statistic; often restricted to pertain to ratios and indices.

determinant—formal definition is beyond the scope of this book; a scalar representing a unique function of a square matrix that involves the precise addition and subtraction of products of elements of the square matrix; important properties of determinants include: (1) if all elements of a column or row are zero, the determinant equals zero; (2) if two columns or rows have equal elements, the determinant equals zero; (3) if one row (or column) can be made equal to another row (or column) by addition, subtraction, multiplication or division of each element by a constant, the determinant equals zero; (4) if the determinant of a matrix equals zero, certain multivariate analyses are not possible.

deviation (score data) vectors—the vectors resulting from subtracting the centroid of a set of data vectors from each data vector.

deviation (score) matrix—a matrix whose columns consist of deviation (score data) vectors of individuals.

deviations—usually, the subtraction of all observed observations from their mean value; also any departures from predicted or theoretical value(s).

diagnosis—after an original discriminant analysis, the classification of new individuals using statistics derived from the original study.

diagonal (of a matrix)—same as principal diagonal of a matrix.

diagonal matrix—a square matrix having nonzero elements only on the diagonal.

dimension (of a matrix or vector)—the "size" of a matrix; the number of rows, r, by the number of columns, c, of a matrix, symbolized $r \times c$.

direction cosines—the coefficients of an eigenvalue normalized so each coefficient represents the cosine of the angle between an original variable axis and the axis defined by the eigenvector.

discontinuous variable—a variable whose measures are limited to whole numbers; e.g., counts; syn.: discrete variable.

discrete variable—a discontinuous variable.

discriminant analysis—a collective term for those multivariate analyses classifying, comparing, contrasting and examining relationships within and between two or more populations.

discriminant function—subject to many interpretations in the literature; here restricted to the first term of the multigroup discriminant function, then becoming a synonym of classification function.

discriminant function analysis—subject to many interpretations in the literature; here applied as a synonym for multigroup discriminant analysis; syn.: classification (function) analysis.

discriminant function constant—the second term in the multigroup discriminant function; syn.: classification constant.

discriminant (function) score—subject to different interpretations in the litrature, here considered a score of an individual on a multigroup discriminant function; syn.: classification score, discrimination score, multigroup discriminant function score.

discriminant space—a space formed from Euclidean space by rotating original variable axes so the angles between any pair of axes have cosines equal to the correlation between the variables; a space typified by a synergistic interplay of the original variables; a space wherein total form, rather than independent variables, is contrasted; see synergistic space.

discrimination score—same as discriminant (function) score.

discrimination vector—a vector whose coefficients represent the transformation of a group centroid to discriminant space; a qualitative appraisal of the nature of a group in discriminant space; syn.: classification vector, group discrimination vector.

dispersion—loosely, any measure of variation.

dispersion matrix—any matrix containing measures of variation, specifically a variance-covariance matrix, a measure of absolute variation, or a correlation matrix, a measure of standardized variation; in manova or discriminant analysis, the pooled within group dispersion matrix.

distance of an individual—the length of a line from the origin of one or more axes in a coordinate system to an individual.

distribution—in statistics, a frequency distribution.

distribution function—same as density function.

divisive (methods)—analyses for arranging individuals into a sequence that proceeds from largest groupings of individuals to successive arrangements and subdivisions within the larger groupings; roughly, the opposite of agglomerative methods.

efficient statistic—a concept pertaining to the relative merits of various possible statistics, e.g., the sample mean, sample median and sample mode as measures of central tendency; the most efficient statistic has the smallest variance; if the most efficient statistic has a variance of σ^2, the measure of the efficiency of any i^{th} other estimator is the decimal fraction σ^2/σ_i^2; in the above example the same mean is the most efficient estimator of the population mean.

eigenvalue—a scalar variable λ for the ith value of which there is a non-zero eigenvector $\mathbf{v}_i$ that satisfies $\mathbf{A}\mathbf{v}_i = \lambda_i\mathbf{v}_i$ where $\mathbf{A}$ is a square matrix; a root solved by a characteristic equation; see characteristic equation; syn.: Hotelling's root, lambda root, lambda value, latent root, latent value.

eigenvalue function—same as characteristic equation.

eigenvector—see eigenvalue and eigenvector function; syn.: characteristic vector, latent vector.

eigenvector function—given the characteristic equation $\mathbf{A} - \lambda\mathbf{I} = 0$, eigenvectors are derived from $(\mathbf{A} - \lambda\mathbf{I})\mathbf{v} = 0$ and define the new axes that provide the features of $\lambda\mathbf{I}$; see characteristic equation.

element (of a matrix or vector)—any of the individual values in a matrix or vector.

error (statistical)—the difference between an observed value and its "true" or "expected" value; a chance deviation, e.g., from the sample mean, rather than a mistake.

error effect—in anova or manova, the term in the model to account for (statistical) error.

error mean square—same as pooled error.

estimation (of a parameter)—a statistical method that will include a population parameter within a range (confidence interval) or circumscribed area (confidence circle) with a certain level of probability, a confidence coefficient.

Euclidean space—in a limited sense ordinary 2- or 3- dimensional space as defined by orthogonal reference axes; often extended to hyperspace; rectangular Cartesian space, including its generalization beyond 3-dimensions.

experimental design—in a statistical sense the organization of a problem from concept through sampling and mensuration with a view toward proper use of a specific statistical analysis; often experimental organization for the use of anova or manova; also pertaining to the anova or manova that satisfied the design; loosely any experimental plan.

explanatory variable—same as independent variable.

FA—abbreviation for factor analysis.

factor—in a mathematical sense, to subdivide into component parts, e.g., 6 is factored into 2×3; in an experimental design, (e.g., a factorial design) a possible source of variation, a set of similar treatments, e.g., several related species, localities, habitats, etc.; in derived variables, any of the involved variables; in multivariate analysis, a vector of structural correlation coefficients or analogous items, an eigenvalue standardized to have unit variance, or loosely any eigenvalue or analogous vector of factor analysis.

factor analysis—a group of multivariate methods for the purpose of studying variation in a sample; e.g., principal component analysis and factor model analysis.

factor loading—loosely, the coefficient of an eigenvector or analogous vector in a factor analysis; in a restricted sense, a (factor) pattern, especially of the common (factor) vectors in factor model analysis.

factor model analysis—generally pertains to subdividing a correlation matrix into theoretical and residual additive fractions and then performing a component or related analysis to examine variation in the theoretical portion; also see theory variance.

factor score—a factor analysis quantity analogous to a component score, but often having unit variance rather than that of an eigenvalue; in principal factor analysis, factor scores are derived by dividing each component score by the square root of the eigenvalue.

factor score coefficients—analogous to a matrix of principal component eigenvectors in which each component coefficient is divided by the square root of its corresponding eigenvalue thereby causing factor scores to have unit variance.

factorial design—an anova or manova having two or more categories of treatment classification, i.e., two or more experimental factors; a multiway anova or manova.

F-distribution—the random sampling distribution of two independent estimates of the variance from a normal distribution; also generalized to the multinormal distribution.

fixed model anova—an anova in which the groups of each factor are determined by the investigator in contrast to being selected at random; e.g., the particular species to be studied are chosen by the worker; syn.: Model I anova, general linear hypothesis model.

fixed variate—same as independent variable.

FMA—abbreviation for factor model analysis.

frequency distribution—a set of values of a variable that includes the frequency of each value or for which the frequency can be calculated.

F-test—a test based upon the ratio of two independent estimates of the variance of a normal population; also as generalized to a multinormal population; syn.: variance ratio.

general component—in PCA an eigenvector with all coefficients of the same sign, implying a general size or growth component.

generalized correlation—the scalar value ranging from zero to unity of the determinant of a correlation matrix; a general measure of independence rather than association among variables, the minimum value occurring with linear dependence and maximum value occurring with zero correlation between variables.

generalized distance—the distance between individuals and/or group centroids in discriminant space.

generalized variance—the scalar value of the determinant of a covariance matrix; a general measure of independence and variation, a minimum value of zero being due to a zero variance or linear dependence among variables and a maximum value owing to zero correlations between variables and equal to the product of all variances.

Gompertz curve—a function of an independent variable x, including constants a, b and k, that predicts the dependent variable y, $y = da^{b^x}$ where $0 < a < 1$, $0 < b < 1$, $y = ka$ when $x = 0$, and y approaches k as x approaches infinity; also used in the form $\log y = \log k + (\log a)b^x$; a growth curve.

Gramian matrix—a symmetric matrix of the form $\mathbf{A}'\mathbf{A}$ or $\mathbf{A}\mathbf{A}'$, e.g., the $SSCP$ matrix, XX', and such matrices as covariance and correlation matrices derived from a $SSCP$ matrix, whose possible determinants are all greater than zero.

grand mean or centroid—the mean or centroid based upon all individuals of all samples.

group—a set of observations or individuals, all of which possess common characteristics, e.g., a sample or population.

group discrimination vector—see discrimination vector.

growth component—in PCA, a general component pertaining to an independent pattern of variation that is consistent with size increase in the group of organisms, etc. sampled.

growth curve—an expression giving population increase as a function of time; also a size increase of one structure as a function of another; also see Gompertz curve and logistic curve.

heteroscedasticity—in reference to two or more population; inequality of population variances in the univariate case or inequality of population dispersions in the multivariate case.

heuristic (analysis)—placing statistical inference secondary to the discovery and refinement of hypotheses by stressing scientific judgment consistent with statistical results.

hit—see classification.

homoscedasticity—in reference to two or more populations, equality of population variances in the univariate case or equality of population dispersions in the multivariate case.

hyperellipse—a generalization of the concept of an ellipse to hyperspace.

hypermultivariate (distribution)—any multivariate distribution that cannot be summarized adequately in 3 or fewer dimensions by multivariate analysis; in a more restricted sense, the inability of canonical analysis of discriminance to summarize most between group differences in the first three canonical axes.

hyperplane—a generalization of the concept of a plane to hyperspace.

hyperspace—extension of the concept of Euclidean space to more than 3-dimensions; in a more general sense, multidimensional Euclidean and discriminant space; loosely, any space of more than 3-dimensions.

hypothesis—generally, a proposition assumed true since its consequences are consistent with general principles and are found to be true, a refinement of syllogistic logic; in statistics, an assumption employed to prove something else, specifically a null hypothesis.

I—symbol for an identity matrix.

identity matrix—a diagonal matrix with diagonal elements of unity; syn.: unit matrix.

independence (statistical)—formally, the probability of two events are independent if the probability of either is the same whether the other is given or not and the probability of the two events occurring simultaneously is the product of the two independent events; in terms of pairs of scalars, vectors, or matrices, their correlations are zero; zero correlation, the basis for random error and lack of interaction.

independence of errors—pertaining to a sample in which the probability for obtaining any observation in the population is identical for individuals in the sample, i.e., error in observations is a random variable; random error.

independent variable—when y is a function of a variable x, each value of x producing a calculated value of y, x is an independent variable and y a dependent variable; syn.: covariate, explanatory variable, fixed variable or variate, predictor, predictor variable, predicated variable.

index—any derived variable supposedly measuring an operation, function and/or specific characteristic of a phenomenon.

individual—a single data record; in the univariate case, a single observation; in the multivariate case, a data vector containing an observation on each of p variables; syn.: case.

inefficient statistic—any possible estimator of a parameter other than the most efficient; see efficient statistic.

interaction—in general, lack of independence resulting in synergism or interference between factors in a multiway anova or between variables and/or factors in any multivariate analysis.

intercept (coefficient)—in a simple linear regression function, the coefficient indicating the value of the dependent variable when the value of x is zero.

intercorrelation—correlation between a pair of variables in contrast to the correlation of a variable with itself.

interference—the condition where the value of one of two associated variables is reduced by a third variable.

inverse matrix—the form of a matrix that when used as a multiplier acts like a divisor, division being impossible in matrix algebra.

ipsative measures—pertaining to a set of a data vectors, the sum of all variables in each being equal to a constant, e.g., data vectors of percentages summing to 100%; ipsative measures cause dependence of the p variables in data vectors and rank less than p, generally $p - 1$, in their cross products, covariance, correlation, etc., matrices, i.e., singular matrices.

isometry—a condition pertaining to equality of dimensions, geometric form, variables, etc.; in growth of organisms, two or more structures growing at exactly the same rate as indicated by having identical coefficients in a formula describing growth; growth in which increase in size is independent of shape changes.

item—a single observation.

iterative method—a procedure that starts with a trial value, performs some operation on it to obtain a second trial value, and continues the procedure until two successive trial values agree to a specified degree of precision, a kind of algorithm basic to many multivariate computations.

Jacobi canonical form—transformation of a square matrix to a diagonal matrix in which the diagonal elements are eigenvalues; a procedure accomplished by solution of a characteristic equation; generally implied by the term canonical form.

kurtosis—pertaining to the ''peaked'' nature of a unimodal distribution; generally curves following a normal distribution are termed mesokurtic, curves lower and flatter than normal are platykurtic and curves higher and sharper than normal are leptokurtic; various statistics crudely measure kurtosis but are applied to test normality.

lambda root or value—same as eigenvalue.

large sample theory—pertaining to various statistics that require large sample sizes to be applied with validity; unfortunately the minimum sample size sufficiently large to apply such statistics generally is unknown; see sample size.

latent root or value—same as eigenvalue.

latent variance—in factor analysis, that part of an individual theoretical variable's variance that is not shared with other variables; see theory variance; syn.: specific variance, unique variance, specificity.

latent vector—same as eigenvector.

leptokurtic—see kurtosis.

level of significance—the magnitude of the Type **I** error, usually symbolized α.

linear—along or pertaining to a straight line; having one dimension; often pertaining to an equation of the first degree in its variable or variables.

linear equation or expression—any algebraic form of the first degree in its variable or variables, not the entire equation; see degree of an equation.

linear function (of a matrix)—a transformation of the type $\mathbf{y} = \mathbf{A}'\mathbf{x}$, an equality, where $\mathbf{y}$ and $\mathbf{x}$ are vectors and $\mathbf{A}$ is a matrix of vectors, all containing elements of the first degree; see degree of an equation.

linear regression—any regression based upon equations of the first degree, simple linear regressions having one independent variable and multiple (linear) regression having two or more.

linearly dependent—the rows or columns, $\mathbf{a}_i$, of a matrix are linearly dependent if a scalar k_i can be applied to each, so $k_1 a_1 + k_2 \mathbf{a}_2 + \ldots + k_p \mathbf{a}_p = 0$; all rows or columns for which the above is not possible are linearly independent.

linearly independent—see linearly dependent.

logistic curve—a function of an independent variable x, including constants a, b and k, that predicts the dependent variable y, $y = k/(1 + e^{a+bx})$ where $b < 0$; $y = k/(1 + e^a)$ when $x = 0$; and y approaches k as x approaches infinity; a growth curve; syn.: Pearl-Reed curve.

Mahalanobis' distance—a statistic originally described by Mahalanobis as equal to the square of the generalized distance for the purpose of measuring the distance between populations but later corrected by Mahalanobis to equal the generalized distance; distance as measured in discriminant space.

main diagonal (of a matrix)—same as principal diagonal.

major axis—in reference to the scatter of points in a coordinate system, the longest axis that can be constructed through those points; in a multivariate normal or other ellipsoidal distribution, the longest axes of the ellipsoid; in PCA, the first component axis as defined by the first eigenvector.

mancova—abbreviation for multivariate analysis of covariance.

manova—abbreviation for multivariate analysis of variance.

marginal distribution—the univariate distribution of any single variable in a vector variable; when a vector variable follows the multivariate normal distribution the marginal distribution of each variable is normal but if all marginal distributions are normal the vector variable need not follow the multivariate normal distribution.

matrix—in general, any rectangular row by column array of numbers or terms, the array written between brackets, parentheses or double lines on either side.

matrix algebra—a generalization of simple (scalar) algebra to operations involving vectors and matrices; the language of multivariate analysis.

MDFA—abbreviation for multigroup discriminant (function) analysis.

MDS—abbreviation of nonmetric multidimensional scaling.

mean—the average or center of gravity of all observations in a sample or a population.

mean vector—same as centroid.

median—in a set of observations sorted in order of magnitude, the value of the middle observation or the average of the two middle observations in a set with an even number of observations; in a normal distribution, the population median equals the population mean.

mesokurtic—see kurtosis.

metric—a standard of measurement; often pertaining to measures leading to a coordinate system in which the distance between any two points can be determined.

minor axis—consider a set of orthogonal axes in which the first and each successive axis is defined on the basis of being the longest possible, i.e., by PCA, if all axes are of unequal length, the axis defining the last dimension is the shortest and the minor axis.

miss—see classification.

mnd—abbreviation for multivariate normal distribution.

mode—the value of a variate in a sample or population which is possessed by the greatest number of individuals; in a normal distribution the population mode is equal to the population mean.

model—the formalized mathematical expression of a theory or law; also loosely applied to any formalized diagram, formula, or statement that agrees with observed data.

monothetic (methods)—in contrast with polythetic methods, procedures involving a sequence of decisions that perform well early in the sequence (lower levels) but suffer later (higher levels) owing to early level decisions.

Monte Carlo (study or method)—loosely, the approximation of features of statistical distributions by repeated random samples; more appropriately the evaluation of mathematical problems by repeated random sampling; many current Monte Carlo studies rely on computers for comprehensive examination of known man-made, artificial populations.

morphometric taxonomy—morphometrics applied to taxonomy; see numerical taxonomy and quantitative taxonomy.

morphometrics—multivariate analysis directed towards the interpretation of the form of phenomena in terms of independent and contrasting patterns of variation; generally a use of heuristics (data analysis) and emphasis of statistical interference for such purposes.

MPCA—abbreviation for multigroup principal component analysis.

MRA—abbreviation for multiple regression analysis.

multigroup discriminant (function) analysis—a method to examine relationships between two or more multivariate populations; classically and still is applied as a device for classification; in morphometrics, only the group discrimination vectors from this analyis are used; syn.: classification function analysis.

multigroup (principal) component analysis—the comparison of group components, using dispersion components as a reference, to ascertain similarities and differences between groups after rejecting the null hypothesis of equality of group dispersions; also acts as a support for the discriminant analysis of groups when the above hypothesis is rejected; also comparing the components of several groups directly without a reference set of components.

multinormal—same as multivariate normal.

multiple correlation analysis—generalization of (simple) correlation analysis involving a single independent variable to two or more independent variables.

multiple correlation coefficient—the correlation between the observed and the predicted values of the dependent variable in multiple regression; a measure of goodness of fit of observed dependent variables to the line of regression that ranges in value from zero (no correlation) to unity (perfect correlation and fit); also see coefficient of determination.

multiple part correlation—consider three sets of data, after covariance adjustment of the second set by the first as in multiple partial correlation, correlation within and between adjusted variables of the second set and the raw variables of the third set.

multiple partial correlation—the correlation between pairs of variables while one or more other variables are held fixed; amounts to the correlation between variables after their being adjusted in the sense of the analysis of covariance by one or more other variables.

multiple range test—an a posteriori statistical procedure used (after a significant anova involving three or more groups) to create subsets, each consisting of groups whose population means are not significantly different from one another.

multiple regression—linear regression of a dependent variable on more than one independent variable.

multivariable—pertaining to the independent variables of multiple regression; since regression stresses the prediction of a single dependent variable, in a strict sense, it is not multivariate or a multivariate method.

multivariate—pertaining to more than one dependent variable in contrast to any regression which has a single dependent variable.

multivariate analysis—a loosely applied term to denote any analysis involving more than one variable; any statistical analysis of multivariate data; often restricted to multivariate statistical analyses involving the assumption of multivariate normality, i.e., parametric multivariate statistics, sometimes, but not in a strict sense including multiple regression and multiple correlation.

multivariate analysis of covariance—generalization of the analysis of covariance to the multivariate case.

multivariate analysis of variance—generalization of the analysis of variance to the multivariate case.

multivariate normal density function—a density function that generates a multivariate normal distribution; see density function; syn.: multinormal density function.

multivariate normal distribution—an ellipsoidal probability distribution generated by the multivariate normal density function; a distribution generally approximated by data vectors of biological measures; syn.: multinormal.

MVA—abbreviation for multivariate analysis.

nodal (axis)—see ordination.

nonmetric multidimensional scaling—a nonparametric method for ordination that involves a nonlinear model.

nonparametric statistics—that branch of statistical analysis which is independent of the distribution of data; syn.: distribution free methods.

nonsingular matrix—a matrix whose determinant is nonzero; only nonsingular matrices possess an inverse so can be involved in all mathematical operations.

normal—pertaining to the normal and, loosely, multinormal distributions.

normal density function—a density function that generates a normal distribution; see density function.

normal distribution—a bell-shaped probability distribution of a single variable that is generated by the normal density function; a distribution approximated by the great majority of biological variables; syn.: Gaussian curve and distribution.

normal form (of a matrix)—same as canonical form.

normal variate—a variate that follows the normal distribution.

normalize—the creation of normalized scores by the conversion of original scores to some standard scale, generally a mean of zero and standard deviation of unity, e.g., a z-score; also the process of transforming variables to variates and eigenvectors to direction cosines; also see standardize.

normalized score—see normalize.

normalized vector—any vector the sum of the square of whose elements is equal to unity; in multivariate analysis, the form of many vectors that produces direction cosines.

norming group sample sizes—pertaining to diagnosis in discriminant analysis, the sample sizes for each group for the prior discriminant analysis that is the basis for a diagnosis problem.

null hypothesis—the particular statistical hypothesis being tested with a predetermined Type I error in contrast to the alternate hypothesis or hypotheses verified by significance; statistical proof occurs in rejecting but not in accepting a null hypothesis; also see statistical inference.

null (matrix or vector)—one having all elements equal to zero.

numerical taxonomy—originally, the use of numerical methods to group taxonomic units based upon their character states; often multivariate procedures that differ from conventional multivariate analyses and operate on similarity matrices, when multivariate methods are used, Q-techniques regularly are applied; in a less specific sense, also morphometric and quantitative taxonomy.

oblique axes—see axis.

observation—the record of a single value of a variable for a single individual; syn.: item.

off diagonal elements (of a matrix)—in a square matrix all elements other than those along the diagonal from the upper left hand to the lower right hand corner; elements of a square matrix other than diagonal elements.

operational taxonomic unit—the lowest rank of taxa employed in a given study, often applied without reference to affinity among such units or to units of questionable taxonomic rank; generally restricted in use to numerical taxonomy.

order (of a matrix or vector)—the number of rows and columns of a matrix stated as the number of rows by the number of columns, symbolically $r \times c$, read "r by c."

ordinate—the vertical or more nearly vertical set of numbers that locate points in 2-dimensional Cartesian space.

ordination—the placement of individuals into a space of 1 to a minimum of $(N-1,p)$ dimensions, where p is the number of variables and N is the number of individuals, into a theoretically continuous sequence reflecting some fundamental property of the individuals, in multivariate analysis, generally only the first 2- or 3-dimensions are examined, e.g., in a component graph, principal coordinate graph, or canonical graph; methodology often involves identifying portrayed axes as serial (individuals are plotted from one extreme to the other without noticeable breaks), nodal (individuals are grouped into two or more clusters), or polynomial (individuals conform to a curve).

orthogonal (axes)—at right angles; also pertaining to variates, vectors, or linear functions defining axes at right angles to one another.

orthogonal matrix—a matrix equal to the inverse of its transpose.

orthogonal vectors—any two vectors having a scalar product of zero; any two vectors in an orthogonal matrix; vectors at right angles to one another.

OTU—abbreviation for operational taxonomic unit.

outlier—in the sense of a population of individuals, an individual so far removed from the mean of the population as to have a remote probability of belonging to that population; often an individual rejected from belonging to a population owing to the above reason.

parameter—a characteristic of a population, often defining the distribution of a population as does the population mean and population variance of a normal population.

parametric statistics—in contrast to non-parametric statistics, stastistical methods involving theoretical distributions, e.g., normal, and parameters defining the population; in general, the statistics in this book.

part correlation—the same as multiple part correlation.

partial correlation—the same as multiple partial correlation.

partial regression coefficient—in multiple regression, the coefficient of an independent variable; termed partial since its value differs from that which would be obtained from a simple regression of it and the dependent variable alone; specifically such coefficients of the raw variable or its deviation scores models.

pattern—a measure of the regression of original variables on some related quantity; in multivariate analysis, generally the regression coefficient for each variable on a score derived from the variable, generally normalized so the regression coefficients are correlation coefficients, i.e., structure; if not normalized, the eigenvector.

PCA—abbreviation of principal component analysis.

PCORD—abbreviation of principal coordinate analysis.

PFC—abbreviation of principal factor analysis.

phenetic (relationship)—in contrast with affinity or relationship between organisms based upon ancestry, the comparisons of overall phenotypic similarities and/or differences as judged by the characters or traits of the organisms without any implication as to their relationships by ancestry.

phenotypic plasticity—the phenomenon of the same geotype being expressed as a different phenotype, generally as a response to different ecological conditions, i.e., environments; also pertaining to the tendency of species to be so characterized; nongenotypic variation.

platykurtic—see kurtosis.

PO—abbreviation for polar ordination.

polar ordination—the oldest and simplest of the better techniques for ecological ordination.

polynomial—an algebraic expression consisting of two or more terms; a polynomial of a single variable can be linear, quadratic, cubic, quartic, etc. according to its degree of 1, 2, 3, 4, etc., the equation of simple linear regression, $y = bx,$ being of the first degree and of polynomial regression, $y = a + b_1x + b_2x^2 + \ldots + \ldots + b_rx^r,$ being of degree r; polynomials of many variables likewise can be linear, etc., the equation for multiple regression being linear since all independent variables are of the first degree.

polynomial (axis)—see ordination.

polynomial regression—any polynomial regression equation of the first degree produces a linear regression (see degree of an equation, polynomial and linear regression); any higher order polynomial regression (with degree higher than one) by definition is nonlinear, e.g., polynomial regressions involving various successive magnitudes in the exponents of a single variable, and produces a curve; generally, restricted to higher order polynomial regressions.

polythetic (method)—in contrast with monothetic methods, procedures involving a sequence of decisions that perform well both early and late in the sequence.

pooled (error)—given g separate samples and a sum of squares of deviations from the mean of each, the sum of the g sum of squares is the pooled sum of squares; when the pooled sum of squares is divided by the sum of the degrees of freedom of the g samples, the result is a pooled error; syn.: error mean square, within (group) mean square.

pooled within group dispersion matrix—over-descriptive term for the dispersion matrix in the sense of the multivariate analysis of variance or discriminant analysis; the matrix formed by pooling the covariance matrices of each group; syn.: within group dispersion matrix.

population—(1) biological: loosely an aggregation of organisms, the individuals of a species or any definable unit of a species; more specifically, the group of potentially interbreeding individuals at a locality, a deme. (2) statistical: the total set of actual or potential observations recognized by a specific characteristic of the observations; a set of measurements that can be defined in reference to any kind of biological population; syn.: universe.

positive definite—pertaining to a matrix of a type whose determinant always is greater than zero; see quadratic form.

positive semidefinite—pertaining to a type of matrix whose determinant can be zero or greater than zero; see quadratic form.

postmultiplier matrix—when two matrices are multiplied, the first is the premultiplier and the second is the postmultiplier, e.g., in the product **AB**, **A** is the premultiplier and **B** is the postmultiplier.

power (of a statistical test)—the probability of rejecting the alternate hypothesis, i.e., accepting a false null hypothesis; the magnitude of the Type **II** error.

precision—pertaining to the conformity of repeated measurements of the same item, the repeatability of measurements in contrast to the closeness of a measurement to its true value, accuracy.

predicated variable—same as independent variable.

predictand (variable)—same as dependent variable.

predicted matrix—in the sense of regression calculating the elements of a matrix (analogous to a dependent variable) from a predictor matrix (analogous to an independent variable).

predicted variable—same as dependent variable.

prediction (statistical)—generally pertaining to a regression equation calculating a value for a dependent variable as a function of one or more independent variables; in no sense is extrapolation permissible beyond the limits of the independent variables, e.g., one cannot predict future events; also includes generalization to predicting one matrix from another.

predictor (variable)—same as independent variable.

premultiplier matrix—see postmultiplier matrix.

principal axes—same as component axes, i.e., axes defined by eigenvectors in PCA.

principal component analysis—method of studying variation in a sample or population by extraction of independent facets of variation from a matrix measuring dispersion.

principal coordinate analysis—a method similar to PCA that generally is more appropriate for ordinations involving qualitative data.

principal diagonal (of a matrix)—in a square matrix, the elements from the upper left hand through the lower right hand corner; syn.: diagonal, main diagonal.

principal factor analysis—analogous to PCA in studying variation in a sample or population but operating from the premise that certain components must be corrected and/or omitted owing to error.

priors—same as a priori probabilities, but generally as used in classification or diagnosis, i.e., arbitrarily probabilities of group membership selected before a discriminant analysis.

proper values—same as eigenvalues.

Pythagorean logic—logic presuming that vectors of multivariate analysis best describe phenomena as consisting of several independent patterns of variation or contrasting pattern of variation; especially pertaining to the nature of the results of multivariate analysis prior to formulating conclusions, in contrast to diverse premises and conclusions about the nature of the data per se prior to its analysis.

Q-technique—multivariate methods that examine relationships between pairs of individuals in a similarity matrix in contrast to relationships between pairs of variables in a dispersion matrix (a *R*-technique), e.g., a *Q*-type principal component analysis of a similarity matrix.

quadratic form—a second degree polynomial that yields a scalar and whose terms are of the second degree for all variables, e.g., in multivariate analysis the vector $\mathbf{v}_i$ of order p_i in $\mathbf{v}_i'\mathbf{R}\mathbf{v}_i$, is "multiplied" by itself; the form of the variance of a transformed variable, e.g., the variance of a component score or canonical variate; quadratic forms having all values greater than zero are positive definite and those with values zero or greater are positive semidefinite.

qualitative variables (data)—an attribute that is not subject to direct numerical measurement, e.g., sex, habitat, presence of a feature, etc.; many such variables can be pseudoquantified, especially sex and other dichotomous variables using a code of zero for one condition and unity for others; if other attributes can be arranged into a logical sequence, they likewise can be coded.

quantitative taxonomy—in the broad sense any taxonomic analysis or judgment based upon mathematical methods; often restricted to multivariate methods, including subdivision into morphometric and numerical taxonomy, a subdivision now hardly recognizable.

quantitative variables (data)—variables in the form of numerical quantities.

quick-and-dirty method—a method involving statistics that are obtained quickly by inspecting data, e.g., the range, median or mode, or comparably simple computations that often are inefficient (dirty) since they tend to announce fewer significant results; sometimes applied to nonparametric methods, comparison of means and their standard errors, conclusions drawn from raw data, etc.

R—generally the symbol for the multiple correlation coefficient.

R^2—symbol for the square of the multiple correlation coefficient, or coefficient of determination.

RA—abbreviation for reciprocal averaging and correspondence analysis.

random error—deviation from a true value that acts as a variate which distorts individual measurements but balances out on the average;

random sample—a sample selected from a population in such a manner that every individual in the population had the same probability of being represented in the sample; a fundamental assumption of any statistical analysis that, in the strictest sense is impossible to satisfy when sampling in nature.

random sampling distribution—the distribution of one or more statistics calculated from all possible samples drawn singly, in pairs, etc., that can be drawn from a population; often samples are of the same size.

random variable—same as variate.

rank (of a matrix)—the greatest number of linearly independent columns (or rows) in a matrix.

ratio—the result of dividing one number by another; often used as a variable.

reciprocal averaging—an iterative algorithm for solution of two or more simultaneous equations in reference to average values for variables in each equation; an algorithm associated here with correspondence analysis.

rectangular axis—see axis.

rectangular matrix—any matrix whose number of columns do not equal the number of rows.

redundancy—in canonical analysis the measure of overlap between variables of two sets in terms of single pairs of canonical axes or all pairs (total redundancy).

regressand (variable)—same as dependent variable.

regression—a method to study the predictability of a so-called dependent variable from a function of one or more other variables; also generalized to the prediction of one matrix from another.

regression coefficient(s)—the coefficient of an independent variable in regression.

regression weight—same as partial regression coefficient.

regressor (variable)—same as independent variable.

reification—examination of the morphometric implications of a vector by identifying its biological or physical implications, e.g., identifying the biological feature implied by a component; also extended to relating various vectors from different multivariate analyses with a view towards identifying their implications.

resemblance—in the loosest sense, a measurement of likeness or dissimilarity of objects by different functions, generally measures of association or distance, based upon characteristics of the objects syn.: similarity, affinity.

resemblance matrix—same as similarity matrix.

residual—same as error.

residual mean square—same as pooled error.

residual matrix—same as pooled within group dispersion matrix; also the derivation of a predicted matrix from an original matrix.

residual matrix—same as pooled within group dispersion matrix; also the deviation of a predicted

robust (method)—pertaining to any statistical method that performs well even when certain statistical assumptions are not satisfied; see assumption (statistical)

rotation (of axes)—the basic device for transforming data vectors to comply with the various models of multivariate analysis, i.e., rotating original orthogonal variables axes to a new set of axes that are orthogonal (except for the transformation to discriminant space); more frequently discussed in reference to further rotation of originally rotated axes to develop so-called simple structure in factor analysis.

R-technique—see Q-technique.

sample—a subset of a population.

sample size—the number of individuals in a selected subunit of individuals from a population, large samples often are assumed to have 30 or more observations so small samples have less than 30 observations, other interpretations exist; in many cases the size distinction is unknown.

scalar (quantity)—a single number in contrast to a matrix, vector, etc.

scalar algebra—simple algebra, a generalization of arithmetic.

serial (axis)—see ordination.

set—a collection of particular things of the same kind that belong or are put together, e.g., points on a line.

shape component—in PCA, a bipolar component pertaining to a pattern of variation that is consistent with the shape of organisms or phenomena in general.

significance (significant)—statistically, the rejection of a null hypothesis; the consequence of the value of a calculated statistic occurring in the critical region; statistical significance is an all-or-none proposition based upon an a priori level of significance so such terms as highly significant are inappropriate.

similarity matrix—an OTU × OTU or an individual × individual matrix whose off diagonal elements are estimates of similarity or resemblance between OTU's and diagonal elements are self comparisons, generally values of unity; includes association matrices but in the strictest sense not matrices of Euclidean distances between OTU's.

simple structure—in factor analysis, the special case where a few common factors, appearing to explain the variation, are rotated to have:

(1) high factor loadings for only a few variables,

(2) near zero loadings for most variables,

(3) high loadings for most variables only in one factor, etc.

simultaneous equations—two or more equations that define simultaneously imposed restrictions on all variables but not necessarily equations having common solutions.

sine (of an angle)—in a right angle triangle, the ratio of the side opposite the angle and the hypotenuse.

singular matrix—any matrix having a determinant equal to zero so no inverse; see ipsative measures.

size component—in PCA, a general component interpreted to imply the increase in size or magnitude of a phenomenon, e.g., a growth component; more specifically a general component containing coefficients of equal magnitude indicating size independent of shape, isometry.

skewness—pertaining to the asymmetry of a frequency distribution and its measurement, generally as a departure from a normal distribution; for unimodal distributions, negative skewness indicates an extension towards lower values and positive skewness towards higher values of the variable.

small sample theory—pertaining to parametric statistics which typically allow smaller sample sizes than do nonparametric statistics; also pertaining to parametric statistics for which random sampling distributions are known in a probability sense in contrast to statistics treated as parameters, especially in statistical tests of hypothesis.

space—any formal mathematical system defined in terms of its objects and/or geometric nature; the system defined by a set; generally pertaining to a p-dimensional space defined by a vector of variables, i.e., a vector space; also see Cartesian space, component space, discriminant space, Euclidean space.

specific variance—same as latent variance.

specificity—same as latent variance.

sphericity—pertaining to axes defining a spheroid; in multivariate analysis, the condition where two or more derived vectors, e.g., component axes, are of equal length hence arbitrarily defined; pertaining to the fact that when three or more axes are of equal length, a sphere or hypersphere is defined by the axes; also see vector.

spurious correlations—apparently meaningful correlation between two actually poorly or uncorrelated variables owing to extraneous factors such as data handling; also correlation between two variables that are influenced by a common multiplicative or added variable, a common result of two ratios or indices that share the same variable or are associated with an extraneous variable; often extended to include many types of nonsense correlations.

square of the multiple correlation coefficient—in the z-score multiple regression model, an estimate of the proportion of the total variance of unity of the dependent variable that is accounted for (= predicted) by the independent variables, a proportion that can assume values from zero to unity; syn.: coefficient of determination.

square matrix—any matrix with the same number of columns and rows.

SS—abbreviation for sum of squares.

SSCP matrix—abbreviation for sum of squares and cross products matrix.

standard partial regression coefficients—partial regression coefficients for the z-score model.

standardize (standardized score)—often used interchangeably with normalize; more specifically, an original score adjusted by some quantity, usually a variance, standard deviation or dispersion; standardization may or may not result in the standardized scores being normalized to a mean of zero and standard deviation of unity.

standardized normal score—loosely, either a normalized or standardized score; more specifically a z-score produced by standardization by the standard deviation to produce a z-score standard deviation of unity and by normalization to cause the mean z-score to be zero; syn.: standardized score, z-score.

standardized score—generally the same as standardized normal score.

standardized variation—in contrast to absolute variation where units of measure influence the magnitude of the variance of each variable, equalizing the variance of each variable by transforming each variable to a set of z-scores; ''equal'' variation of all variables.

statistic—a summary value calculated from a sample; usually an estimator of a population parameter.

statistical analysis—same as statistics.

statistical inference—making judgments about the nature of one or more populations on the basis of random samples and certain assumptions about the population(s) that lead to knowing the probability of the method providing correct judgments on alternate hypotheses; the method is dependent on random samples and assumptions being valid; knowing the probability of rejecting a null hypothesis applies to the method and not the individual study, see Type **I** error; also see statistical proof.

statistical proof—the statistical procedure that establishes the truth of a statement; generally, the result of rejecting a test of a null hypothesis; statistical proof does not occur whenever a null hypothesis is accepted; also see statistical inference.

statistically significant—same as significant.

statistics—numerical data pertaining to individuals; the science of collecting, analyzing and interpreting such data; inferences from samples to populations based on probability (statistical inference).

stepwise discriminant analysis—methods that add or subtract one variable at a time for the avowed purpose of obtaining the ''best'' set of discriminators; not recommended here owing to inherent problems related to those of stepwise regression and the fact that variables are evaluated according to their independent rather than synergistic contribution.

stepwise regression—methods that add or subtract one independent variable at a time to a multiple regression equation and that supposedly, but generally do not, provide the ''best'' set of independent variables.

structure—same as correlation structure.

submatrix—any row by column portion of a matrix that also is a matrix; in multivariate analysis most submatrices are symmetric.

subset—a set that is part of a larger set, e.g., a local population of a species.

subvector—a portion of a vector containing contiguous elements, generally a portion that defines a recognizable set; a subset of a vector.

sum of squares—the sum of the squares of the deviations of observations from their mean.

sum of squares and cross products matrix—given p variables, a $p \times p$ symmetric matrix containing sum of squares along the diagonal and cross products as off diagonal elements, all in terms of the p variables; syn.: cross products matrix, sum of squares matrix.

sum of squares matrix—same as a sum of squares and cross products matrix.

supressor variable—in multiple regression, an independent variable having little correlation with the dependent variable but having a sizeable regression coefficient owing to high correlation with other independent variables; so named since it reduces the magnitude of the multiple correlation coefficient.

syllogism—a logical, often true statement having three propositions (a major premise, a minor premise and a conclusion) and being true if the premises are true; an unfortunate past example was that birds and insects fly, bats are not insects, so bats are birds; another form relates implications such as, *if a* implies *b* and *b* implies *c, then a* implies *c;* also, such as, *if* the value of a mean and variance are given *then* the standard deviation and coefficient of variation can be provided.

symbatic vectors—vectors having little angular departure but different in that they reflect different biological and/or physical phenomena.

symmetric matrix—any square matrix equal to its transpose, e.g., a correlation or covariance matrix.

synergistic space—a space in which all variables act as a single unit greater than the sum of the effects of the individual variables; a space of total form rather than its component parts, e.g., discriminant space; a space whose units of total form often are difficult to appreciate in reference to individual variables.

taxon—the group of organisms that represent a particular taxonomic category, e.g., the organisms comprising a species, a genus, etc.

taxonomic distance—same as Euclidean distance but often standardized to a maximum value of unity between groups.

test of (null) hypothesis—evaluation of the probability of a null hypothesis being untrue; includes a null hypothesis, alternate hypothesis or hypotheses, assumptions, level of significance, critical region, calculation of the appropriate statistic(s) from one or more samples, interpretation of the statistic(s) as significant or not, and conclusions and/or inferences; syn.: test of significance.

test of significance—same as test of hypothesis.

theory matrix—used as a synonym for both common (factor) matrix and predicted matrix; but, in factor analysis, more often the former.

theory variance—in factor analysis where an original variable is partitioned into a theoretical and residual fraction, the variance of the theoretical variable which often is further partitioned into variation shared by the variables (a common variance) and variation not shared (a latent variance).

total (effect)—in anova or manova the deviation of individual observations or data vectors from the grand mean or grand centroid.

trace (of a matrix)—in a square matrix, the sum of the diagonal elements.

transformation—the changing of a variable or expression to a mathematical form of another type; in multivariate analysis the changing of one variable to another that involves a change from one set of reference axes to another set; also logarithmic and other transformations of data to attempt satisfying statistical assumptions.

transpose (of a matrix or vector)—the matrix resulting by interchanging the rows and columns of a matrix; the arrangement of the elements of a vector in a row.

treatment effect—same as among groups effect.

triangular matrix—a matrix portraying elements only above or below the diagonal, but sometimes the diagonal elements as well; often a device to present unique elements of a symmetric matrix.

Type *I* error—the probability of rejecting a true null hypothesis; an a priori selected error that is an integral part of any test of hypothesis.

Type *II* error—the probability of accepting a false null hypothesis; generally considered indirectly to a test of hypothesis by increasing the sample size to minimize this error.

unbiased estimate—a statistic, the mean of whose random sampling distribution is equal to a population parameter, e.g., the mean but not the median or mode.

unique variance—same as latent variance.

unit matrix—same as identity matrix.

unit vector—a vector with one element of unity and all other elements of zero.

univariate (statistics)—statistics involving a single dependent variable for each individual; statistical methods other than those of multivariate analysis.

universe—same as statistical population.

unreliability—in factor analysis, the residual variance.

variable—generally any quantity which varies; a varying feature measured on each individual of a sample, or such a feature or characteristic of a population; often used in lieu of variate, especially here since the relative frequency of variables is rarely known in practice.

variance—a measure of variation of a variable; in a population, the mean of the squares of the deviations from the population mean; in a sample, the sum of squares divided by the degrees of freedom, one less than the sample size.

variance analysis—same as analysis of variance.

variance-covariance matrix—consider a symmetric matrix containing p columns and p rows that correspond for p variables, a $p \times p$ matrix, in which the off diagonal elements are covariances and diagonal elements are variances; a dispersion matrix measuring absolute variation within and between variables; syn.: covariance matrix.

variance explained—same as variance extracted.

variance extracted—loosely, the amount of variance predicted by a function; in multiple regression, measured by the square of the multiple correlation coefficient; in factor analysis, the achieved communality; in canonical analysis, extended to redundancy; syn.: variance explained.

variance of an individual—in multivariate analysis, the sum of square of squares, either of the component scores of an individual in PCA or of canonical variates of an individual in CA.

variance of a variable—in multivariate analysis the sum of the products of each eigenvalue and the square of the variable's eigenvector coefficient associated with the eigenvalue.

variance ratio—same as F-test.

variate—a quantity that can assume any of a set of values with a specified relative frequency (probability); generally applied to a variable following the normal distribution; syn.: random variable.

vector—here, a column of numbers called elements (a column vector) elsewhere, also a row of such elements (a row vector); data vectors are of the same type; normalized eigenvectors contain elements, each the cosine of the angle between an original variable axis and the axis represented by the eigenvalue; in geometry, the coordinates of a point that defines a line from the origin to the point, an axis; the length or absolute value of a vector is the square root of the sum of the squares of its elements.

vector of a variable—the row vector defined by the consecutive N values of a single variable in a sample of N individuals; a row of a data matrix.

vectors of variables—here, same as data vector.

weight—in multivariate analysis, a coefficient.

within (group)—in anova or mancova, pertaining to the sample for a group.

within group dispersion matrix—same as pooled within group dispersion matrix; the dispersion matrix of discriminant analysis and manova.

within (group) mean square—same as pooled error.

z-scores—a score derived by transforming an original data set to a set having a mean of zero and standard deviation of unity; a normalized score from raw data; since normalization involved division by the standard deviation, also properly called a standardized score; also properly termed a standardized normal score.

References

Anderson, T. W. 1958. An introduction to multivariate statistical analysis. Wiley, New York.
————. 1963. Asymptotic theory for principal component analysis. *Annals of Mathematical Statistics,* 34:122-148.
Atchley, W. R. and E. H. Bryant. 1975. Multivariate statistical methods: among groups covariation. Halsted Press, New York.
————, et al. 1976. Statistical properties of ratios. I. Emperical results. *Systematic Zoology,* 25:137-148.
Beals, E. W. 1973. Ordination: mathematical elegance and ecological naiveté. *Journal of Ecology,* 61:23-35.
Blackith, R. E. 1957. Polymorphism in some Australian grasshoppers. *Biometrics,* 13:183-196.
————. 1960. A synthesis of multivariate techniques to distinguish patterns of growth in grasshoppers. *Biometrics,* 16:28-40.
————. 1961. Multivariate statistical methods in human biology. *Medical Documentation,* 5:26-28.
————. 1965. Morphometrics. In *Theoretical and mathematical biology.* T. H. Waterman and H. J. Morowitz, eds. Blaisdell Publ. Co., New York.
———— and R. A. Reyment. 1971. Multivariate morphometrics. Academic Press, New York.
Bray, J. R. and J. T. Curtis. 1957. An ordination of the upland forests of southern Wisconsin. *Ecological Monographs,* 27:325-349.
Bryant, E. H. and W. R. Atchley. 1975. Multivariate statistical methods: within groups covariation. Halsted Press, New York.
Cacoullos, T. (ed.). 1973. Discriminant analysis and applications. Academic Press, New York.
Calhoun, R. E. and D. L. Jameson. 1970. Canonical correlation between variation in weather and variation in size in the Pacific tree frog, *Hyla regilla,* in southern California. *Copeia,* 1970:124-134.
Cassie, R. M. 1972. A computer programme for multivariate statistical analysis of ecological data. *Journal of Experimental Marine Biology and Ecology,* 10:207-241.
Cooley, W. W. and P. R. Lohnes. 1971. Multivariate data analysis. Wiley, New York.
Corruccini, R. S. 1977. Correlation properties of morphometric ratios. *Systematic Zoology,* 26:211-214.
Dempster, A. P. 1969. Elements of continuous multivariate analysis. Addison-Wesley, Menlo Park, California.
Dixon, W. J. (ed.). 1973. BMD biomedical computer programs. 3rd ed. University of California Press, Berkeley.
Eades, D. C. 1965. The inappropriateness of the correlation coefficient as a measure of taxonomic resemblance. *Systematic Zoology,* 14:98-100.
Everitt, B. 1974. Cluster analysis. Wiley, New York.
Farris, J. S. 1977. On the phenetic approach to vertebrate classification. In *Major patterns in vertebrate evolution.* M. K. Hecht, P. C. Goody, and B. M. Hecht, eds. NATO Adv. Stud. Inst. Ser., Ser A Life Sci., Plenum Press, New York.
Fasham, M. J. R. 1977. A comparison of nonmetric multidimensional scaling, principal components and reciprocal averaging for the ordination of simulated coenoclines, and coenoplanes. *Ecology,* 58:551-561.
Fisher, R. A. 1936. The use of multiple measurements in taxonomic problems. *Annals of Eugenics,* 7:179-188.
Gauch, H. G., Jr. 1973. The relationship between sample similarity and ecological distance. *Ecology,* 54:618-622.
————, R. H. Whittaker and T. H. Wentworth. 1977. A comparative study of reciprocal averaging and other ordination techniques. *Journal of Ecology,* 65:157-174.
Geisser, S. 1964. Posterior odds for multivariate normal classifications. *Journal of the Royal Statistical Society,* 26:69-76.
Gittins, R. 1969. The application of ordination techniques. In *Ecological aspects of the mineral nutrition of plants.* I. H. Rorison, ed. *British Ecological Symposium,* 9:37-66. Blackwell Scientific Publications, Oxford.
Gower, J. C. 1966. Some distance properties of latent root and vector methods used in multivariate analysis. *Biometrika,* 53:325-338.

————. 1971. A general coefficient of similarity and some of its properties. *Biometrics,* 27:857-871.

———— and G. J. S. Ross. 1969. Minimum spanning trees and single linkage cluster analysis. *Applied Statistics,* 18:54-64.

Harman, H. H. 1967. Modern factor analysis. 2nd ed. University of Chicago Press.

Harris, R. J. 1975. A primer of multivariate statistics. Wiley, New York.

Hartigan, J. A. 1975. Clustering algorithms. Wiley, New York.

Hazel, J. E. 1970. Binary coefficients and clustering in biostratigraphy. *Bulletin Geological Society of America,* 81:3237-3252.

Hill, M. O. 1973. Reciprocal averaging: an eigenvector method of ordination. *Journal of Ecology,* 61:237-249.

————. 1974. Correspondence analysis: a neglected multivariate method. *Applied Statistics,* 23:340-354.

————, R. G. H. Bunce and M. W. Shaw. 1975. Indicator species analysis, a divisive polythetic method of classification, and its applications to a survey of native pinewoods in Scotland. *Journal of Ecology,* 63:597-613.

Holland, D. A. 1969. Component analysis: an aid to the interpretation of data. *Experimental Agriculture,* 5:151-164.

Hope, K. 1969. Methods of multivariate analysis with handbook of multivariate methods programmed in Atlas Autocode. Gordon and Breach, Sci. Publ., New York.

Hotelling, H. 1931. The generalization of Student's ratio. *Annals of Mathematical Statistics,* 2:360-378.

————. 1933. Analysis of a complex of statistical variables into principal components. *Journal of Educational Psychology,* 24:417-441.

————. 1935. The most predictable criterion. *Journal of Educational Psychology,* 26:139-142.

————. 1936a. Simplified calculation of principal components. *Psychometrika,* 1:27-35.

————. 1936b. Relations between two sets of variates. *Biometrika,* 28:321-377.

Huxley, J. S. 1932. Problems of relative growth. Methuen and Co., London.

Ito, K. and W. J. Schull. 1964. On the robustness of the T_0^2 test in multivariate analysis of variance when variance-covariance matrices are not equal. *Biometrika,* 51:71-82.

James, R. C. and E. F. Beckenback (eds.). 1968. James and James mathematical dictionary. Van Nostrand Reinhold, New York.

Jeffers, J. N. R. 1967. Two case studies in the application of principal components analysis. *Applied Statistics,* 16:225-236.

Jolicoeur, P. 1959. Multivariate geographical variation in the wolf, *Canis lupus* L. *Evolution,* 13:283-299.

————. 1963a. The degree of generality of robustness in *Martes americana. Growth,* 27:1-27.

————. 1963b. The multivariate generalization of the allometry equation. *Biometrics,* 19:497-499.

———— and J. E. Mosimann. 1960. Size and shape variation in the painted turtle, a principal component analysis. *Growth,* 24:339-354.

Kaiser, H. F. 1960. Comments on communalities and the number of factors. Paper read at an informal conference, *The communality problem in factor analysis.* Washington University.

Kendall, D. G. 1971. Seriation from abundance matrices. In *Mathematics in the archaeological and historical sciences.* F. R. Hodson, D. G. Kendall and P. Tatu, eds. Edinburgh University Press.

Kendall, M. G. 1957. A course in multivariate analysis. Charles Griffin and Co., London.

———— and W. R. Buckland. 1957. A dictionary of statistical terms. Hafner Publ. Co., New York.

Klecka, W. R. 1975. Discriminant analysis. In *SPSS statistical package for the social sciences.* N. H. Nie et al. 2nd ed. McGraw-Hill, New York.

Kruskall, J. B. 1964a. Multidimensional scaling by optimizing goodness of fit to a nonmetric hypothesis. *Psychometrika,* 29:1-27.

————. 1964b. Nonmetric multidimensional scaling: a numeric method. *Psychometrika,* 29:115-129.

Lee, P. J. 1971. Multivariate analysis for the fisheries biology. *Fisheries Research Board of Canada, Technical Report No.* 244. Freshwater Institute, Winnipeg, Manitoba.

Lefebvre, J. 1976. Introduction aux analyses statistique multidimensionales. Masson, New York.

Mahalanobis, P. C. 1930. On tests and measures of group divergence. *Journal and Proceedings of the Asiatic Society of Bengal,* 26:541-588.

————. 1936. On the generalized distance in statistics. *Proceedings of the National Institute of Science, India,* 2:49-55.

Mather, P. M. 1976. Computational methods of multivariate analysis in physical geography. Wiley, New York.

Moore, C. S. 1965. Inter-relations of growth and cropping in apple trees studied by the method of component analysis. *Journal of the Horticultural Sciences,* 40:133-149.

Morrison, D. F. 1967. Multivariate statistical methods. McGraw-Hill, New York.

Mosimann, J. E. 1958. An analysis of allometry in the chelonian shell. *Review Canadienne de Biologie*, 17:137-228.

———. 1970. Size allometry: size and shape variables with characterizations of the log-normal and generalized gamma distributions. *Journal of the American Statistical Association*, 65:930-945.

———. 1975a. Statistical problems of size and shape. I. Biological applications and basic theorems. In *Statistical distributions in scientific work*. Vol. II. G. P. Patil et al. (eds.). D. Reidel Publ. Co., Dordrecht, Holland.

———. 1975b. Statistical problems of size and shape. II. Characterizations of the lognormal, gamma and Dirichlet distributions. In *Statistical distributions in scientific work*. Vol. II. G. P. Patil et al. (eds.). D. Reidel Publ. Co., Dordrecht, Holland.

Mulaik, S. A. 1972. The foundations of factor analysis. McGraw-Hill, New York.

Nie, N. H. et al. 1975. SPSS statistical package for the social sciences. 2nd ed. McGraw-Hill, New York.

Noy-Meir, I. 1973a. Data transformation in ecological ordination. I. Some advantages of non-centering. *Journal of Ecology*, 61:329-341.

———. 1973b. Divisive polythetic classification of data by optimized divisions on ordination components. *Journal of Ecology*, 61:753-760.

———, D. Walker and W. T. Williams. 1975. Data transformation in ecological ordination. II. On the meaning of data standardization. *Journal of Ecology*, 63:779-800.

Orloci, L. 1975. Multivariate analysis in vegetation research. Dr. W. Junk, Publishers, The Hague.

Overall, J. and C. J. Klett. 1972. Applied multivariate analysis. McGraw-Hill, New York.

Parks, J. M. 1969. Multivariate facies maps. In *Computer applications in petroleum exploration*. Computer Contribution 40:6-12. State Geological Survey, Lawrence, Kansas.

Pearce, S. C. 1965a. The measurement of a living organism. *Biometr.-Praxim.*, 6:143-152.

———. 1965b. Biological statistics: an introduction. McGraw-Hill, New York.

Pearson, K. 1901. On lines and planes of closest fit to systems of points in space. *Philosophical Magazine, ser.* 6, 2:559-572.

Press, S. J. 1972. Applied Multivariate analysis. Holt, San Francisco.

Rao, C. R. 1952. Advanced statistical methods in biometric research. Wiley, New York.

———. 1964. The use and interpretation of principal component analysis in applied research. *Sankhayā, ser. A*, 26:329-358.

Rohlf, F. J. 1972. An empirical comparison of three ordination techniques in numerical taxonomy. *Systematic Zoology*, 21:271-280.

Seal, H. 1964. Multivariate statistical analysis for biologists. Methuen, London.

Searle, S. R. 1966. Matrix algebra for the biological sciences. Wiley, New York.

Shepard, R. N. 1962a. The analysis of proximities. Multidimensional scaling with an unknown distance function. I. *Psychometrika*, 27:125-140.

———. 1962b. The analysis of proximities. Multidimensional scaling with an unknown distance function. II. *Psychometrika*, 27:219-246.

———. 1974. Representation of structure in similarity data: Problems and prospects. *Psychometrika*, 39:373-421.

Sneath, P. H. A. and R. R. Sokal. 1973. Numerical taxonomy. W. H. Freeman and Co., San Francisco.

Snedecor, G. W. and W. G. Cochran. 1967. Statistical methods. 6th ed. Iowa State University Press, Ames.

Sokal, R. R. and F. J. Rohlf. 1969. Biometry, the principles and practices of statistics in biological research. W. H. Freeman and Co., San Francisco.

——— and P. H. A. Sneath. 1963. Principles of numerical taxonomy. W. H. Freeman and Co., San Francisco.

Tatsuoka, M. M. 1971. Multivariate analysis: techniques for educational and psychological research. Wiley, New York.

Tukey, J. W. 1962. The future of data analysis. Annals of Mathematical Statistics, 33:1-67.

Wilks, S. S. 1932. Certain generalizations in the analysis of variance. *Biometrika* 2:471-494.

Wright, S. 1954. The interpretation of multivariate systems. In *Statistics and mathematics in biology*. O. Kempthorne et al. (eds.). The Iowa State University Press, Ames.

Young, D. N. 1971. Autecology of *Postelsia palmaeformis*. M. S. dissertation, Biological Sciences Department, California Polytechnic State University, San Luis Obispo.

Index

Errata for *MORPHOMETRICS*
the multivariate analysis of biological data

The following typographical errors appear in *Morphometrics*. Please make the necessary corrections in your copy.

1. Omission of . . . on either side of a formula to indicate a determinant or an absolute value. Unless otherwise stated the correct form is shown.

   p. 17—$|\mathbf{S}^2 - \lambda \mathbf{I}| = 0$ (in 2 cases)

   p. 19—$\Pi\lambda_i = |\mathbf{A}|$; $|\mathbf{A}| \geqq 0$ (2 cases); $|\mathbf{A}| = 0$ (3 cases)

   p. 24—$|\mathbf{R}| = 3/8$; $|\mathbf{S}^2| = 0$; $|\mathbf{R}| = 0$

   p. 25—all 4 **R** to $|\mathbf{R}|$ and all but the first $\mathbf{S}^2$ to $|\mathbf{S}^2|$ (7 cases) Note that this does not include the $\mathbf{S}^2$ in the line after, "2. The minimum . . ." This line will be replaced (see below).

   p. 26—$\mathbf{x_1}'$ to $|\mathbf{x_1}'|$ in lines 11, 14, 16, 23
   $\mathbf{x_2}'$ to $|\mathbf{x_2}'|$ in lines 11, 15, 16, 23
   all $\mathbf{S}^2$ to $|\mathbf{S}^2|$ (4 cases)

   p. 27—all $\mathbf{S}^2$ to $|\mathbf{S}^2|$ (4 cases)
   all **A** to $|\mathbf{A}|$ (5 cases)

   p. 29—$|\mathbf{S}^2 - \lambda \mathbf{I}| = 0$

   p. 30—$|\mathbf{R}_{11}^{-1}\mathbf{R}_{12}\mathbf{R}_{22}^{-1}\mathbf{R}_{21} - \lambda \mathbf{I}| = 0$

   p. 31—$|\mathbf{W}^{-1}\mathbf{B} - \lambda \mathbf{I}| = 0$

   p. 53—same correction as on p. 29

   p. 71—$(\widetilde{\mathbf{S}}^2 - \lambda \mathbf{I}) = 0$ to $|\widetilde{\mathbf{S}}^2 - \lambda \mathbf{I}| = 0$

   p. 106—add determinant lines to characteristic equation just below center of page (formula resembles that on p. 30).

   p. 121—add determinant lines to two characteristic equations in center of page and in the first change $\mathbf{R}_{22}^{*}$ to $\mathbf{R}_{22}^{*-1}$

   p. 136—$a_{ijk} = 1 - |X_{ik} - X_{jk}|/(range\ of\ X_k)$

   p. 152— m
   $\sum_{i=1}^{m} |x_{\varrho i} - x_{\varrho+1, i}| < m$

   p. 154—$|\mathbf{L} - \lambda \mathbf{I}| = 0$ and $|\mathbf{V} - \lambda \mathbf{I}| = 0$

p. 166—last term of formula to $\lambda |d_{im} - d_{jm}|$.

p. 168—in last paragraph, line 2, to $|\beta| \leq 1$

p. 178—in formula M = add a perpendicular line on each side of $\mathbf{S}^2$ and of $\mathbf{S}_k^2$

p. 192—in formula 4, read $\delta = |\mu_1 - \mu_2|$

p. 193—formula 3 read $D = \dfrac{|d|}{s} = \dfrac{|\bar{X_1} - \bar{X_2}|}{s}$

p. 211—$q_j = |\mathbf{S}_j^{-2}|/\Sigma|\mathbf{S}_j^{-2}|$

p. 221—read $|\mathbf{W}^{-1}\mathbf{B} - \lambda\mathbf{I}| = 0$

p. 243—$|\mathbf{A} - \lambda\mathbf{I}| = 0$

2. Other errors

p. 25—line 2 following, "2. The minimum value . . ." change from, "reflects the degree of independence . . .", to, "the minimum value of any variance is zero."

p. 25—fifth line up from bottom, change $\mathbf{R}_{12}$ to $|\mathbf{R}|$

p. 77—3rd line under *Sample Size,* change more to most

p. 78—par. 1, l. 4, change not to no

p. 100—line 9, change to = $\mathbf{f'H'} + \mathbf{d'U'} + \mathbf{c'E'}$ with variances

p. 135—3rd formula, change 2nd Σ symbol to $\displaystyle\sum_{m=1}$

p. 143—par. 2, l. 1, change ordination to ordinations

p. 145—formula, change + after d_{ij}^2 to =

p. 149—line after 1st formula, change the sum to The sum

p. 151—Table 8.3 header, line 2, change Table 8.1 to Table 8.2

p. 152—under 9, formula, change 2nd = to −

—line 8, change to . . . . Tables 8.3 and 8.4.

p. 173—par. 3, l. 9, change (Blackith, 1967) to (Blackith, 1965)

p. 180—in 2nd formula, delete / /

p. 187—par. 1, l. 6, change becomes to become

p. 192—in formula 1 and 2 add / just before the 2

—in formula 6, add / just before σ

p. 197—line 1, change . . . . . $'ia$- to y_v

p. 207, 208, 210—confusing alignment of the discrimination vector ℓ might cause it to appear as a subscript rather than a vector. See p. 209 for correct presentation.

p. 217—there are several instances of the use of bold capital letters where there should be bold low case letters: $\mathbf{V}$ in $\mathbf{V}_i'\mathbf{V}_i = 1$ and $\mathbf{V}_i'\mathbf{S}^2\mathbf{V}_i$ plus the following two $\mathbf{V}_i$ should all be $\mathbf{v}$
$\mathbf{C}$ in $\mathbf{C}_i'\mathbf{R}\mathbf{C}_i = 1$ $\mathbf{C}_i = \mathbf{S}_{diag_i}\mathbf{a}_i$ should be $\mathbf{c}$

p. 224—1st formula, for R_j read R_{cj}^2

3. Other known errors are not critical

MRINAL K. DAS